AF657507

Recent Mycological Researches

Recent Mycological Researches

Editor
S.C. SATI
Department of Botany
Kumaun University
Nainital

I.K. International Publishing House Pvt. Ltd.
NEW DELHI • BANGALORE

Published by

I.K. International Publishing House Pvt. Ltd.
S-25, Green Park Extension, Uphaar Cinema Market
New Delhi – 110 016 (India)
E-mail : info@ikinternational.com

ISBN 81-88237-80-9

10 9 8 7 6 5 4 3 2

© 2006 I.K. International Publishing House Pvt. Ltd.

All rights reserved. No part of this publication may be reproduced, stored in a retrieval system, or transmitted in any form or any means: electronic, mechanical, photocopying, recording, or otherwise, without the prior written permission from the publisher.

Published by Krishan Makhijani for I.K. International Publishing House Pvt. Ltd., S-25, Green Park Extension, Uphaar Cinema Market, New Delhi – 110 016 and Printed by Rekha Printers Pvt. Ltd., Okhla Industrial Area, Phase II, New Delhi – 110 020.

Dedicated to
My revered teacher

Late Prof. R.D. Khulbe
(1949–2002)

Who devoted his life to
The Himalayan Mycology and Society

Contents

Prof. R. C. Pant
VICE-CHANCELLOR

KUMAUN UNIVERSITY
NAINITAL - 263001

Off. : (05942) 235068
Resi. : (05942) 236855
Fax : (05942) 235576
e-mail : vc@kumuni.ren.nic.in

FOREWORD

Man has been acquainted with fungi since times immemorial, as one of the practical example is alcohol fermentation and many macro-fungi as mushrooms and toadstools were recognized by Aristotle in 300 BC. It is estimated that there are now more then one lakh species of fungi known from different habitats. A lot of research is going on these fungi in India and abroad. The Recent Mycological Researches is an excellent collection of papers from India and Nepal covering the diverse topics like Fungal flora / diversity, Fungal diseases, management and control, Fungal physiology and ecology, Fungal biotechnology and symbiosis.

It gives me a great pleasure to write a foreword for this monograph compiled by Dr. S. C. Sati of the Botany Department, Kumaun University, Nainital, which is one of the premier institutions, in the field of Mycology and Plant Pathology in Uttaranchal. Mycological research at Nainital was initiated by Prof. K. S. Bhargava FNA, the first Head of Department and continued by many mycologists including Late Prof. R. D. Khulbe. As a result, Kumaun University has produced a good number of Ph.Ds. and publications in mycology. It is hearten that Prof. Khulbe's student Dr. S. C. Sati is now carrying the torch of mycological research in this region. It is fitting that Dr. Sati has commemorated this treatise in honor of his teacher.

I wish a grand success to this endeavor of Dr. Sati. I am sure that it will fulfill the need of numerous research students of mycology, teachers and will be a useful contribution to the science.

R.C. Pant 18.6.05

Preface

Microbes including fungi constitute an important component of biodiversity. They are our closest neighbours and comprise one of the biggest kingdom in the living world. There are more than one lakh species of fungi known which grow and multiply in all diverse habitats in nature. Due to indiscriminate exploitation and habitat destruction a substantial part of the world biodiversity is being concomitantly regularly lost. Who knows many of the microbes could be extinct well before we are ready to recognize them? Unfortunately a little attention has been paid towards this direction. Their management and exploitation for the betterment of mankind is must. Though there are many national and international journals that give information about fungi and microbes but the recent knowledge is almost in scattered state. Hence, it is desirable to keep this at one place to provide up-to-date information about mycological research especially being done in India.

Recent Mycological Researches is dedicated to my teacher late Prof. R.D. Khulbe (Nov. 1949–Jan. 2002) as a tribute, who contributed best of his time to mycological studies in Kumaun Himalaya. It is an excellent collection of 26 original research/ review articles of eminent mycologists /personalities from different corners of India including Nepal. It has following four major sections to cover all possible aspects of mycological research:

1. Fungal flora/diversity
2. Fungal diseases, management and control
3. Fungal physiology, ecology and pollution
4. Fungal biotechnology and symbiosis

It is doubtless to say that this monograph would be very useful to all scientists, teachers, students interested in mycology and mycological literature. A line of critical comment on this book is always welcome from the readers/ users of this book.

I wish to express my gratitude to all the persons who have contributed their valuable articles to make the present monograph more meaningful. I would also like to thank my colleagues and research students for their support and encouragement. Thanks are also due to my family members for keeping me free from all domestic burdens to complete the work and to Dr. M. Belwal for his help.

S.C. Sati

Late Prof. R.D. Khulbe
(1949–2002)

Prof. Raghuvar Dutt Khulbe was the eldest son of Sri C. S. Khulbe, Member of INA and grandson of Sri B. R. Khulbe, a renowned astrologist of his times. Born on 18th Nov., 1949 in a small village Chaura near Dwarahat, in Almora district of Uttaranchal, Kumaun, he did his early schooling from Vinayak Inter College, Vinayak (Ranikhet) and Intermediate from Dwarahat (Almora). He had a brilliant academic record. After obtaining B. Sc. from Bareilly College, Bareilly he did his M. Sc. (Botany) from D. S. B. Govt. College, Nainital in 1971. He joined research and did his Ph. D. under the erudite guidance of Prof. K. S. Bhargava, FNA, the founder Head of Botany Department of the then D.S.B. Govt. College, Nainital on the 'Watermolds of Nainital and its suburbs' in 1977 from Agra University. Later he also earned his D.Sc. on "Aquatic Phycomycetes" in 1987 from Kumaun University, Nainital.

Prof. Khulbe began his professional life after appointment as lecturer in the Botany Department, D. S. B. Govt. College, Nainital (now Kumaun University) in 1971. He became Reader in 1985, thereafter Professor in 1995. He had nearly 150 research papers to his credit in various national and international journals on aquatic fungi, plant pathology as well as water pollution. He had guided 15 students, viz., Dr. S. C. Sati, Dr. G .S. Mer, Dr. B. L. Verma, Dr. (Km.) Kamla Pandey, Dr. M.C. Sati, Dr. A.P. Dhyani, Dr. (Smt.) Prathiba Harbola, Dr. R.S.Negi, Dr. Chandra Joshi, Dr. Anjala Durgapal, Dr. Rajender Kaur, Dr. Deepa Bisht, Dr. Deependra Nayal, Dr. M. C. Pandey and Dr. Archana Bhatt for Ph.D. in different disciplines of mycology, plant pathology and water pollution. He organized a National Seminar on *Perspectives in Aquatic Biology* in 1988, which was attended by a number of aquatic biologists including Prof. K.S. Bhargava, Lucknow; Prof. S.P. Singh; Prof. B.M. Johri, New Delhi; Prof. R. Dayal, B.H.U.; Prof. C. Manoharachary, Hyderabad; Prof. M.N. Madhyastha, Karnataka; Prof. A.K. Verma, J.N.U.; Prof. M.R. Vijayaraghwan, New Delhi and Dr. G.C. Srivastava, Gorakhpur. In the year 1990 he visited Regensburg (Germany) to participate in IV International Mycological Congress.

He authored two monographs—*Parasitic Watermolds* (1994) and *A Manual of Aquatic Fungi* (2001), which are considered one of the monumental contributions in the field of Indian aquatic mycology. He also edited two books *Perspectives in Aquatic Biology* (1989) and *Microbial Activity in Himalaya* (1992) and wrote one book on *Shak Shabjion Ke Kavak Rog* in Hindi for the benefit of village farmers.

Prof. Khulbe had been Fellow of Indian Botanical Society, Fellow of Phytopathological Society of India, besides, life members of various scientific societies like ISCA, ISNA, NAS etc. He was the founder Organizer of Swami Vivekanand Cultural and Scientific Environmental Organization, Nainital. He executed a number of research projects, namely, Aquatic Fungi in Relation to Fish Diseases in Kumaun Himalaya, C.S.I.R., New Delhi; Bio-degradation of Leaf Litter in Different Forest Ecosystems in Kumaun Himalaya; A part of the Project on an integrated Ecological Study of Eastern Kumaun Himalaya with Emphasis on Natural Resources, D.S.T., New Delhi; Seed Transmission and Control of some Important Seed-borne Pathogens of Major Vegetables Grown in Kumaun Hills, I.C.AR. New Delhi; Studies on Watermolds in relation to Agricultural Crops of Tarai, Nainital, C.S.T., U.P., Lucknow; Studies on Water Pollution Nainital and Bhimtal Lakes of Kumaun Himalaya, Ministry of Environment and Forests, Govt. of India, New Delhi; Aquatic Fungi of Kumaun Himalayas: in vitro Culture, Exploration of Pathological and Biotechnological Potentials, D.S.T., New Delhi. Fungal diseases of fishes in Nanaksagar, Nainital; Itiology, Pathology, Immunology and Control, I.C.A.R., New Delhi; Mycotic Infection and their Management in Fresh water Fishes of Central Himalaya, Deptt. of Biotechnology, Govt. of India. and established a strong tradition of research in the field of Himalayan Mycology.

Prof. Khulbe discovered 6 species of aquatic fungi viz., *Dictyuchus pisci* Khulbe & Sati, *Saprolegnia bhargavii* Khulbe & Verma, *Protoachlya oryzae* Khulbe, *Apodachlya minima* Coker and Leitner var. *hypogyna* (Khulbe) Mer and Sati, *Pythium vexans* var. *minuta* Mer and Khulbe and *Scoliolegnia centrica* Khulbe. This line of investigation is now also being continued successfully by one of his students (Dr. S.C. Sati) at the same center.

Prof. Khulbe, besides a modest mycologist was a good teacher. He had also been quite popular for his homeopathy and social service to the society. He had been closely associated with the Extension Program and Public Awareness Program of the Central Himalayan Environmental Association (CHEA) for a long time and held the honorary post of Joint Secretary till his death. He was suffering from congenital heart valve problem, which got aggravated during the last few years. Not caring for this he had gone on a visit to CHEA extension area involving a long climb. On return he suffered from pneumonia, external as well as internal herpes respectively. This coupled with his heart problem led to breathing his last on the ill-fated morning of 2nd Jan, 2002 at Sushila Tiwari Memorial Forest Hospital, Haldwani (Nainital). Just one day before his death on the eve of New Year, I rang him to know about his health, he replied, "O K, I am fine now but some problem of herpes is still there and will be cured soon". Despite of many good wishes conveyed to him on 1st Jan 2002, on the eve of New Year, could not prevent him from the cruel hands of death. Prof. Khulbe is survived by his wife Vijaya, two daughters (Manisha, an Assistant Professor of Electronic Engineering and Anjani, student of B. Sc.Ag., both at Pantnagar) and one son (Kapil, M. Sc. Botany, working for his Ph. D.).

Prof. R. D. Khulbe contributed best of his time to our department and maintained a prestigious position for Mycology and Plant Pathology to the University. The untimely tragic departure of Prof. Khulbe has created lacunae in the department. The passing away of Prof. Khulbe is undoubtedly an irreparable loss in the field of Mycology and Plant Pathology of our department as well as to whole Uttaranchal. In his death the scientific society has lost a dedicated researcher, teacher, mycologist, plant pathologist, homeopathic doctor and social worker of Kumaun Himalaya. It is our humble attempt to dedicate this monograph to him as homage for his novel contribution in the field of Himalayan Mycology and Plant Pathology.

Section A

Fungal Flora / Diversity

1

Fungal Flora of Two Altitudinally Different Streams of Kumaun Himalaya

M. Belwal and S.C. Sati

Department of Botany, Kumaun University, Nainital (India)

SUMMARY

In the present study, two altitudinally different streams of Kumaun Himalaya were surveyed and 57 species of waterborne conidial fungi belonging to 34 genera were isolated from different habitats including water foam, submerged leaf litter samples and plant roots. Out of these 57 species, 35 were present in Jeoli stream and 54 species were recorded from Jageshwar stream. Seventeen species were also recorded as endophytes of riparian plant roots. Four species viz, *Diplocladiella appendiculata, D. longibrachiata, Flabellospora acuminata* and *Flabellocladia tetracladia* were found to be new to Indian aquatic fungi, whereas 11 species were found to be new to this region.

Keywords: Microbial diversity, Aquatic hyphomycetes, Occurrence

The freshwater bodies, a part of the aquatic ecosystem, harbour a great microbial diversity comprising different groups of microbial communities, fungal community being one of them. It was Ingold (1942) who described water-borne conidial fungi in detail and named them 'Aquatic hyphomycetes' as they complete their life cycle on submerged substrate in well aerated waters. These fungi are known as dominant microorganisms during the first stages of leaf litter decomposition process in running freshwater bodies (Findlay and

Arsuffi, 1989; Barlocher and Kendrick, 1976; Kaushik and Hynes, 1971; Suberkropp and Klug, 1976).

Waterborne conidial fungi are biologically important as they provide inoculum potential, responsible for the decomposition of organic matter in running freshwater bodies. These fungi constitute a major component of freshwater ecosystem and they help in promoting the productivity of freshwater streams through nutrient release by the decomposition of leaf litters (Barlocher and Kendrick, 1976). Manoharachary (1985) reported that decomposed detritus and submerged decaying leaf material was more palatable and nutritious to aquatic invertebrates than unprocessed leaves. In a study on the decomposition of oak, pine and mixed (pine-oak) litter in high altitude Himalayan stream, Tiwari (1992) and Sati and Pant (2000) found nearly 30% weight loss within one month.

The occurrence and distribution of waterborne conidial fungi have been studied by a number of workers all over world (Alasoadura, 1968, Anastasiou, 1964; Charmier, 1985; Conway, 1970; Descals and Webster, 1982 a, b; Hudson and Ingold, 1960; Ingold, 1956, 1965, 1975, 1976; Ingold and Ellis, 1952; Iqbal *et al.*, 1973; Jooste and Merwe, 1990; Khan, 1986; Kuthubutheen, 1987; Marvanova, 1975; Marvanova and Bandoni, 1987; Marvanova and Hywel-Jones, 2000; Nawawi, 1975, 1987; Nawawi and Kuthubutheen, 1987; Nilsson, 1964; Sridhar and Kaveriappa, 1984; Tubaki, 1957; Webster and Descals, 1979; Webster, 1959; Webster *et al.*, 1976, 1994; Willoughby and Archer, 1973; Willoughby and Minshall, 1975; Wolf 1972).

In India, Ingold and Webster (1973) for the first time reported three species of waterborne conidial fungi from Madras, followed by Subramanian and Bhatt (1981) who studied waterborne conidial fungi from different streams of the Western Ghats. Later on, some more Indian mycologists explored the Southern part of India for these fungi (Chandrashekhar *et al.*, 1991; Madhusudan Rao and Manoharachary, 1982; Manoharachary and Murthy, 1981; Soosamma *et al.*, 2001; Sridhar, 1984; Sridhar and Kaveriappa, 1982; Sridhar *et al.*, 1992; Sreekala and Bhatt 2002). These fungi are also being studied in the Kumaun Himalaya by Dr. S.C. Sati and his co-workers who have explored the waterborne conidial fungi from various localities (Sati *et al.*, 1989, 1992, 2002; Sati and Tiwari 1990, 1992, 1997; Tiwari and Sati 1991, 1993).

MATERIALS AND METHODS

In the present investigation, waterborne conidial fungi occurring in running freshwater bodies of Nainital and Almora, districts, located in the outer part of Central Himalaya, ie Kumaun Himalaya (29° 5¢ N to 31° 25¢ N latitude and 77° 48¢ to 81° 0¢E longitude) were studied. Submerged, dark brown, decaying plant material, especially leaf litter samples of different plant species viz *Acer oblongum, Barberries* sps., *Betula alanoides, Cedrus deodara, Lyonia ovalifolia, Mangifera indica, Myrica indica, Nerium sps, Pinus roxburghii, Polygonum nepalensis,*

Pyrus pashia, Quercus leucotrichophora, Rhododendron arboreum, Syzigium cumuni and foam samples accumulated at water barriers were collected from various habitats at periodic intervals from two altitudinally different streams, Jageshwar (1775m.asl) and Jeoli (1175 m.asl) of Kumaun Himalaya and brought to the laboratory.

The leaf litter was washed thoroughly with running tap water for 2–3 hours and then cut into small pieces and placed for incubation at 15 ± 2°C in sterile petridishes (9 cm diam.) containing sterile water. After two days, the incubated leaves were periodically examined under the microscope to detect the conidia of waterborne conidial fungi, following Ingold (1975), Webster and Descals (1981), Descals and Webster (1982), and Marvanová and Descals (1985). The foam samples were fixed in 5% FAA on the spot, to avoid conidial germination and examined under the microscope by placing drops on glass slides. Semi-permanent slides of these conidial fungi have been deposited in the Kumaun University Mycological Slide (KUMS) Collection at the Department of Botany, Nainital.

RESULTS AND DISCUSSION

Altogether, 57 species belonging to 34 genera of water-borne conidial fungi viz *Acaulopage, Alatospora, Anguillospora, Articulospora, Campylospora, Camposporium, Clavariopsis, Clavatospora, Dimorphospora, Diplocladiella, Dwyaangam, Flabellocladia, Flabellospora, Flagellospora, Helicomyces, Helicosporium, Heliscina, Heliscus, Isthmotricladia, Lateriramulosa, Lemonniera, Lunulospora, Magdalannaea, Naidella, Pestalotiopsis, Setosynnema, Speiropsis, Tetracladium, Tetraploa, Tricladium, Tripospermum, Triscelophorus, Tetrachaetum* and *Volucrispora* were recovered from different submerged leaf litters and water foam accumulated at stream barriers. A brief account of their occurrence and distribution of each species in two studied streams is given below:

1. ***Acaulopage tetraceros*** Derchsler, Mycologia, 27:195, 1935 (Fig. 1).

Drechsler described this species as predaceous on *Amoebae* and other protozoa, capturing their prey by adhesion to mycelium. In this study it was isolated from decaying submerged leaf litter of *Pyrus pashia* during September 2001 and found restricted to Jageshwar site only (KUMS 101). It was also isolated from the living roots of *Murraya koenegii* and *Lyonia ovalifolia* and discovered as a new root endophyte from Jageshwar site only. This was a new record in Uttaranchal.

2. ***Alatospora acuminata*** Ingold, Trans. Brit. Mycol. Soc., 25: 339, 1942 (Fig. 2)

This species was found in Jageshwar throughout the study period while from Jeoli site it was reported during September 2000. It was isolated from to the decaying submerged leaf litter of *Pyrus pashia, Pinus roxburghii, Rhododendron arboreum, Cedrus deodara, Acer oblongum, Betula alanoides, Lyonia oval·folia,*

Syzigium cumuni and *Mangifera indica,* and from water foam (KUMS 201). It was also recovered from both the study sites on the living roots of unidentified grass and *Equisetum* sp. as a new root endophyte.

3. ***Alatospora flagellata*** (J. Gönczöl) Marvanová,Trans. Brit. Mycol. Soc. 75: 227, 1980 (Fig. 3).

The present species was commonly found in both the study sites where it was collected on submerged leaf litter of *Pyrus pashia, Pinus roxburghii, Rhododendron arboreum, Acer oblongum* and *Polygonum nepalensis,* as well as from water foam in September 2000 (KUMS 202).

4. ***Alatospora pulchella*** Marvanová, Archiv fur Protistenk., 119 :71, 1977 (Fig. 4).

This species was restricted to Jageshwar site only and was collected on submerged leaf litter of *Pyrus pashia, Pinus roxburghii, Rhododendron arboreum, Acer oblongum, Betula alanoides* and *Lyonia ovalifolia,* as well as from foam samples in July 2000 (KUMS 203). It was also isolated from the living roots of unidentified grass and found to be a new root endophyte from Jageshwar.

5. ***Anguillospora crassa*** Ingold, Trans. Brit. Mycol. Soc., 41: 365, 1958 (Fig. 5).

The present species was commonly found in both the sites where it was isolated from submerged decaying leaf litter of *Polygonum nepalensis,* as well as from water foam on October 2000 (KUMS 301). It was also isolated from the living roots of *Equisetum* sp. and *Barberis* and found as a new root endophyte.

6. ***Anguillospora filliformis*** Greathead, South Africa Bot. 27:202, 1961 (Fig. 6).

The present species was restricted in its occurrence to Jageshwar stream only. It was isolated from decaying submerged leaf litter of *Pyrus pashia* and *Betula alanoides* as well as from water foam in August 2000 (KUMS 302).

7. ***Anguillospora furtiva*** Webster & Descals, Can. J. Bot., 71: 1647, 1998 (Fig. 7).

The present species was restricted in its occurrence and found only in Jageshwar stream. It was isolated from submerged leaf litter of *Quercus leucotrichophora* and *Barberis* as well as from foam in August 2000 (KUMS 303).

8. ***Anguillospora longissima*** (Sacc and Therry) Ingold, Trans. Brit. Mycol. Soc., 41:367,1958 (Fig. 8).

This species was common in its occurrence in both the sites. It was isolated from submerged decaying leaf litter of *Pyrus pashia, Rhododendron arboreum, Cedrus deodara, Acer oblongum, Lyonia ovalifolia, Syzigium cumuni* and *Mangifera indica* as well as from water foam in May 2000 (KUMS 304). It was also recovered from the living roots of an unidentified grass as a common root endophyte.

9. *Articulospora tetracladia* Ingold, Trans. Brit. Mycol. Soc. 25: 376, 1942 (Fig. 9).

The present species was found to be common in its occurrence for Jageshwar and Jeoli. It was isolated from the submerged decaying leaf litter of *Pyrus pashia, Rhododendron arboreum, Acer oblongum, Betula alanoides, Lyonia ovalifolia* and *Syzigium cumuni,* in July 2000 (KUMS 401).

10. *Camposporium pellucidum* (Grove) Hughes, Mycol. Pep. 36: 9, 1951 (Fig. 10).

The present species was commonly found at both the sites. The species was isolated from water foam as well as submerged decaying leaf litter of *Acer oblongum* and *Syzigium cumuni,* in July 2000 (KUMS 501).

11. *Campylospora chaetocladia* Ranzoni, Farlowia, 4: 353, 1953 (Fig. 11).

The present species was found as common in its occurrence in both the sites. It was solated from decaying submerged leaf litter of *Rhododendron arboreum, Nerium* sp. and *Polygonum nepalensis,* as well as from water foam during June 2000(KUMS 601). It was also recovered from the living roots of *Murraya koenegii* as a new root endophyte.

12. *Campylospora purvula* Kuzuha, J. Jap. Bot. 48:220, 1973 (Fig. 12).

The present species was found restricted in its occurrence on submerged decaying leaf litter of Jeoli, but found as a common root endophyte on the living roots of the Pterdophytic plant, *Lyonia ovalifolia* and *Barberis* sp. in both the sites. It was isolated during June 2000 from the submerged leaf litter of *Polygonum nepalensis* and *Bauhinia* sp. as well as from water foam (KUMS 602). It created a new record in Uttaranchal.

13. *Clavariopsis aquatica* de Wildeman, Ann. Soc. Belge. Microsc., 19: 197, 1895 (Fig. 13).

The present species was found to be a frequently occurring species from Jageshwar and Jeoli. It was isolated from submerged decaying leaf litter of *Pyrus pashia, Pinus roxburghii, Rhododendron arboreum, Cedrus deodara, Acer oblongum, Betula alanoides* and *Polygonum nepalensis,* as well as from foam (KUMS 701). It was also isolated from the living roots of Pteridophytic plant as a root endophyte.

14. *Clavariopsis azlanii* Nawawi, Trans. Brit. Mycol. Soc. 88:428, 1987 a (Fig. 14).

This species was restricted in its occurrence for Jeoli. It was isolated from submerged decaying leaf litter of *Syzigium cumuni* and *Mangifera indica,* as well as from water foam in September 2000 (KUMS 702). It was found to be new record to Uttaranchal.

15. ***Clavatospora tentacula*** (Umphlett) Nilsson, Symb. Bot. Upsal. 18:89, 1964 (Fig. 15).

It was found to be a restricted species in its occurrence for Jeoli site. It was isolated from the submerged decaying leaf litter of *Bauhinia* sp., *Syzigium cumuni* and *Mangifera indica*, as well as from foam in May 2000 (KUMS 801).

16. ***Dimorphospora foliicola*** Tubaki, J. Hattori Bot. Lab., 20: 150, 1958 (Fig. 18).

The present species was found to be restricted in its occurrence for Jageshwar site only and collected from decaying submerged leaf litter of *Rhododendron arboreum, Cedrus deodara, Acer oblongum, Betula alanoides* and *Lyonia ovalifolia* in September 2000 (KUMS 1001).

17. ***Diplocladiella appendiculata*** Nawawi, Mycotaxon 28(2): 298, 1987 b (Fig. 16).

The present species was found to be restricted in its occurrence for Jageshwar stream only and isolated from decaying submerged leaf litter of *Polygonum nepalensis* only as well as from water foam during July 2000. This species was found to be new to Indian aquatic fungi (KUMS 1101).

18. ***Diplocladiella longibrachiata*** Nawawii and Kuthubutheen, Mycotaxon, 24: 217, 1985 (Fig. 17).

The present species was commonly found at both the sites. While it was isolated on foam samples only from the Jeoli site from Jageshwar it was isolated from both submerged decaying leaf litter of *Polygonum nepalensis* as well as foam samples in 2001(KUMS 1102). It created a new record in Indian aquatic fungi.

19. ***Diplocladiella scalaroides*** Arnaud, Bull. Timest. Soc. Mycol. France, 89:295, 1954 (Fig. 19).

The present species was found to be common in its occurrence in both the sites. It was collected from foam samples only during July 2000 (KUMS 1103).

20. ***Dwyaangam cornuta*** Descals, Trans. Brit. Mycol. Soc. 78:408, 1982 (Fig.20).

The present species was found to be restricted in its occurrence in Jageshwar from foam samples, during July 2000 (KUMS 1201). It created a new record in Uttaranchal.

21. ***Flabellocladia gigantea*** Nawawi, Trans. Brit. Mycol. Soc. 85: 174, 1985 (Fig. 21).

The present species was found in Jageshwar site only. It was collected from the water foam samples only during September 2000 (KUMS 1301).

22. *Flabellocladia tetracladia* (Nawawi) Nawawi, Trans. Brit. Mycol. Soc. 85:175, 1985 b. (Fig. 22).

The present species was found to be restricted to Jageshwar site only. It was isolated from the decaying submerged leaf litter of *Quercus leucotrichophora* and *Lyonia ovalifolia* as well as from water foam samples in October 2000 (KUMS 1302). This species was found to be a new addition to Indian aquatic fungi.

23. *Flabellospora acuminata* Descals & Webster, Trans Brit. Mycol. Soc. 78: 411, 1982 (Fig. 23).

The present species was commonly found in both the sites. It was isolated from both decaying submerged leaf litter of *Quercus leucotrichophora Q. floribunda* and *Polygonum nepalensis* and water foam samples, during August 2000 (KUMS 1401).

24. *Flagellospora penicillioides* Ingold, Trans Brit Mycol. Soc. 27:44, 1944 (Fig. 24).

This was a common species in both the sites. It was isolated from decaying submerged leaf litter of *Pyrus pashia, Pinus roxburghii, Rhododendron arboreum, Cedrus deodara, Acer oblongum, Betula alanoides, Lyonia ovalifolia, Syzigium cumuni* and *Mangifera indica,* as well as from water foam samples, in May 2000 (KUMS 1501).

25. *Helicomyces roseus* Link, Freunde Berlin Mag. Neuestern. Entdeek, Gesammten Naturk 3:21, 1809 (Fig. 25).

It was found as a common species in both the sites. It was collected from foam as well as from decaying submerged leaf litter of *Rhododendron arboreum* and *Polygonum nepalensis,* in August 2000 (KUMS 1701).

26. *Helicosporium lumbricoides* (Sac.) Linder, Ann. Misso. Bot. Gad, 16: 282, 1929 (Fig. 26).

The present species was commonly found in both the sites and isolated from the decaying submerged leaf litter of *Polygonum nepalensis,* as well as from water foam samples, during August 2000 (KUMS 1601).

27. *Heliscina campanulata* Marvanova, Trans. Brit. Mycol. Soc. 75: 21, 1980 (Fig. 27).

The present species was found restricted to Jageshwar site only and isolated from decaying submerged leaf litter of *Pyrus pashia,* and *Lyonia ovalifolia* as well as from foam samples during February 2001(KUMS 1801).

28. *Heliscus lugdunensis* Ingold, Trans. Brit. Mycol. Soc. 25: 357, 1942 (Fig. 28).

The present species was commonly found as common in both the sites. It was isolated from water foam as well as from decaying submerged leaf litter of

Pyrus pashia, Pinus roxburghii, Cedrus deodara, Acer oblongum, Syzigium cumuni and *Bauhinia* sp. in July 2000 (KUMS 1901). It was also isolated from the living roots of Pteridophytic plant and *Lyonia ovalifolia* and found a new root endophyte in both the study sites.

29. *Isthmotricladia gombakiensis* Nawawi, Trans Brit. Mycol. Soc. 64: 243, 1975a. (Fig. 29).

The present species was found to be restricted in its occurrence to Jageshwar. Collected from the water foam as well as decaying submerged leaf litter of *Polygonum nepalensis* during July 2001(KUMS 2001), it created a new record in Uttaranchal.

30. *Lemonniera alabamensis* Sinclair & Morgan-Jones, Mycotaxon, 9: 469–481, 1979 (Fig. 31).

The present species was found restricted to Jageshwar only. It was collected from decaying submerged leaf litter of *Polygonum nepalensis* and *Nerium* sp., in December 2000(KUMS 2101).

31. *Lemonniera aquatica* de Wildeman, Ann. Soc. Belge. Microsoc. 19: 197, 1895 (Fig. 32).

The present species was found restricted to Jageshwar site only. It was isolated from water foam samples and submerged decaying leaf litter of *Nerium* sps., in September 2000 (KUMS 2102).

32. *Lemonniera cornuta* Ranzoni, Farlowia, 4: 353, 1953 (Fig. 33).

The present species was commonly found in both the sites. It was collected from foam as well as decaying submerged leaf litter of *Mangifera indica, Pyrus pashia, Pinus roxburghii, Rhododendron arboreum, Acer oblongum, Cedrus deodara, Betula alanoides* and *Lyonia ovalifolia* on September 2000(KUMS 2103). It was also isolated from the living roots of *Lyonia ovalifolia* and an unidentified grass, as a new root endophyte.

33. *Lemonniera pseudofloscula* Dyko, Trans. Brit. Mycol. Soc., 69: 89, 1977 (Fig. 34).

The present species was found commonly. It was frequently isolated from water foam samples as well as from submerged decaying leaf litter of *Rhododendron arboreum, Betula alanoides, Lyonia ovalifolia and Polygonum nepalensis* in November 2000 (KUMS 2104). This species was also isolated from the living roots of an unidentified grass and *Lyonia ovalifolia* where it was to be a found new root endophyte.

34. *Lemonniera terrestris* Tubaki, J. Hatt. Bot. Club. 20: 165, 1958 (Fig. 35).

The present species was found commonly. It was isolated from decaying submerged leaf litter of *Pyrus pashia, Pinus roxburghii, Rhododendron arboreum, Acer oblongum, Betula alanoides, Mangifera indica* and *Lyonia ovalifolia* in September 2000(KUMS 2105). It was also isolated from the living roots, *Lyonia ovalifolia* and found to be a new root endophyte.

35. *Lunulospora curvula* Ingold, Trans. Brit. Mycol. Soc. 25: 404, 1942 (Fig. 36).

The present species was commonly found in and it frequently occurred on water foam as well as on decaying submerged leaf litter of *Pyrus pashia, Pinus roxburghii, Rhododendron arboreum, Cedrus deodara, Acer oblongum, Betula alanoides, Lyonia ovalifolia, Syzigium cumuni and Mangifera indica* in May 2000 (KUMS 2201). This species was also isolated from the living roots of *Eupatorium adenophyllum* and Pteridophytic plant and found to be a new root endophyte.

36. *Lunulospora cymbiformis* Miura, J. Jap. Bot. 47: 65, 1972 (Fig. 37).

It was found common. It was isolated from water foam samples as well as submerged decaying leaf litter of *Pyrus pashia, Pinus roxburghii Cedrus deodara, Acer oblongum, Betula alanoides, Lyonia ovalifolia, Rhododendron arboreum, Syzigium cumuni* and *Mangifera indica* in May 2000 (KUMS 2202).

37. *Lateriramulosa uni-inflata* Matsushima, New Guinea, Kobe, 78 pp., 1971 (Fig. 30).

The present species was found restricted to Jageshwar only and was recovered from foam samples only during July 2000 (KUMS 2301).

38. *Magdalannaea monogramma* G. Arnaud, Bull. Soc. Mycol. Fr., 68: 209, 1952 (Fig. 38).

This species was found to be a rare species and occurred at Jageshwar site only. It was recovered from water foam samples, only during July 2000 (KUMS 2401).

39. *Naidella fluitans* Marvanova & Bandoni, Mycologia, 79: 578, 1987 (Fig. 39).

The present species was found to be restricted in its occurrence for Jageshwar site only. It was isolated from decaying submerged leaf litter of *Polygonum nepalensis* and *Nerium* species, in July 2000 (KUMS 2501).

40. *Pestalotiopsis submersus* Sati and Tiwari, Nova Hedwigia, 56: 543–547, 1993 (Fig. 40).

The present species was found commonly. It was isolated from water foam samples as well as from decaying submerged leaf litter of *Pinus roxburghii, Rhododendron arboreum* and *Syzigium cumuni* on July 2001 (KUMS 2601). It was also isolated from the living roots of *Equisetum*, Fern and *Lyonia ovalifolia* and found to be a new root endophyte.

41. ***Setosynnema isthmosporum*** Shaw & Sutton, Bot. J. Linn. Soc. 91: 25, 1985 (Fig. 41).

The present species was found commonly in both the sites. It was isolated from water foam as well as decaying submerged litter of *Pinus roxburghii* and *Polygonum nepalensis*, in June 2000 (KUMS 2701).

42. ***Speiropsis pedatospora*** Tubaki, J. Hattori Bot. Lab. 20:171, 1958 (Fig. 42).

The present species was found restricted in its occurrence to Jageshwar site. It was isolated from water foam samples as well as from decaying submerged leaf litter of *Polygonum nepalensis* and *Rhododendron arboreum*, during June 2000 (KUMS 2801). It was created a new record in Uttaranchal.

43. ***Speiropsis scopiformis*** Kuthu. & Nawawi, Trans. Brit. Mycol. Soc. 89: 584, 1987 (Fig. 43).

It was found to be a common species. It was isolated from water foam samples as well as decaying submerged leaf litter of *Pyrus pashia, Rhododendron arboreum, Betula alanoides, Syzigium cumuni* and *Mangifera indica* on July 2000 (KUMS 2802).

44. ***Tetrachaetum elegans*** Ingold, Trans. Brit. Mycol. Soc, 25:377, 1942 (Fig. 46).

The present species was found commonly. It was solated from water foam samples as well as from the decaying submerged leaf litter of *Pyrus pashia, Pinus roxburghii, Rhododendron arboreum, Cedrus deodara, Acer oblongum, Betula alanoides, Lyonia ovalifolia, Syzigium cumuni, Mangifera indica* and *Syzigium cumuni* in May 2000 (KUMS 2901). It was also isolated from the living roots of *Lyonia ovalifolia* and Pteridophytic plant species and found to be a common root endophyte.

45. ***Tetracladium apiense*** Sinclair and Eicker, Trans. Brit. Mycol. Soc. 76: 515, 1981 (Fig. 47).

The present species was found commonly. It was isolated from water foam samples as well as from the decaying submerged leaf litter of *Polygonum nepalensis* and *Mangifera indica* on September 2000 (KUMS 3001).

46. ***Tetracladium marchalianum*** de Wildeman, Ann. Soc. Belge. Microsc.,17:35, 1893 (Fig. 48).

It was found a most commonly occurring species. Isolated from water foam samples and decaying submerged leaf litter of *Pyrus pashia, Pinus roxburghii, Rhododendron arboreum, Cedrus deodara, Acer oblongum, Betula alanoides, Lyonia ovalifolia, Syzigium cumuni* and *Mangifera indica*, in May, 2000 (KUMS 3002).

This species was also isolated from the living roots of unidentified grass and *Equisetum* and found to be a new root endophyte.

47. *Tetracladium setigerum* (Grove) Ingold, Trans. Brit. Mycol. Soc., 25:339, 1942 (Fig. 49).

The present species was also found to be a commonly occurring species in both the study sites. It was isolated from the water foam samples and submerged decaying leaf litter of *Barberries, Lantana* sp., *Polygonum nepalensis* and *Syzigium cumuni,* in May 2000 (KUMS 3003). It was also isolated from the living roots of unidentified grass and *Equisetum* as a new root endophyte.

48. *Tetraploa aristata* Berkeley & Broome, Ann. & Mag. Nat. Hist. Ser.2: 459, 1850 (Fig. 44).

The present species was found commonly. It was isolated from water foam as well as from decaying submerged leaf lifter of *Barberis* sp. during September 2000(KUMS 3101). It created a new record in Uttaranchal.

49. *Tricladium aciculum* Nawawi, Trans. Brit. Mycol. Soc. 85: 177, 1985 (Fig. 50).

The present species was found to be restricted in its occurrence to Jageshwar site only. It was isolated from decaying submerged leaf litter of *Nerium* and *Quercus leucotrichophora,* as well as from foam in January 2001(KUMS 3201).

50. *Tricladium chaetoladium* Ingold, Trans. Brit. Mycol. Soc. 63: 624, 1974 (Fig. 51).

The present species was also found to be restricted to Jageshwar site only. It was isolated from water foam samples as well as from decaying submerged leaf litter of *Pyrus pashia, Pinus roxburghii, Rhododendron arboreum, Cedrus deodara, Acer oblongum, Betula alanoides* and *Lyonia ovalifolia* throughout the study period (KUMS 3202).

51. *Tricladium indicum* Sati and Tiwari, Mycol Res. 96: 229–232, 1992 (Fig. 52).

The present species was found to be restricted in its occurrence to Jageshwar site only. It was isolated from water foam samples and decaying submerged leaf litter of *Myrica indica* and *Nerium* sp., during July 2000 (KUMS 3203).

52. *Tricladium splendens* Ingold, Trans. Brit. Mycol. Soc. 25: 388, 1942 (Fig. 53).

The present species was found to be restricted occurrence to Jageshwar site only and was isolated from water foam samples as well as decaying submerged leaf litter of *Rhododendron arboreum,* only in July 2000 (KUMS 3204).

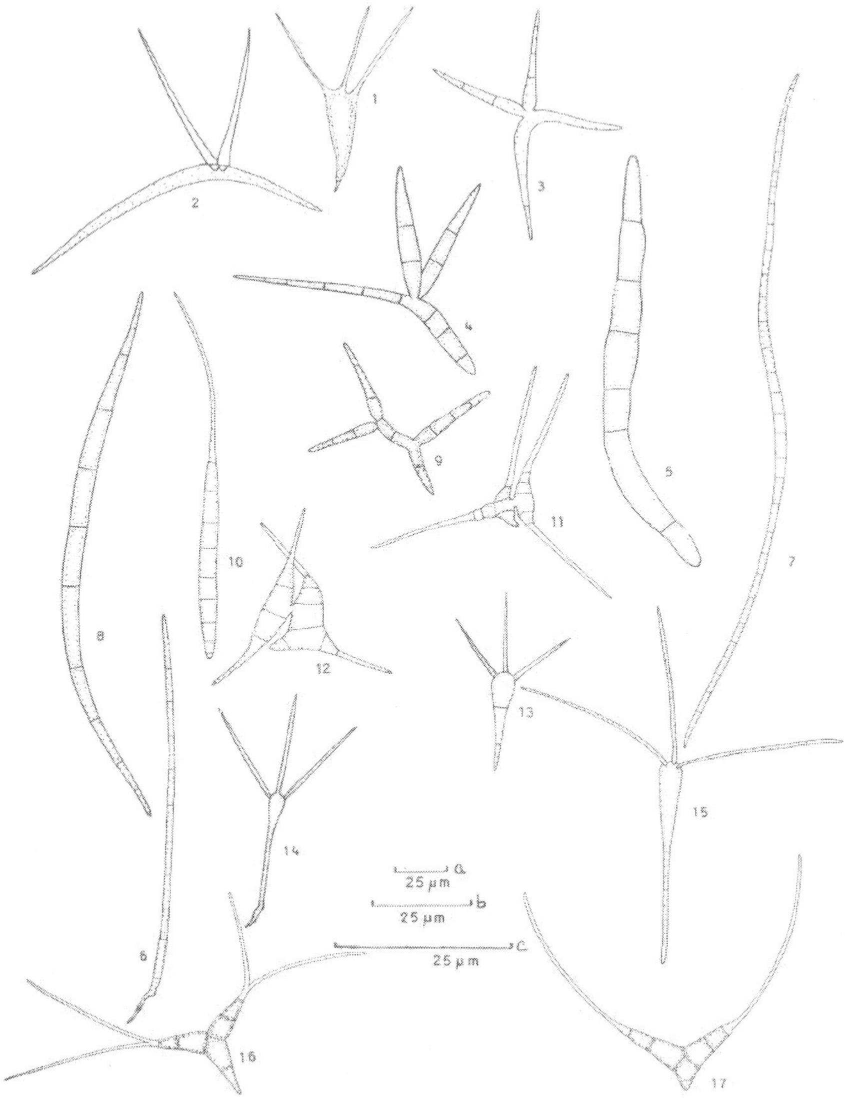

Fig. 1-17. 1-*Acaulopage tetraceros* Derchsler, 2- *Alatospora acuminata* Ingold, 3- *Alatospora flagellata* (J. Gönczöl) Marvanová, 4- *Alatospora pulchella* Marvanová, 5- *Anguillospora crassa* Ingold, 6- *Anguillospora filliformis* Greathead, 7- *Anguillospora furtiva* Webster & Descals, 8- *Anguillospora longissima* (Sacc.and Therry) Ingold, 9- *Articulospora tetracladia* Ingold, 10- *Camposporium pellucidum* (Grove) Hughes, 11- *Campylospora chaetocladia* Ranzoni, 12- *Campylospora purvula* Kuzuha, 13- *Clavariopsis aquatica* de Wildeman, 14- *Clavariopsis azlanii* Nawawi, 15- *Clavatospora tentacula* (Umphlett) Nilsson, 16- *Diplocladiella appendiculata* Nawawi & 17- *Diplocladiella longibrachiata* Nawawii and Kuthubutheen

(Fig. 13, 14 in scale a; 1,5,6,7,8,11,15,16,17 in scale b and 2,3,4,9,12 in scale c)

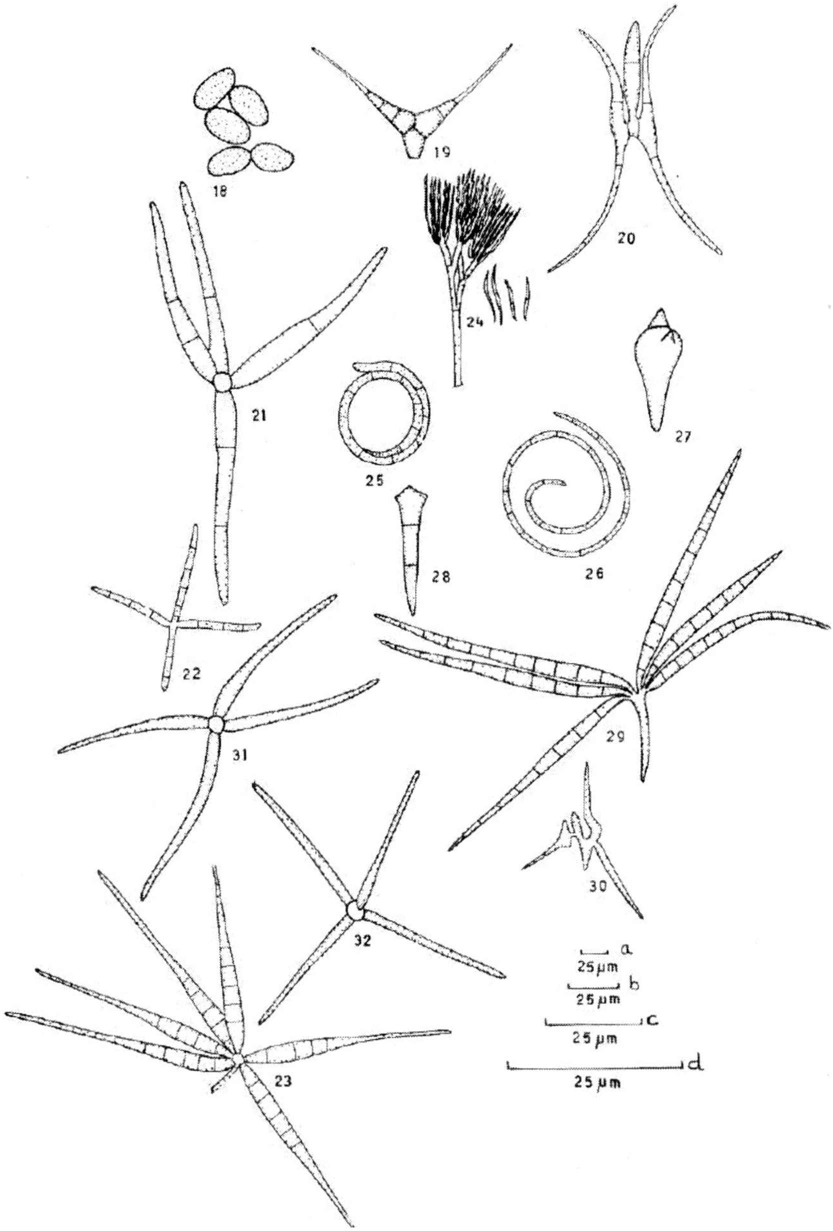

Fig. 18-23. 18-*Dimorphospora foliicola* Tubaki, 19- *Diplocladiella scalaroides* Arnaud, 20- *Dwyaangam cornuta* Descals 21- *Flabellocladia gigantea* Nawawi, 22- *Flabellocladia tetracladia* (Nawawi) Nawawi, 23- *Flabellospora acuminata* Descals & Webster, 24- *Flagellospora penicillioides* Ingold, 25- *Helicomyces roseus* Link, 26- *Helicosporium lumbricoides* (Sac.) Linder, 27- *Heliscina campanulata* Marvanova, 28- *Heliscus lugdunensis* Ingold, 29- *Isthmotricladia gombakiensis* Nawawi, 30- *Lateriramulosa uni-inflata* Matsushima, 31- *Lemonniera alabamensis* Sinclair & Morgan-Jones & 32- *Lemonniera aquatica* de Wildeman

(24 in scale a; 22, 25 in scale b; 18,19,20,21,23,27,28,29,31,32 in scale c and 26, 30 in scale d)

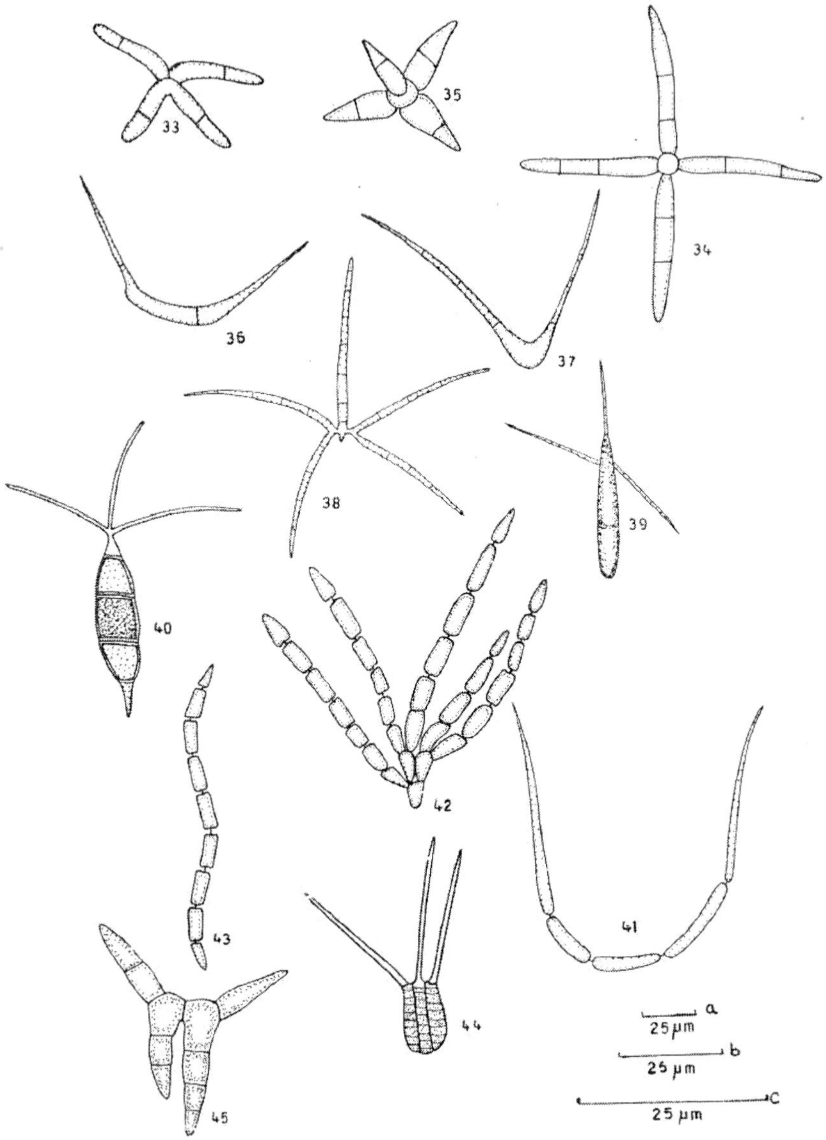

Plate C

Fig. 33-45. 33- *Lemonniera cornuta* Ranzoni, 34- *Lemonniera pseudofloscula* Dyko, 35- *Lemonniera terrestris* Tubaki, 36- *Lunulospora curvula* Ingold, 37- *Lunulospora cymbiformis* Miura, 38- *Magdalannaea monogramma* G. Arnaud, 39- *Naidella fluitans* Marvanova & Bandoni, 40- *Pestalotiopsis submersus* Sati and Tiwari, 41- *Setosynnema isthmosporum* Shaw & Sutton, 42- *Speiropsis pedatospora* Tubaki, 43- *Speiropsis scopiformis* Kuthu. & Nawawi, 44- *Tetraploa aristata* Berkeley & Broome, 45- *Tripospermum cameleopardus* Ingold, Dann and Mc Dougall

(Fig. 44 in scale a;33,34,35,36,37,38,39,40,41,42,43 in scale b; and 45 in scale c)

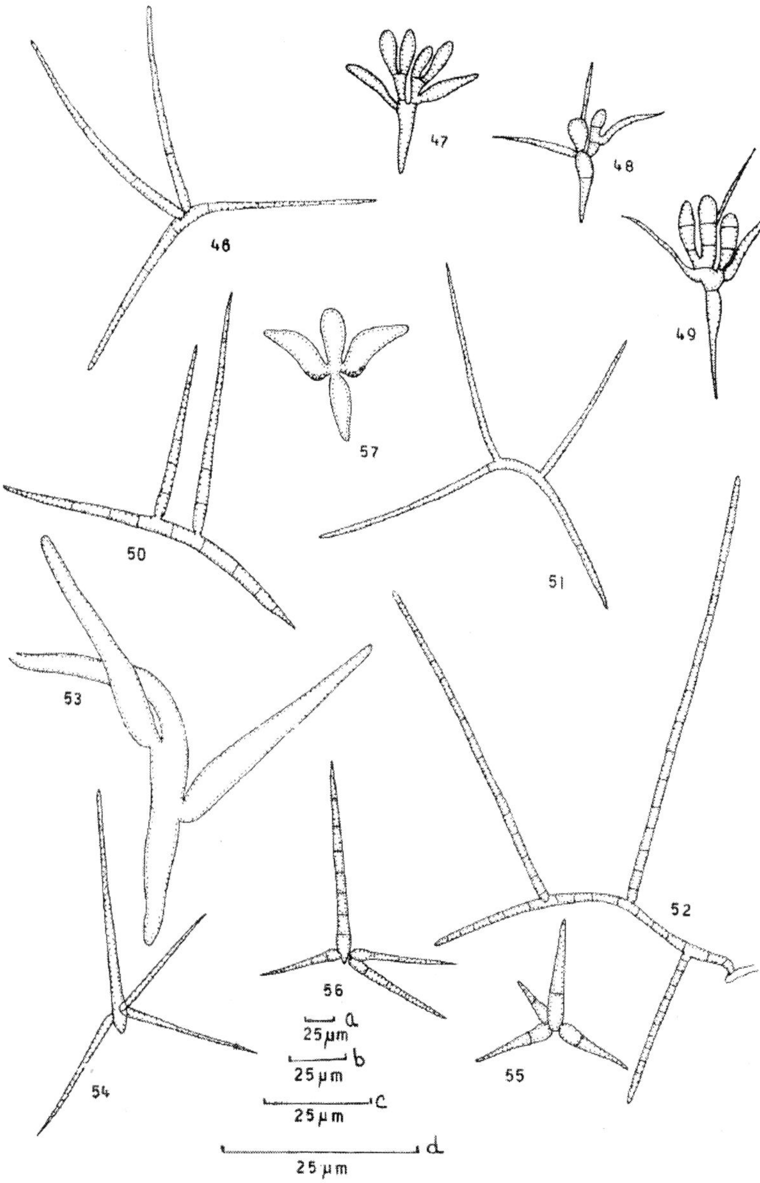

Fig. 46-57. 46- *Tetrachaetum elegans* Ingold, 47- *Tetracladium apiense* Sinclair and Eicker, 48- *Tetracladium marchalianum* de Wildeman, 49- *Tetracladium setigerum* (Grove) Ingold, 50- *Tricladium aciculum* Nawawi, 51- *Tricladium chaetoladium* Ingold, 52- *Tricladium indicum* Sati and Tiwari, 53- *Tricladium splendens* Ingold, 54- *Triscelophorus acuminatus* Nawawi, 55- *Triscelophorus konajensis* Sridhar & Kaveriappa, 56- *Triscelophorus monosporus* Ingold, 57- *Volucrispora ornithomorpha* (Trotter) Haskins

(Fig. 52 in scale a; 46 in scale b; 47,48,49,50,51,53,54,55,56 in scale c and 57 in scale d)

53. *Tripospermum cameleopardus* Ingold, Dann and Mc Dougall, Trans. Brit. Mycol. Soc. 51: 51–56, 1968 (Fig. 45).

It was found as a common species in both the sites. This species was frequently collected from foam samples, as well as from submerged decaying leaf litter of *Quercus leucotrichophora* and *Lantana* sp. in January 2001(KUMS 3301). It created a new record in Uttaranchal.

54. *Triscelophorus acuminatus* Nawawi, Trans. Brit. Mycol. Soc.64: 31, 1975 (Fig. 54).

The present species was found commonly in both the sites. It was isolated from water foam samples and decaying submerged leaf litter of *Pyrus pashia, Pinus roxburghii, Rhododendron arboreum, Cedrus deodara, Acer oblongum, Betula alanoides, Lyonia ovalifolia, Syzigium cumuni* and *Mangifera indica* in October 2000 (KUMS 3401).

55. *Triscelophorus monosporus* Ingold, Trans. Brit. Mycol. Soc., 26:148, 1943 (Fig. 56).

The present species was found commonly. It was isolated from water foam samples, as well as from decaying submerged leaf litter of *Pyrus pashia, Pinus roxburghii, Rhododendron arboreum, Cedrus deodara, Acer oblongum, Betula alanoides, Lyonia ovalifolia, Syzigium cumuni* and *Mangifera indica* in May 2000 (KUMS 3402).

56. *Triscelophorus konajensis* Sridhar & Kaveriappa, Indian Phytopathology, 40:102–105, 1987 (Fig. 55).

The present species was found to be common in both Jageshwar and Jeoli. It was solated from decaying submerged leaf litter of *Polygonum nepalensis* and *Rhododendron arboreum* during July 2000 (KUMS 3403). It created a new record in Uttaranchal.

57. *Volucrispora ornithomorpha* (Trotter) Haskins, Can. J. Microbiol.4: 279, 1958 (Fig. 57).

The present species was found to be restricted to Jageshwar site only, where it was isolated from water foam samples only in November 2000 (KUMS 3501).

ACKNOWLEDGEMENT

The authors are thankful to Prof. S.P. Singh, Head, Department of Botany, Kumaun University, Nainital for providing lab facilities to work.

REFERENCES

Alasoadura, S.O. 1968. Some aquatic Hyphomycetes from Nigeria. Trans.Brit.Mycol. Soc, 51: 535–540.

Anastasion, C.J. 1964. Some aquatic fungi imperfecti from Hawaii. Pacif. Sci., 18: 202–206.

Barlocher, F. and Kendrick, B. 1976. Hyphomycetes as intermediaries of energy flow in streams. In "Recent Advances in Aquatic Mycology" (E.B. Gareth Jones Ed.) 435–446.

Chandrashekhar, K.R., Sridhar, K.R. and Kaveriappa, K.M. 1991. Aquatic hyphomycetes of a sulphur spring. Hydrobiologia, 218: 151–156.

Charmier, A.C. 1985. Cell wall degrading enzymes of aquatic hyphomycetes: a rivew. Botanical Journal of Linnean Society, 91: 67–81.

Conway, K.E. 1970. The aquatic Hyphomycetes of central New York, Mycologia, 62: 516–530.

Descals, E. and Webster, J. 1982a. Taxonomic studies on "Aquatic Hyphomycetes" III Some new species and a new combination. Trans. Brit. Mycol. Soc., 78:405–437.

Descals, E. and Webster, J. 1982b. Taxonomic studies on "Aquatic Hyphomycetes" IV. Pure culture and typification of various species Trans. Brit. Mycol. Soc. 79: 45–64.

Drechsler, Mycologia, 27: 195, 1935.Findley, S.E.G. and Arsuffi, T.L. 1989. Microbial growth and detritus transformations during decomposition of leaf litter in a stream. Freshwater Biol., 21: 261–269.

Hudson, H.J. and Ingold, C.T. 1960. Aquatic Hyphomycetes from Jamaica. Trans. Brit. Mycol. Soc., 73: 109–116.

Ingold, C.T. 1942. Aquatic Hyphomycetes of decaying alder leaves. Trans. Brit. Mycol. Soc., 25: 339–417.

Ingold, C.T. 1956. Stream spora in Nigeria. Trans. Brit. Mycol. Soc., 39: 108–110.

Ingold, C.T. 1965. Hyphomycete spores from mountain torrents. Trans. Brit. Mycol. Soc, 48: 453–458.

Ingold, C.T. 1975. *An illustrated guide to aquatic and water borne Hyphomycetes (Fungi imperfecti) with notes on their biology.* Freshwater Biol. Assoc. Scient. Publ. No. 30 England 96 pp.

Ingold, C.T. 1976. The morphology and biology of freshwater fungi excluding Phycomycetes in "Recent advances in aquatic Mycology" (Ed. E.B. Gareth Jones) Publ. Paul Elek Ltd. 335–357.

Ingold, C.T. and Ellis, E.A. 1952. On some hyphomycetes spores, including those of *Tetracladium maxilliformis,* from Wheatfen. Trans. Brit. Mycol. Soc., 35: 158–161.

Ingold, C.T. and Webster, J. 1973. Some aquatic Hyphomycetes from India. Kavaka, 1: 5–9.

Iqbal, S.H., Sultana K. and Farzana, S. 1973. Stream spora of Pakistan. Biologia, 19: 79–84.

Jooste, W.J. and Merwe, W.J.J. van der. 1990. Ultrastructure of conidiogenesis and conidia of *Anguillospora pseudolongissima* and *Flagellospora penicillioides.* South African Journal of Botany, 56: 319–325.

Kaushik, N.K. and Hynes, H.B.N. 1971. The fate of dead laves that fall into streams. Arch Hydrobiol., 68 : 465–515.

Khan, M.A. 1986. Secondary conidial production in *Articulospora tetracladia.* Pakistan Journal of Botany, 18: 175–177.

Kuthubutheen, A.J. 1987. A new species of *Phalangispora* and further observation on *P. constricta* from Malaysia. Trans. Brit. Mycol. Soc., 89: 414–420.

Madhusudan Rao, M. and Manoharachary, C. 1982. Processing of leaves by fungi in aquatic ecosystem. Indian Phytopathology, 35 : 654–656.

Manoharachary, C. (1985). Aspects and Prospects of water borne conidial fungi from India. Ad. Biol. Res., 4: 160–183.

Manoharachary, C. and Murthy, A.B.S. 1987. Foam spora from Andhra Pradesh, India. Curr. Sci., 50: 378–379.

Marvanova, L. 1975. Concerning *Gyoerffyella* Kol. Trans. Brit. Mycol Soc., 65: 555–565.

Marvanova, L. and Bandoni, R.I. 1987. *Naidella fluitans* gen. et. sp. nov, a conidial basidiomycetes. Mycologia, 79: 573–886.

Marvanova, L. and Descals, E. 1985. New and critical taxa of aquatic hyphomycetes. Bot. J. Linn. Soc., 91: 1–23.

Marvanova, L. and Hywel-Jones, N.L. 2000. *Sigmoidea contorta* sp. nov. and two rare hyphomycete species from streams in Thailand. Cryptogamie, Mycol., 21: 13–16.

Nawawi, A. 1975. *Triscelophorous acuminatus* sp. nov. Trans. Brit. Mycol. Soc, 64: 31, 243, 345–348.

Nawawi, A. 1987a. *Clavariopsis azlanii* sp. nov. A new aquatic hyphomycete from Malaysia. Trans. Brit. Mycol. Soc, 88: 428–432.

Nawawi, A. and Kuthubutheen, A.J. 1987. *Triscelosporium verrucosum* gen. et sp. nov., a dematiaceous aeroaquatic hyphomycetes with tetraradiate conidia. Trans. Brit. Mycol. Soc., 29: 285–290.

Nilsson, S. 1964. Freshwater Hyphomycetes. Taxonomy, Morphology and Ecology. Symb. Bot. Upsal., 18:1–30.

Sati, S.C. and Tiwari, N. 1990. Some aquatic hyphomycetes of Kumaun Himalaya, India. Mycotaxon, 39: 407–414.

Sati, S.C. and Tiwari, N. 1992. A new species of *Tricladium* from Kumaun Himalaya, India. Mycol. Research, 96: 229–232.

Sati, S.C. and Tiwari, N. 1997. Glimpses of conidial aquatic fungi in Kumaun Himalaya in *"Recent Researches in Ecology, Environment and pollution"* X: 17–37. Eds. S.C. Sati, J. Saxena and R.C. Dubey. *Today and Tomorrow printers and Publishers,* New Delhi.

Sati, S.C., Mer, G.S. and Tiwari, N. 1989. Occurrence of waterborne conidial fungi on *Pinus roxburghii* needles, Curr. Science, 58: 918–919.

Sati, S.C., Tiwari, N., and Mer, G.S. 1992. Colonization of aquatic hyphomycetes on pine needles. Indian Phytopath., 45: 106–107.

Sati, S.C., Tiwari, N. and Belwal, M. 2002. Conidial aquatic fungi of Nainital, Kumaun Himalaya, India. Mycotaxon, 81: 445–455.

Sati, S.C. and Pant, N. 2000. Microbial decomposition of leaf litter in a fast flowing fresh water stream of Kumaun Himalaya in *"Microbes, Agriculture, Industry and Environment"* (Eds. D.K. Maheshwari *et. al.*) Pub. Bishan Singh Mahendra Pal Singh, Dehradun, 197–207.

Soosamma, M., Lekha, G., Sreekala, K.N. and Bhatt, D.J. 2001. A new species of Trinacrium from submerged leaves from India. Mycologia, 93: 1200–1202.

Sreekala, K.N. and Bhatt, D.J. 2002. *Dendrospora yessemredda* sp. nov. from freshwater foam. In " Frontiers in Microbial Biotechnology and Plant Pathology". (Eds. C. Manoharacharya, et al.) Scientific Publishers, Jodhpur. 295–298.

Sridhar, K.R.1984. Studies on waterborne fungi of Dakshina Kannada and Kodagu regions. Ph.D. Thesis, Mangalore University.

Sridhar, K.R. and Kaveriappa, K.M. 1982. Aquatic fungi of the Western Ghat forests in Karnataka, Indian Phytopathol, 35:293–296.

Sridhar, K.R. and Kaveriappa, K.M. 1984. Seasonal occurrence of waterborne fungi in Konaje stream (Mangalore), India Hydrobiologia, 119:101–105.

Sridhar, K.R., Chandrashekhar, K.R. and Kaveriappa, K.M. 1992. Research on the Indian Subcontinent. In "The Ecology of Aquatic Hyphomycetes (ed. Bärlocher, F.) Springer-Verlag, Heidelberg. pp.182–211.

Suberkropp, K., and Klug, M.J. 1976. Fungi and bacteria associated with leaves during processing in a woodland stream. Ecology, 57: 707–719.

Subramanian, C.V. and Bhatt, D.J. 1981. Conidia from freshwater foam samples from the Western Ghats, Southern India. Kavaka, 9: 45–62.

Tiwari, N. and Sati, S.C. 1991.Contribution to Indian aquatic hyphomycetes. Two new additions. Acta Botanica Indica, 19: 256 –258.

Tiwari, N. 1992. Taxonomy and species composition of Hyphomycetes in forested streams and their colonisation pattern on tree leaves in Nainital, Central Himalaya. Ph.D. Thesis, Kumaun University, Nainital.

Tiwari, N. and Sati, S.C. 1993. Some Waterborne conidial fungi from a freshwater stream of Nainital, Kumaun Himalaya. Abst. In 80th Indian Science Congress, p 29.

Tubaki, K. 1957. Studies on Japanese Hyphomycetes III Aquatic group. Bull. Nat. Sci. Mus. Tokyo, 41: 294–268.

Webster, J. 1959. Experiments with spores of aquatic Hyphomycetes I. Sedimentation and impaction on smooth surfaces. Ann. Bot. N.S., 23:595–611.

Webster, J. and Descals, E. 1979. The teleomorph of waterborne Hyphomycetes from freshwater. In *"The Whole Fungus"* Vol. 2 (B. Kendrick, Ed.) Ottawa, National Museum of Canada and Kananaskis Foundation. 419–451.

Webster J. and Descals E. 1981. Morphology, distribution and ecology of conidial fungi in freshwater habitats. In the *Biology of Conidial Fungi* (Cole. G.T. and Kendrick, B. eds) N.Y. Academic Press, 295–355.

Webster J., Moran S.T. and Davey R.A. 1976. Growth and sporulation of *Tricladium chaetocladium* and *Lunulospora curvula* in relation to temperature. Trans. Brit. Mycol. Soc., 67: 491–499.

Webster J., Marvanova L. and Eicker A. 1994. Spores from foam from South African rivers. Nova Hedwigia., 57: 379–398.

Willoughby L.G. and Archer J.F. 1973. The fungal spora of a freshwater stream and its colonisation pattern on wood. Fresh Wat. Biol., 3: 219–239.

Willoughby L.G. and Minshall G.W. 1975. Further observation on *Tricladium giganteum.* Trans. Brit. Mycol. Soc., 65: 77–82.

Wolf C.C. 1972. Taxonomic considerations in freshwater Hyphomycetes. Virginia. Polytechnic Institute and State University, Ph.D. 1971.

2

Diversity and Occurrence of Watermoulds in Kumaun Himalaya

P.C. PALIWAL AND S.C. SATI

Department of Botany, D.S.B. Campus, Kumaun University, Nainital, India

SUMMARY

This article deals with a consolidated account of the occurrence and distribution of watermoulds, isolated from water bodies, soils of agricultural or non-agricultural fields, diseased roots and seedlings as well as with diseased fish and their eggs. A total of 91 species of watermoulds, distributed over fifteen genera, which have been reported from different habitats of Kumaun Himalaya are discribed hereunder.

Keywords: Watermoulds, Diversity, Occurrence

Although the term watermoulds suggests that these fungi occur exclusively in water, investigations have shown their occurrence in other habitats too (Mer, 1982; Sati, 1997). The watermoulds are primarily saprophytic in nature (posess the ability to colonise organic animal matter and plant remains) but they can also parasitise algae, fungi, field crop, plant and animals.

A considerable amount of taxonomic and ecological studies have been carried out all over the world by a number of investigators including Coker (1923), Coker and Mathews (1937), Middleton (1943), Johnson (1956,1971), Sparrow (1960), Scott (1961), Waterhouse (1942), Seymour (1970), Robertson (1980) and Dick (1990). In India, a systematic study of watermoulds was initiated by Butler (1907), followed by Chaudhari (1931), Hamid (1942),

Bhargava (1944), Dayal (1961), Rao (1963), Manoharachary (1981), Kiran and Dayal (1983), Murida (1988) and Mishra and Dwivedi (1987). In Kumaun Himalaya, Khulbe (1977, 2001), Sati (1981, 1997), Mer (1982), Verma (1984), Mer and Khulbe (1983, 1984) made an important contribution to watermoulds.

The Kumaun Himalaya, the eastern region of Uttaranchal comprises six districts viz Almora, Bageshwar, Champawat, Nainital, Pithoragarh and US Nagar. It is situated between 28°44¢ and 30 44¢ N latitude and 78°45¢ and 85°1¢ E longitude. It has a distinct geographical entity, ranging from an elevation 204–7436 m and an area of 21035 km^2 with monsoon pattern of rainfall. Thus, it offers a unique opportunity of studying the watermoulds under tropical, subtropical and temperate climatic conditions.

The present paper deals with a comprehensive study of the occurrence and distribution of watermoulds in Kumaun Himalaya, as carried out by previous workers. Altogether, 91 species of watermoulds belonging to fifteen genera of Blastocladiales, Saprolegniales, Leptomitales, Lagenidiales and Peronosporales which have been recovered from different habitats by different investigators are listed hereunder. The species new to India are marked with an asterisk (*), while those new to science, are marked with a double asterisk (**).

Allomyces anomalus Emerson, Llyodia 4:133, 1941.

Collected from forest soil of Haldwani, Ganai, Nainital, Pithoragarh and Ranikhet (Khulbe, 1985) and the waters of Nanak Sagar (Kaur, 1996).

Allomyces arabuscula Butler, Ann. Bot. London 28:1027,1911.

Collected from soil of Ayarpata and Tiffintop, Nainital (Sati and Khulbe, 1980) and fish eggs (Sati, 1981).

Allomyces moniliformis Coker and Braxton, J. Elisha Mitchell Sci. Soc. 42:139, 1926.

Collected from soil and water samples of Haldwani (Chowdhry and Agrawal, 1980).

Allomyces macrogynous Emerson

Isolated from soil of Nainital and Haldwani (Sati, 1990).

Achlya americana Humphrey, Trans. Amer. Phil. Soc. 17: 116, 1892 (1893).

Collected from the waters of Nainital lake (Khulbe, 1977; Mer and Khulbe, 1984) and fish (Sati, 1981).

Achlya apiculata de Bary, Bot. Z. 46:635, 1988.

Collected from the waters of Nainital (Sati, 1992).

Achlya conspicua Coker, The Saprolegniaceae p. 131, 1923.

Collected from Gomati and Kosi river waters (Khulbe, 1985).

Achlya difusa Harvey ex Johnson, J. Elisha Mitchell Sci. Soc. 58:29,1942.

Isolated from rotten seedlings of wheat at Khatima, Nainital (Verma, 1987).

Achlya flagellata Coker, The Saprolegniaceae, p. 116, 1923.

Isolated from Bhimtal lake water (Khulbe, 1977), fish eggs (Sati, 1981) and from soil of tomato, wheat and pea fields near Nainital (Khulbe, 1983 a).

Achlya imperfecta Coker, The Saprolegniaceae, p. 118–119, 1923.

Collected from the waters of Nainital and Bhimtal lake (Khulbe, 1977); fish (Sati, 1981) and soil and rice seedling (Mer, 1982).

Achlya klebsiana Pieters, Bot. Gaz. 60: 486, 1915.

Collected from Naukuchiyatal lake water (Khulbe, 1977); soil of Haldwani (Chowdhry and Agrawal, 1980); seedlings of *Oryza sativa*, sal forest soil of Haldwani and Ramnagar and oak forest soil of Nainital (Khulbe, 1985).

**Achlya oryzae* Ito et Nagai, Jr. Fac. Agri. Hokkaido Imper. Univ. 32,17: 1931.

Collected from the soil of Ramtal, Garurtal and Naukuchyatal (Mer et al., 1980); rhizophere and non-rhizosphere soils of tomato, paddy, pea and wheat near Nainital (Khulbe, 1983a); pine forest soils of Almora, Dwarahat and Nainital and oak forest soils of Naina peak (Khulbe, 1985).

Achlya **orion** Coker and Couch, The Sparolegniaceae p. 112,1923.

Collected from soils of Ramtal, Sattal (Mer 1982); from fish (Sati, 1981) and rhizosphere and non-rhizosphere soils of tomato, paddy, pea and wheat near Nainital (Khulbe, 1983 a).

Achlya prolifera Nees v. Esenbeck Nova Acta Acad. Loop. Carol. 11: 514, 1823.

Collected from soils of Ramtal, Sattal (Mer, 1982); from coldwater fish and fish eggs (Sati, 1981); from pine forest soil of Nainital and sal forest soil of Haldwani (Khulbe, 1985) and paddy fields of Khatima and Nainital (Verma, 1984).

Achlya rodrigueziana Wolf, Mycologia 33:274, 1941.

Collected from cultivated field soils of Nanakmatta (Verma, 1984) and waters of Tumariya dam (Nayal, 1999).

Achlya sp.

Collected from fish and fish eggs (Sati, 1981) and water and soil of Sattal (Mer, 1982)

Aphanomyces helicoides Minder, Kryptogamenphyll. Mark Branderburg, 5:559, 1985.

Collected from fish (Khulbe, 1985) and water (Kaur, 1996).

Aphanomyces keratinophilous (Ook and Kob.) Seymour and Johnson, Mycologia 65 (6): 1312–1318,1973.

Collected from the waters of Khurpatal and Nainital lake (Khulbe, 1985) and the soil of Khurpatal (Sati, 1992).

Aphanomyces laevis de Bary, Johrb. Wiss. Bot. 179, 1860.

Collected from the water of Nainital lake (Khulbe, 1977) fish and fish eggs (Sati, 1981) and forest soils of Kilburry, Nainital and cultivated fields (Khulbe, 1985, 1991).

Aphanomyces scaber de Barry, Johrb, Wis. Bot. 2 : 178, 1860.

Isolated from Nainital lake water (Khulbe and Bhargava, 1977) and infected fry of fish (Kaur, 1996).

Aphanomyces sp. (Sterile)

Collected from fish and fish eggs (Sati, 1981) and soil, and water of Ramtal (Mer, 1982).

****Brevilegnia diclina*** Harvey, J. Elisha Mitchell Sci. Soc. 42: 243, 1927

Collected from fish eggs (Sati, 1981) and soil of Ramtal (Mer, 1982).

Brevilegnia linearis Coker and Braxton, Jr Elisha Mitchell Sci. Soc. 42:214–215, 1927.

Collected from the waters of Nainital lake (Khulbe, 1977) and forest soil of Kilburry, Ramgarh and Nainital (Khulbe, 1985).

Brevilegnia megasperma Harvey, J. Elisha Mitchell Sci. Soc. 45: 322, 1930.

Collected from the soil of Khurpatal (Sati, 1991).

Brevilegnia subclavata Couch, J. Elisha Sci. Mitchell Soc., 42:229, 1927.

Collected from the soil of Ramtal (Mer, 1982); cultivated soil of Khatima (Verma, 1984), pine forest of soil Khairna, oak forest soil of Ramgarh and forest soil of Haldwani (Khulbe, 1985).

Dictyuchus anomalus Nagai, J. Fac. Agri. Hokkaida, Imp. Univ. 32: 1–43, 1931.

Collected from the paddy fields of Nanakmatta (Verma, 1984).

Dictyuchus monosporus Leitgeb, N. Amer. Flora, 2(1): 1–76, 1937.

Collected from soil, roots and seedlings of rice in Khatima, Nainital (Verma, 1984) and the waters of Kosi river (Paliwal, 2004).

** ***Dictyuchus pisci*** Khulbe and Sati, Bibliotheca Mycologica 91:609–617, 1983.

Collected from the infected fish of a fish pond at Jeolikot, Nainital (Khulbe and Sati, 1983).

Dictyuchus pseudodictyon Coker and Baxton, J. Elisha Mitchell Sci. Soc. 42: 139–149, 1926

Collected from the soil of Khurpatal (Sati, 1990)

Dictyuchus sterile Coker, The Saprolegniaceae p151,pol. 52, 1923

Collected from fish (Sati, 1981); soil of Ramtal (Mer, 1982) and waters of Kosi river (Paliwal, 2004).

*****Geolegnia inflata*** Coker and Harvey, J. Elisha Mitchell Sci. Soc. 41: 154, 1925.

Collected from sal forest soil near Haldwani (Khulbe, 1982).

*****Geolegnia septisporangia***

Collected from the soil of Khurpatal, Nainital (Sati, 1990)

Isoachlya anisospora (Cooker and Matthews) var. indica, Saxena and Bhargava, Curr. Sci. 13:79, 1944.

Collected from the waters of Khurpatal lake (Khulbe, 1977).

Isoachlya monilifera (de Bary) Kauffman, Amer. J. Bot. 8: 231, 1921.

Collected from Sariyatal lake water (Khulbe, 1977).

Isoachlya toruloides Kauffman and Coker, Amer. J. Bot. 8: 231, 1921.

Collected from Nainital lake water (Khulbe, 1980).

*****Leptolegnia caudata*** de Bary, Bot. Zeit., 46: 631, 1888.

This species was collected from the infected eggs of fish from Bhimtal lake and the water of Ramtal (Mer et al., 1981).

******Protachlya oryzae*** Khulbe, Curr Sci. 52(9): 428, 1983.

The species was isolated from the infected roots of *Oryza sativa* from Bhimtal (Khulbe, 1983 b).

*****Protachlya paradoxa*** Coer, The Saprolegniaceae, p. 91, pl. 26–28, 1923.

Collected as a parasite of fish (Sati and Khulbe, 1980).

******Saprolegnia bhargavi*** Khulbe and Verma, Nova Hedwigia 44(1–2): 215–219, 1987.

Collected from the cultivated field soil of Khatima (Khulbe and Verma, 1987).

*****Saprolegnia crustosa*** var. III Maurizio,Mitt. d. Dentsch. Fiscerei-Vereims 7:52, 1899.

Collected from Bhimtal lake water (Khulbe, 1977).

Saprolegnia delica Coker, The Saprolegniaceae, p. 30, 1923.

Collected from water of Bhimtal lake (Khulbe, 1977).

Saprolegnia diclina Humphery, Trans. Amer. Phil. Soc. 17:109, 1982 (1893).

Isolated from the water of Khurpatal lake (Khulbe, 1977) fish eggs (Sati, 1981) and soil of RamTal (Mer, 1982).

Saprolegnia ferax (Gruith) Thuret, Ann. Sci. Nat. Bot III, 14: 214, 1850.

Collected from Nainital lake water (Khulbe, 1977); forest soil of Ramtal (Sati and Khulbe, 1980) and tomato field soil at Nanakmatta (Verma, 1984).

****Saprolegnia glomerata*** (Tiesenhousen) Lund, Kgl. Danske. Selsk. Skrift, Naturv, Math. Afd. IX 6(I):14, 1934.

Collected from the soil of Ramtal (Mer, 1982) and cultivated field soil of Sattal (Khulbe, 1985) and Kosi river water (Paliwal, 2004).

Saprolegnia hypogyna (Pringsheim) de Bary, Jahrb. F. Wiss. Bot. 9:191, 1873.

Collected from lake water of Nainital (Khulbe, 1977).

Saprolegnia kauffmaniana Pieters, Bot. Gaz. 60: 488, 1915.

Collected from Nainital lake water (Khulbe, 1977).

Saprolegnia lapponica Gaumann, Botaniska Notiser, p. 156, 1918.

Collected from Nainital lake water (Khulbe, 1977).

****Saprolegnia mixta*** var. asplundii Gaumann, Botaniska Notiser., p. 155, 1918.

Isolated from Nainital (Khulbe, 1977) and fish (Sati, 1981).

Saprolegnia monoica, Pringsheim , Johrb. f. wiss. Bot. 1: 292, 1858.

Collected from the lake water of Khurpatal (Khulbe and Bhargava, 1977) fish (Sati, 1981) and cultivated field soils of Nanakmatta (Verma, 1984).

Saprolegnia parasitica, Coker, The Saprolegniaceae, p. 57–58, 1923.

Collected from the water of Nainital lake (Khulbe, 1977) fish and fish eggs (Sati, 1981) and Ramtal (Mer, 1982).

****Saprolegnia subterranea*** (Dissman) Seymour, Nova Hedwigia, 19: 59–62, 1970.

Isolated from Ramtal (Sattal) lake soils (Mer, 1982) and forest soil of Nainital (Khulbe, 1985).

Saprolegnia sp. (Sterile)

Collected from Nainital lake water (Khulbe, 1977) fish and fish eggs (Sati, 1981) and soil of Ramtal (Mer, 1982).

*****Scoliolegnia centrica*** Khulbe, Sydowia Annales Mycologici Ser. II. 36: 150–153, 1983.

Collected from oak forest soil of Naina peak and Kilbury near Nainital (Khulbe, 1983c).

Thraustotheca clavata (de Bary) Humphrey, Trans. Amer. Phil. Soc., 17:131, 1892 (1893).

Collected from Bhimtal lake water (Khulbe, 1977) fish (Sati, 1981) and soil of Ramtal (Mer, 1982) cultivated soil of Dwarahat and forest soil of Kilburry and Nainapeak (Khulbe, 1985).

****Apodachlya minima*** Coker and Leitner var. hypogyna (Khulbe) Mer and Sati, Curr. Sci. 56(17): 906–907.

Collected from Sattal (Nainital) lake water (Khulbe, 1985) and soil of Naina peak, Nainital (Mer and Sati, 1987).

Olpidiopsis achlyae Mc Larty, Bull. Torrey Bot. Club, 68: 62, 1941.

Collected from the soil and water of Nainital (Sati, 1991), (it was found to be parasitic on the hyphae of *Achlya* spp.)

Olpidiopsis brevispinosa Whiffen, Amer. J. Bot. 29: 610.

Collected from Sariyatal lake, Nainital (Khulbe, 1977) (this isolate was found to be parasitic in the hyphae of *Pythium monospermum*).

Olpidiopsis fusiformis Corun, Ann. Sci. Nat. Bot. V 15: 147, 1852.

Collected from Khurpatal lake and found to be parasite on *Achlya flagellta* (Khulbe, 1977).

Olpidiopsis luxurians Barrett, Ann. Bot. London, 26: 231, 1912.

Collected from Nainital lake it was found parasitic in *Aphanomyces helicoides* von Minden (Khulbe, 1977).

Olpidiopsis pythii (Butler) Karling, Simple Holocarpic Biflagellate Phycomycetes, p. 47, 1942.

Collected from Naukuchiyatal lake (Khulbe, 1977) it was found to be collected as parasitic in the hyphae of *Pythium monospermum.*

Olpidiopsis saprolegniae var. ***levis*** Coker, The Saprolegniaceae p. 185, 1923.

Collected from Nainital lake it was found to be parasitic on *Saprolegnia ferax* Humphrey (Khulbe, 1977) and from Ramtal water; it was found to be parasitic on *Saprolegnia diclina* (Mer 1982).

Pythium afertile Kanouse and Humphrey, Papers Mich. Acad. 8: 129–140, 1927.

Collected from Nainital lake water (Khulbe, 1977), fish eggs (Sati, 1981), soil of Ramtal lake (Mer, 1982) and infected roots and soil from Nagara-tarai (Verma, 1984).

Pythium aphanidermatum (Edson) Fitzpatrick, Jour. Agri. Res. 4: 279–292, 1915.

Isolated from soil of Khatima (Verma, 1984) and cabbage roots of Nainital and Haldwani region (Sati and Tiwari, 1992).

Pythium anguillulaeacti Sadeback, Bot. Zbl. 29 (10) : 318–319, 1887.

Collected from diseased roots of brinjal from Bamori, Haldwani (Bhatt, 2000).

Pythium artotrogus (Mont.) de Bary,

Collected from soil (Mer et al., 1984) and pea roots of Nainital region (Sati, 1990).

Pythium debaryanum Hesse, Inaug Diss. Halle, 14–34, 1874.

Isolated from rice roots and soils from Khatima (Verma, 1984).

Pythium dissotocum Dreschsler, Jour. Wash. Acad. 20: 398–418, 1930.

Isolated from Ramtal soil and from diseased roots and seedlings of *Trigonell foenum-graceum* L. near Nal Damyantital (Mer 1982).

Pythium echinulatum Matthews, Univ. N. Carol. Press. 136 p 1931.

Collected from Ramtal soil (Mer, 1982) diseased roots and soils from Khatima (Verma, 1984) and the waters of Nanaksagar dam (Kaur, 1996).

Pythium elongatum Matthews, Studies on the Genus *Pythium* (Univ. N. Corol. Press): 106, 1931.

Isolated from the forest soil of Kilbury Nainital (Khulbe, 1983 a).

Pythium gracile Schenk, Verh. Phys. Med. Ges. Wizbz. 9:13–20, 1859.

Isolated from Ramtal water (Mer, 1982) and from fish gills (Sati and Khulbe, 1983).

Pythium graminicolum Subramanium Agri. Res. Inst. Pusa (India) Bull. 177: 1–7, 1928.

Isolated from diseased roots of bell-pepper from Bamori, Haldwani (Bhatt, 2000).

Pythium hypogynum Middleton, Mem. Torrey Bot. Club, 20: 69, 1943.

Isolated from diseased wheat roots and soil from Khatima (Verma, 1984) and the water of Nanaksagar dam (Kaur, 1996).

Pythium inflatum Matthews, Univ. N. Carol. Press, 45: 1931.

Isolated from diseased tomato roots and soils from Nagara-tarai (Verma, 1984).

Pythium iwayamai Ito, Trans. Sapporo. Nat. Hist. Soc. 14: 11–33. 1935.

Isolated from diseased roots and collar region of tomato and brinjal from Russy, Nainital and Bamori, Haldwani (Bhatt, 2000).

Pythium marsipium Drechsler, Phytopathology, 31: 478–507, 1941.

Collected from diseased roots of brinjal from Bithoria, Haldwani (Bhatt, 2000).

Pythium middletonii Sparrow, Aqu. Hypo. (Univ. of Mich Press) p. 1187, 1960.

Isolated from Ramtal water and soil (Mer, 1982) and mustard roots (Sati and Tiwari, 1992).

Pythium monospermum Pringsheim Jahrb, Wiss. Bot. I 288–290, 298–299, 1858.

Collected from water (Mer et. al 1984) and forest soil of Nainital (Khulbe, 1985).

Pythium multisporum Poitras, Mycologia, 41: 171–173, 1949.

Isolated from diseased roots and collar region of bell pepper from Russy, Nainital (Bhatt, 2000).

Pythium myriotylum Drechsler, Jour. Wash. Acad. Sci. 20: 404–405, 1930.

Isolated from diseased roots of tomato from Bamori, Haldwani (Bhatt, 2000).

Pythium paroecandrum Drechsler, J. Wash. Acad. Sci., 20: 406–407, 1930.

Isolated from diseased roots of cabbage and cauliflower from Nanakmatta (Verma, 1984).

Pythium polytylum Dreschler, Jour. Wash. Acad. Sci. 20: 398–418,1930.

Isolated from the soil of Nanakmatta (Tarai) Nainital (Verma 1984).

Pythium proliferum de Bary, Jb. Wiss. Bot. 2: 182–189, 1860.

Collected from water (Khulbe, 1977) and diseased roots of chilli from Bamori, Haldwani (Bhatt, 2000).

Pythium pulchrum Minden, Mykol. Unternuch, Ber, R. Falck 1: 146–255, 1916.

Isolated from Ramtal soil and infected seedlings of *Hibiscus esculenta* L near Nal Damyanti tal (Mer, 1982).

Pythium rostratum Butler, Memo.of the Dept. of Agri., India Bot. Series 1: 84, 1907.

Isolated from roots of wheat, rice, cabbage, cauliflower and soil (Mer et al., 1984).

Pythium spinosum Sawada apud Sawada and Chen, Trans. Nat. Hist. Soc. Formosen, 16:199–200, 1926.

Collected from rice fields of Sattal (Mer et al., 1984) and roots of cabbage (Sati and Tiwari, 1992).

Pythium splendens Braun Jour. Agri. Res. 30: 1043–1062, 1925.

Isolated from the diseased roots of chilli and potato from Bithoria, Haldwani and Russy, Nainital (Bhatt, 2000).

****Pythium torulosum*** Coker and Patterson, J. Elisha Mitchell Sci. Soc. 42: 247, 1927.

Isolated from the diseased roots of wheat and soil from Khatima (Verma, 1984) and the water of Nanaksagar dam (Kaur, 1996).

Pythium ultimum Trow, Ann. Bot., 15: 300–301, 1901.

Isolated from the roots of wheat from Nanakmatta (Verma, 1984).

Pythium undulatum Peterson, Bot. Tiddskr. 29: 345–440, 1909.

Collected from the waters of Nainital lake (Khulbe, 1977), fish (Sati,1981), soil (Mer et al., 1984) and rice roots from Nanakmatta (Verma, 1984).

*****Pythium vexans*** var ***minuta*** Mer and Khulbe, Curr. Sci. 54 (14): 735–736, 1983.

Isolated from the soils of Ramtal and Naina peak Nainital (Mer and Khulbe, 1983).

Pythium vexans de Bary, J.R. Agric. Soc. 12 (Ser, 12); 225–1856.

Collected from water and soil Nanakmatta (Kaur, 1996).

REFERENCES

Bhatt, A. 2000. Studies on *Pythium* species in relation to their parasitic association with Solaneous vegetables of Haldwani and Nainital (Kumaun Himalaya). Ph. D. Thesis, Kumaun University, Nainital.

Bhargava, K.S. 1944. Fertilisation in *Isoachlya anisopora* (de Bary) Coker var. *indica* Sak. Et. Bhar. Curr. Sci. 13: 237.

Butler, E.J. 1907. An account of the genus *Pythium* and some Chytridiaceae. Mem. Dept. Agric. India. (Bot. Sci.) 15: 1–160.

Chaudhari, H. 1931. On the fungus parasite of *Spirogyra offinis*. Proc. 18th Indian Sci. Cong.: 261.

Chowdry, P.N. and Agarwal, G.P. 1980. Studies in seasonal variation on aquatic fungi from Delhi. Indian Phytopathology 33 (4): 614–615.

Coker, W.C. 1923. The Saprolegniaceae with notes on other watermoulds. Univ. of North Carolina, Carolina press Chapel Hill, North Carolina, USA: p. 116, 118–119, 131, 151–152, 201.

Coker, W.C. and Matthews, V.D. 1937. Blastocladiales, Monoblepharidales and Saprolegniales, North American Flora 2(1): 1–76.

Dick, M.W. 1990. Key to *Pythium*. Dept. of Botany, Univ. of Reading, U.K.: 64.

Dayal, R. 1961. Development of zoosporangium and liberation of zoospores in *Achlya aplanes* Maurizio var *indica*. Cytologia 26(3–4): 372–377.

Hamid, A. 1942. Indian watermoulds III. Proc. Ind. Acad. Sci. 15 (B): 206–215.

Johnson, T.W.Jr. 1956. The genus Achlya; morphology and taxonomy. Univ. of Michigan Press, Ann. Arbor: 180.

Johnson, T.W.Jr. 1971. Aquatic fungi of Iceland-*Pythium*. Mycologia 63: 517–536.

Kaur, R. 1996. Studies on watermoulds in relation to fish diseases and decomposition in Nanak Sagar Nainital. Ph. D. Thesis, Kumaun University, Nainital.

Khulbe, R.D. 1977. Taxonomic and Ecological studies of water moulds in Nainital and its suburbs. Ph.D.Thesis, Agra Univ., Agra: 113.

Khulbe, R.D. 1980. Occurrence of watermoulds in some lakes of Nainital, India. Hydrobiol. 74: 77–80.

Khulbe, R.D. 1982. *Geolegnia inflata* Coker and Harvey, a new record for Indian aquatic fungi from Nainital, Geobios (New Reports), 1: 35–36.

Khulbe, R.D. 1983a. Aquatic fungal flora in rhizosphere and non rhizosphere solis in some cultivated fields in Nainital. Sci. and Cult., 49: 105–106.

Khulbe, R.D. 1983b. *Protoachlys oryzae* sp. nov. from Nainital Current Sci., 52(9): 428.

Khulbe, R.D. 1983c. *Scolilegnia centrica*—a new watermould from India. Sydowia, Annales Mycologici Ser. II. 36: 150–153.

Khulbe, R.D. 1985. Studies in Aquatic Phycomycetes of Nainital in relation to Taxonomy, Physiology, Ecology and Pathology. D.Sc. Thesis, Kumaun Univ., Nainital: 335.

Khulbe, R.D. 2001. A manual of aquatic fungi (Chytridiomycetes & Oomycetes). Daya Publishing Housing House, Delhi: 255.

Khulbe, R.D. and Bhargava, K.S. 1977. Distribution and seasonal periodicity of watermoulds in some lakes in Nainital hills, India. Hydrobiologia 54(1): 67–72.

Khulbe, R.D. and Sati, S.C. 1983. A new species of *Dictychus* and key to the other species. Bibliotheca Mycologia 91; 609–617.

Khulbe, R.D. and Verma, B.L. 1987. *Saprolegnia bhargavii* sp. nov. from India. Nova Hedwigia, 44: 215–219.

Kiran, U. and Dayal, R. 1983. Fresh watermoulds from Varanasi (India) VI. Some Oomycetes. Proc. Nat. Acad. Sci. India. 55(B): 50–52.

Manoharachary, C. 1981. The taxonomy and ecology of freshwater Phycomycetes form India. Indian Rev. Life Sci. 1: 3–21.

Mer, G.S. 1982. Taxonomic and Physiological studies of water moulds of Sattal, Nainital. Ph. D. Thesis.

Mer, G.S. and Khulbe, R.D. 1983. *Pythium vexans* var. *minuta* from Kumaun Himalaya, India. Curr. Sci. 56: 735–736.

Mer, G.S. and Khulbe, R.D. 1984. Water moulds of Sattal (Nainital), India. Sydowia, Annales Mycologic. Ser. II. 37: 195–207.

Mer, G.S. and Sati, S.C. 1987. An unusual strain of *Apodachlya minima* Coker and Leitner, Curr. Sci., 56: 906–907.

Mer, G.S., Sati, S.C. and Khulbe, R.D. 1980. Occurrence, distribution and seasonal periodicity of some aquatic fungi of Sattal (Nainital), India. Hydrobiol. 76: 200–205.

Mer, G.S., Sati, S.C. and Khulbe, R.D. 1981. An addition to Indian aquatic fungi. Ind. Phytopatho. 34: 387–388.

Mer, G.S., Verma, B.L. and Khulbe, R.D. 1984. Aquatic fungi of Kumaun Himalaya., India, *Pythium* Pringhsheim. Sydowia 37: 208–221.

Middleton, J.T. 1943. The taxonomy, host range and geographical distribution of the genus *Pythium*. Mem. Torrey Bot. Club. Mem. 20: 1–171.

Mishra, R.C. and Dwivedi, R.S. (1987). Aquatic moulds from Gujar lake, Jaunpur-II. J. Ind. Bot. Soc. 66:203–208.

Murida, M.S. 1983. Cytological studies of certain members of the family, Saprolegniaceae. Part I. J. Bot. Soc. 17 (5): 301–310.

Nayal, D.S. 1999. Studies on Aquatic fungal flora of Tumriya Dam and Bhimtal lake with special reference to fish disease. Ph. D. Thesis, Kumaun University, Nainital.

Paliwal, P.C. 2004. Studies on Water Pollution of Kosi river of Kumaun Himalaya. Ph. D. Thesis, Kumaun University, Nainital.

Rao, V.G. 1963. An account of the genus *Pythium* Pringhsheim in India. Mycopathologia. Myco. Appl. 2(1): 45–59.

Robertson, G.I. (1980). The genus *Pythium* in New Zealand. New Zealand J. of Botany.18: 73–102.

Sati, S.C. 1981. Aquatic fungi of Kumaun in relation to fish infection. Ph.D. Thesis. Kumaun University, Nainital: 111.

Sati, S.C. 1990. An integrated study on taxonomy, ecology and pathology of zoosporic fungi with reference to Nainital, Kumaun Himalaya, U.G.C. Project Report.

Sati, S.C. 1991. Aquatic fungi parasitic on temperate fishes of Kumaun Himalaya India. Mycoses. 34: 437–441.

Sati, S.C. 1991a. Two fungal parasites of the eggs of *Tor tor* Ham. in Nainital lake of Kumaun Himalaya. Science and Culture, 57: 96–97.

Sati, S.C. 1991 b. Aquatic fungi parasitic on temperate fish of Kumaun Himalaya, India. Mycoses, 34: 437–441.

Sati, S. C. 1992. Watermoulds parasitic on coldwater fish of Nainital in Kumaun Himalaya, India, Abstract in XV International Botanical Congress, Tokyo, Japan, (2384) p. 324.

Sati, S.C. 1997. Diversity of aquatic fungi in Kumaun Himalaya: Zoosporic Fungi. In "Recent Researches in Ecology, Environment and Pollution" Vol. X: (eds. S. C. Sati, J. Saxena and R.C. Dubey), Today and Tomorrow's Printers and Publ., New Delhi: 1–16.

Sati, S.C. and Khulbe, R.D. 1980. Occurrence of aquatic fungi in some soils of Nainital Hills. Geobios 7: 42–43.

Sati, S.C. and Khulbe, R.D. 1983. *Pythium gracile* Schenk, as parasite on fish gills. Ind. Phytopath. 36: 587–588.

Sati, S.C. and Mer, G.S. 1980. Fungal flora of Kumaun region-Aquatic Mastigomycetes. Ind. Phytopath. 41(1): 39–43.

Sati, S.C. and Tiwari, N. 1992. Studies on five species of *Pythium* parasitic on mustard and cabbage in India. Mycopathologia, 119: 97–100.

Seymour, R.L. (1970). The genus *Saprolegnia*. Nova Hedwigia. Z. Kryptogamemkd 19: 1–124.

Scott, W.W. 1961. A monograph of the genus *Aphanomyces*. Vir. Agr. Exp. Sta. Tech. Bull. 115: 95.

Sparrow, F.K.Jr. 1960. Aquatic Phycomycetes (2nd ed.). University of Michigan Press Ann Arbor: 1187.

Verma, B.L. 1984. Studies on lower Phycomycetes in some agricultural fields of Tarai, Nainital with special reference to Saprolegniaceae and Pythiaceae. Ph.D., Thesis, Kumaun Univ., Nainital.

Verma, B.L. 1987. Zoosporic fungi in the wheat fields of Kumaun Himalaya. Madras Agri. Jour.74(1): 1–5.

Waterhouse, G.M. 1942. Some watermoulds of the Hogsmill river collected from 1937–1939. Trans. Brit. Mycol. Soc. 25: 315–325.

3

Aeromycospora of Agra City

S.V.S. CHAUHAN AND ANTIMA KULSHRESTHA

Department of Botany, School of Life Sciences
Dr. B.R. Ambedkar University, Agra-282002 (U.P.), India

SUMMARY

Agra, situated at the confluence of the Yamuna has semi-arid tropical climate with scrub type jungle vegetation. In the present investigation, an aeromycological survey of Agra city was performed at different sites. The role of different environmental parameters in a variety of air spora composition was worked out. A total number of 102 fungal types were recorded and among them *Alternaria* and *Aspergillus* were found to be the major fungal spore types. The maximum number of fungal population in outdoor areas were encountered from Dayal Bagh (37415 spores / m^3, 33370 spores / m^3) and Taj Mahal area (35595 spores / m^3, 33650 spores / m^3 in the year 2000 and 2001 respectively), followed by vegetable and fruit markets, Paliwal Park, Sikandara, Red Fort, Motilal Nehru Park, Mankameshwar temple and Ravali temple. From the indoor environments the fungal spores were better represented in Medical College (103936 spores / m^3) than in District Hospital (70952 spores / m^3).

Keywords: Aeromycospora, Agra, *Alternaria, Aspergillus*

Air carries innumerable spores, mycelia of fungi and pollen. Fungal spores contribute to a major portion of air spora. The study of airborne fungal spores is important for biodeterioration and the diseases that they cause in plants and animals, including man. The seasonal incidence and concentration of fungal spores varies, both in quality and quantity, from place to place, depending on the local flora, topography of the landscape and human interference. Seasonal and diurnal variations of airborne fungal spores have been studied in Jabalpur (George and Verma, 2000), Burdwan (Chakraborty *et al.*, 2000), Nagpur (Kakde *et al.*, 2001), Baroda (Arya *et al.*, 2001) and Lucknow (Khandelwal, 2001).

The present investigation has been undertaken to study the Aeromycospora of Agra city. The study is indispensable as Agra is well known for its beautiful historical monuments and tourism. The tourists can take an idea from it, if they are conscious about microbial pollution.

MATERIALS AND METHOD

Study Area

The city of Agra is situated at 27°10′ and 27°13′ Northern latitude and 77°54′ and 78°4′ longitude at the eastern edge of the extensive plains of the confluence of the Yamuna river. The climate is characterised by a dry and increasingly hot season from March to June, a warm monsoon period from July to October and dry and a cold winter from November to February. The monthly temperature of January ranged between 7.4–21.1°C and 5.3–20.6°C in 2000 and 2001, respectively. It increased gradually in the subsequent months upto May, where it was recorded between 27.8°C–41.6°C and 27.2–40.5°C, in 2000 and 2001, respectively. There was a decline afterwards with minor fluctuations in the months of August, September and October. Relative humidity remained low during the summer months when it was recorded between 46–47% in the month of May. It showed a gradual increase from July onwards, ie, from 83–91% in July, and 80–88% in December. The average rainfall of the city was 384.9 mm. (Table1).

The study was carried out both in field as well as in laboratory conditions. The observations were made from 1st January, 2000 to 31 December, 2001 from outdoor areas.

Sampling Sites

The samples were collected from different sites for extramural (outdoor) and intramural (indoor) aeromycological study (Fig. 1).

Outdoor Environments

Historical monuments and their adjacent areas: (1) Taj Mahal (2) Red Fort (3) Sikandara and (4) Dayalbagh.

Market : (5) Vegetable and fruit market.

Gardens : (6) Moti Lal Nehru Park (7) Paliwal Park.

Temples : (8) Ravali (9) Mankameshwar

Indoor Environments

Hospitals : (10) District Hospital (11) S.N. Medical College.

Sampling Method

Air spora of different sites were studied by the Rotorod sampler for a period of two years. The Rotorod sampler was run for half an hour once a month at a constant height of 1 metre above ground level at each site between 10–11 am. The exposed tape strips were mounted on slides by using glycerine jelly as mountant.

Scanning

It was carried out as described by Tilak and Srinivasu (1967).

Identification

Identification of fungal spores was done on colony characteristics, growth and morphology of fungal spores under microscope (10x–40x), and was subsequently confirmed after consulting relevant literature by Gilman (1585), Burnett and Hunter (1972), Ellis and Ellis (1985), Tilak (1989) and Sutton *et al.* (1998).

Exposed Gravity Petriplates Method

For the confirmed idenfication of fungal spores, petriplates containing Potato Dextrose Agar (PDA) medium and Sabourauds Dextrose Agar (SDA) medium were exposed horizontally at the same time, when the sampler was operated at each site for a period of 15 minutes. After exposure, the petriplates were incubated at room temperature. The fungal colonies developed in petriplates were counted, isolated by subculturing and then identified.

RESULTS AND DISCUSSION

The results of the experiments obtained during the period of investigation revealed a wide range of airborne fungal components. Apart from fungal spores, miscellaneous bioparticles like pollen grains, fungal hyphae, algal parts and insect scales were also seen. As many as 102 types of fungal spores were identified. Their taxonomic classes and percentage contribution in the aeromycospora is given in Table 3.1.

Maximum concentration of fungal population from outdoor environments was encountered from Dayalbagh and its adjacent areas (37415 spores/m^3 33370 spores/m^3), followed by Taj Mahal and adjacent areas (35595 spores/m^3, 33650 spores/m^3), vegetable and fruit markets (27755 spores/m^3, 22645 spores/m^3), Paliwal Park (25180 spores/m^3, 28450 spores/m^3), Sikandara and adjacent area (20595 spores/m^3, 22745 spores/m^3), Red Fort (15810 spores/m^3, 18315 spores/m^3), Motilal Nehru Park (12540 spores/m, 16525 spores/m^3), Mankameshwar temple (14805 spores/m^3, 10310 spores/$m^{3)}$ and Ravali temple (9410 spores/m^3, 7535 spores/m^3), in the year 2000 and 2001 respectively. From the indoor

Table 3.1: Spore Spectra Representing Various Taxonomic Classes in the Air

S. No.	Name of the spores classes	Total No. of Fungal types represented from each class	Contribution (%)
1	Myxomycetes	1	0.98
2	Phycomycetes	4	3.92
3	Ascomycetes	30	29.41
4	Basidiomycetes	4	3.92
5	Dueteromycetes	63	61.76
	Total	102	99.99

Table 3.2: Sampling Sites and Fungal Air Spora Composition

Site no.	Sampling sites	Fungal Population Spores/m^3	
		Year 2000	Year 2001
1	Taj Mahal and adjacent area	35595	33650
2	Red fort and adjacent area	15810	18315
3	Sikandara and adjacent area	20585	22745
4	Dayalbagh and adjacent area	37415	33370
5	Vegetable and Fruit market	27755	22645
6	Motilal Nehru Park	12540	16525
7	Paliwal Park	25180	28450
8	Ravali temple	9410	7535
9	Mankameshwar temple	14805	10310
10	District Hospital	70959	–
11	S.N. Medical College	103936	–

environments, the fungal spores were better represented in Medical College (103936 spores/m^3) than in District Hospital (70952 spores/m^3) (Table 3.2).

OUTDOOR ENVIRONMENTS

Historical Monuments

Taj Mahal and Adjacent Areas

At this site, a total of 23 fungal types were recorded (Table 3.3). *Aspergillus* was found to be the dominant type as it contributed 23.25% to the total air spora. Out of all species of *Aspergillus*, in maximum concentration *A. flavus*. After *Aspergillus*, *Epicoccum purpurascens* (12.42%) was found to be dominant, followed by *Alternaria alternata* (12.32%), *Curvularia lunata* (7.95%), *Cladosporium herbarum* (6.85%), *Penicilliun citrinum* (5.46%), *Trichoderma* sp. (4.77%) and

Table 3.3: Number, types, percentage contribution and seasonal frequency of colonies trapped on exposed Petriplates at Taj Mahal and adjacent areas during the investigation period between January, 2000 to December, 2001.*

S.No.	Fungal Types	No. of colonies	Contribution (%)	Seasonal Frequency		
				Rain	Winter	Summer
Phycomycetes						
1.	*Cunnighamella* sp.	6	0.59	+	–	–
2.	*Rhizopus arrhizus*	32	3.18	+ +	+	+
3.	*Rhizopus microsporus*	14	1.39	+	–	+
Deuteromycetes						
4.	*Alternaria alternata*	124	12.32	++	+ + +	+ +
5.	*Aspergillus flavus.*	100	9.94	+ + +	+	+ + +
6.	*Aspergillus niger*	90	8.94	+ + +	+ +	+ + +
7.	*Aspergillus terreus*	30	2.98	+ +	+	+
8.	*Aspergillus ustus*	14	1.39	+	–	+
9.	*Aureobasidium pollulans*	12	1.19	–	+ +	–
10.	*Candida* sp.	24	2.38	+	–	+ +
11.	*Cladosporium herbarum*	69	6.85	+ + +	+ + +	+
12.	*Curvularia lunata*	80	7.95	+ +	+ + +	+
13.	*Collectotrichum* sp.	15	1.49		+	–
14.	*Epicoccum purpurascens*	125	12.42	+ +	+ + +	+
15.	*Fusarium oxysporium*	30	2.98	+	+	+ +
16.	*Geotrichum* sp.	16	1.59	–	+	–
17.	*Helminthosporium*	44	4.37	+ + +	–	+
18.	*Nigrospora* sp.	24	2.38	+ +	+	–
19.	*Penicillium citrinum*	55	5.46	+ +	+ +	+
20.	*Penicillium purpurogenum*	13	1.29	+	+	–
21.	*Trichoderma* sp.	48	4.77	+ +	–	++
22.	*Trichothecium roseum*	25	2.48	+	+ +	–
23.	*Ulocladium* sp.	2	0.19	–	+	–
	Unidentified colonies	14	1.39			
	Total	1006	99.91			

– absent, + Present, + + moderately present, + + + abundantly present.
* Average data of two years

Helminthosporium sp. (4.37%), while *Ulocladium* sp. (0.19%) was found in least concentration, (Table 3.3).

Though the colonies of *Aspergillus* were recorded throughout the year, but their abundance was more in summer and the rainy season. The colonies of *E. purpurascences, A. alternata, P. citrinum and C. lunata* were more frequent in winter and the rainy months, while *Aureobasidium* pollulans, *Geotrichum sp.* and *Ulocladium sp.* were restricted to the winters. (Table 3.3).

Red Fort and Adjacent Area

It is clear from Table 3.4 that 21 fungal types belonging to 17 genera were recorded from the Red Fort area. In all, 674 fungal colonies were trapped. *Aspergillus* sp. was abundantly recorded at this site and contributed to 41.37% of the total air spora. Among all species of *Aspergillus, A. niger* was recorded most frequently with 29.67% contribution throughout the year, particularly in summer and the rainy season. After *A. niger, Cladosporium herbarum* occupied second rank and contributed 18.10% followed by *Alternaria alternata* (7.12%), *Epicoccum purpurascens* (5.19%), *A. restrictus* (4.74%) and *Curvularia lunata*

Table 3.4: Number, types, percentage contribution and seasonal frequency of colonies trapped on exposed Petriplates at Red Fort and adjacent area during the investigation period between January, 2000 to December, 2001.*

S.No.	Fungal Types	No. of Colonies	Contribution (%)	Seasonal Frequency		
				Rain	Winter	summer
Phycomycetes						
1.	*Mucor* sp.	4	0.59	+	–	+
2.	*Rhizopus arrhizus*	9	1.33	+	–	–
Ascomycetes						
3.	*Chaetomium* sp.	19	2.81	+ +	–	+
Deuteromycetes						
4.	*Alternaria alternata*	48	7.12	+	+ +	+ +
5.	*Alternaria* sp.	20	2.96	–	+ +	–
6.	*Aspergillus flavus*	25	3.70	+	+	+ +
7.	*Aspergillus niger*	200	29.67	+ + +	+ +	+ + +
8.	*Aspergillus restrictus*	32	4.74	+ +	+ +	–
9.	*Aspergillus terreus*	22	3.26	+ +	+	+
10.	*Aureobasidium pollulans*	7	1.03	–	+	+
11.	*Cladosporium herbarum*	122	18.10	+ +	+ +	+
12.	*Curvularia lunata*	26	3.85	+ +	+ +	+
13.	*Candida* sp.	9	1.33	–	–	+
14.	*Drechslera* sp.	5	0.74	–	+	+
15.	*Epicoccum purpurascens*	35	5.19	+	+ +	+
16.	*Fusarium oxysporum*	30	4.45	+	–	+ +
17.	*Helminthosporium* sp.	10	1.48	+	–	+
18.	*Nigrospora* sp.	14	2.07	+	+ +	–
19.	*Penicillium citrinum*	24	3.56	+	+ + +	–
20.	*Scolecobasidium* sp.	6	0.89	–	+	+
21.	*Trichothecium* sp.	7	1.03	–	+	–
	Total	674	99.9			

– absent, + present, + + moderately present + + + abundantly present.
*Average data of two years.

(3.85%), while *Scolecobasidium* sp. contributing 0.89% and *Mucor* sp. (0.59%), were recorded at minimum concentrations. *Alternaria alternata, C. herbarum* and *P.citrinum* were abundantly recorded in the winter months. On the other hand, *A. niger, A. flavus, Chaetomium* sp. and *R. arrhizus* were recorded in the rainy season. *Alternaria sp., Aureobasidium pollulans* and *Trichothecium* sp. were trapped only in the winter months.

Sikandara and Adjacent Areas

Table 3.5 shows that 556 fungal colonies were recorded from Sikandara total number of 25 fungal colonies belonging to 19 genera were recorded. *Aspergillus niger* was the dominant type and contributed 15.82% to the air spora, followed by *Alternaria alternata* (13.10%), *Cladosporium herbarum* (11.69%), *Aspergillus flavus* (10.79%), *Curvularia lunata* (6.74%), *Helminthosporium* sp. (5.03%) and *Epicoccum purpurascens* (4.85%).

Only one member of Ascomycetes and Basidiomycetes ie *Chaetomium* sp. (2.51%) and *Schizophyllum commune* (0.35%) were recorded respectively from this site. *S. commune* was trapped in the month of September. This fungus was recorded for the first time in Agra city. *Aureobasidium pollulans, Geotrichum* sp. and *Nigrospora* sp. were restricted to the winter season, while *Rhizopus arrhizus, S. commune, Pestalotiopsis* sp. and *Trichoderma* sp. were recorded only in the rainy season. On the other hand, *Bipolaris* sp. and *Candida* sp. were trapped in summer months only (Table 3.5).

Dayalbagh and Adjacent Areas

It is clear from Table 6 that 1105 fungal colonies were recorded from Dayalbagh area. A total of 32 fungal species belonging to 23 genera were recorded. *Aspergillus flavus* (13.84%) was the most dominant type, followed by *A. niger* (8.95%), *Alternaria alternata* (8.59%), *Cladosporium herbarum* (8.05%), *Curvularia lunata* (6.87%) and *Epicoccum purpurascens* (6.24%). The colonies of these fungi were recorded throughout the year. On the other hand, *Arthrinium* sp. was trapped in least numbers (0.54%) during the rainy and winter season.

The colonies of *Cunnighamella* sp. and *Helminthosporium* sp. were recorded only in the rainy season, *Aureobasidium pollulans, Collectotrichum* sp. and *Geotrichum* sp. in the winter season; and *Aspergillus candidus* and *Fusarium oxysporum* were found only in the summer season. *A. alternata, Alternaria* sp., *Aspergillus restrictus, C. herbarum, C. lunata, E. purpurascens, P. citrinum* and *Trichothecium roseum* were moderately recorded in the winter season (Table 3.6).

Market

Vegetable and Fruit Markets

It is clear from Table 3.7 that a total of 1201 fungal colonies belonging to 33 fungal types were trapped from vegetable and fruit market. *Aspergillus* species contributed 25.29% to the total air spora. Among all species of *Aspergillus,*

Table 3.5: Number, types, percentage contribution and seasonal frequency of colonies trapped on exposed Petriplates at Sikandara and adjacent area during the investigation period between January, 2000 to December, 2001.*

S.No.	Fungal types	No. of colonies	Contribution (%)	Seasonal Frequency		
				Rainy	Winter	Summer
Phycomycetes						
1.	*Rhizopus arrhizus*	7	1.25	+	–	–
2.	*Rhizopus microporus*	10	1.79	+	–	+
Ascomycetes						
3.	*Chaetomium* sp.	14	2.51	+	–	+
Basidiomytes						
4.	*Schizophyllum commune*	2	0.35	+	–	–
Deuteromycetes						
5.	*Alternaria alternata*	73	13.10	++	+++	+
6.	*Alternaria* sp.	24	4.31	+	+++	–
7.	*Aspergillius flavus*	60	10.79	++	+	+++
8.	*Aspergillius niger*	88	15.82	+++	+	+++
9.	*Aspergillius restrictus*	22	3.95	+	++	–
10.	*Aspergillius terreus*	11	1.97	+	+	+
11.	*Aureobasidium pollulans*	4	0.71	–	+	–
12.	*Bipolaris* sp.	2	0.35	–	–	+
13.	*Candida* sp.	5	0.89	–	–	+
14.	*Cladosporium herbarum*	65	11.69	+	+++	+
15.	*Curvularia lunata*	36	6.47	+	++	+
16.	*Epicoccum purpurascens*	27	4.85	+	++	–
17.	*Fusarium oxysporum*	15	2.69	+	–	++
18.	*Fusarium moniliforme*	5	0.89	+	–	+
19.	*Geotrichium* sp.	7	1.25	–	+	–
20.	*Helminthosporium* sp.	28	5.03	++	+	+
21.	*Nigrospora* sp.	13	2.33	–	++	–
22.	*Penicillium citrinum*	12	2.15	–	+	+
23.	*Pestalotiopsis* sp.	2	0.35	+	–	–
24.	*Pithomyces* sp.	12	2.15	+	+	–
25.	*Trichoderma* sp.	3	0.53	+	–	–
	Undefined colonies	9	1.61	–	–	–
	Total	556	99.78			

– absent, + present, ++ moderately present, +++ abundantly present.
* Average data of two years.

A. niger (11.99%) contributed maximum spora and was recorded throughout the year, more so in the rainy and summer season. *Rhizopus* species occupied second rank. Three species of *Rhizopus* were recorded, among which *R. arrhizus* was the most abundant (8.49%).

Table 3.6: Number, types, percentage contribution and seasonal frequency of colonies trapped on exposed Petri plates at Dayalbagh and adjacent area during the investigation period between January, 2000 to December, 2001.*

S.No.	Fungal types	No. of colonies	Contribution (%)	Seasonal Frequency		
				Rain	Winter	Summer
Phycomycetes						
1.	*Cunnighamella* sp.	8	0.72	+	–	–
2.	*Mucor* sp.	12	1.08	+	–	+
3.	*Rhizopus arrhizus*	34	3.07	+ +	+	+
4.	*Rhizopus schipperae*	7	0.63	+	–	–
Ascomycetes						
5.	*Chaetomium*	12	1.08	+	+	–
Deuteromycetes						
6.	*Alternaria Alternata*	95	8.59	+ +	+ + +	+
7.	*Alternaria* sp.	32	2.89	+	+ +	–
8.	*Arthrinium* sp.	6	0.54	+	+	–
9.	*Aspergillus candidus*	12	1.08	–	–	+
10.	*Aspergillus flavus*	153	13.84	+ + +	+	+ + +
11.	*Aspergillus niger*	99	8.95	+ +	+	+ + +
12.	*Aspergillus restrictus*	39	3.52	+ +	+	–
13.	*Aspergillus ustus*	24	2.17	+	–	+ +
14.	*Aspergillus sydowii*	8	0.72	+	+	–
15.	*Aureobasidium pollulans*	12	1.08	–	+	–
16.	*Cladosporium herbarum*	89	8.05	+ +	+ + +	+
17.	*Colletotrichum* sp.	7	0.63	–	+	–
18.	*Curvularia lunata*	76	6.87	+ +	+ +	+
19.	*Epidermophyton* sp.	29	2.62	+	+	–
20.	*Epicoccum purpurascens*	69	6.24	+ +	+ +	+
21.	*Fusarium oxysporum*	12	1.08	–	–	+
22.	*Fusarium moniliforme*	9	0.81	+	–	+
23.	*Geotrichum* sp.	13	1.17	–	+	–
24.	*Helminthosporium* sp.	30	2.71	+ +	–	–
25.	*Nigrospora* sp. 12	1.08	+	+	–	
26.	*Ochroconis* sp.	18	1.62	+	+	–
27.	*Penicillium citrinum*	47	4.25	+	+ +	–
28.	*Penicillium purpurascens*	24	2.17	+	+	–
29.	*Pithomyces* sp.	21	1.90	+	+	–
30.	*Stemphylium* sp.	9	0.81	+	+	–
31.	*Trichoderma* sp.	29	2.62	+ +	+	–
32.	*Trichothecium roseum*	36	3.25	+	+ +	–
	Unidentified colonies	22	1.99	–	–	–
	Total	1105	99.83			

– absent, + present, ++ moderately present, +++ abundantly present. * *Average data of two years.*

Epicoccum purpurascens (8.74%), *Alternania alternata* (7.91%), *Curvularia lunata* (7.07%) and *Penicillium citrinum* (4.57%) were recorded throughout the year, with abundance in the winter season. No distinct periodicities have been observed in the occurence of *Nigrospora* sp., *Helminthosporium* sp. and *Trichothecium* sp. On the other hand, *Arthrinium* sp., *Aspergillus restrictus, Geotrichum* sp., *Ochroconis* sp. and *Stemphylium* sp. were recorded only in the winter season (Table 3.7).

Garden

Motilal Nehru Park

Table 3.8 shows that colonies of 26 fungal species were recorded. *Aspergillus niger* (17.73%) was the dominant species, followed by *Alternaria alternata* (12.31%), *Cladosporium herbarum* (10.18%), *Aspergillus flavus* (8.37%), *Epicoccum purpurascens* (6.56%), *Fusarium moniliforme* (4.43%), *Helminthosporium* sp. (4.26%) and *Rhizopus arrhizus* (4.10%), while *Syncephalastrum* sp. (0.49%), *Geotrichum* sp. and *Scolecobasidium* sp. each contributed (0.65%) and *Ulocladium* sp. (0.49%) showed the minimum contribution.

Aureobasidium pollulans, Geotrichum sp., *Nigrospora* sp., *Scolecobasidium* sp., and *Trichothecium roseum* were restricted to winters. On the other hand, *Ulocladium* sp. and *Syncephalastrum* sp. were recorded only in the rainy and summer seasons, respectively (Table 3.8).

Paliwal Park

It is clear from Table 3.9 that 27 fungal types belonging to 19 genera were recorded from Paliwal Park. A total of 804 fungal colonies were trapped. In this site also, *Aspergillus* was recorded as the dominant type. Among all species, *Aspergillus niger* was found to be dominant, with 22.38% contribution, followed by *Cladosporium herbarum* (8.95%), *Aspergillus terreus* (5.59%), *Alternaria alternata* (5.47%), *Curvularia lunata* (5.34%), *Penicillium citrinum* (4.22%), while *Ulocladium* (0.24%) showed minimum contribution.

Most of the fungal colonies were trapped during the months of October to February. *Aspergillus* sp. were recorded more or less throughout the year, but colonies of *Cladosporium herbarum* were frequently observed in the winter season. *Rhizopus* sp. and *Mucor* sp. were recorded in the summer and rainy season. *Geotrichum* sp. and *Trichothecium roseum* were observed only in the winter season. On the other hand, *Helminthosporeum* sp., *Trichoderma* sp. and *Ulocladium* sp. were recorded in rainy season only (Table 3.9).

Temples

Ravali Temple

It is clear from Table 3.10 that 14 fungal species belonging to 12 genera were trapped from Ravali temple. *Aspergillus* species were recorded abundantly and

Table 3.7: Number, types, percentage contribution and seasonal frequency of colonies trapped on exposed Petriplates at Vegetables and Fruits Market during the investigation period between January,2000 to December, 2001.*

S.No.	Fungal types	No. of colonies	Contribution (%)	Seasonal Frequency		
				Rain	Winter	Summer
Phycomycetes						
1.	*Cunnighamella* sp.	10	0.83	+	–	–
2.	*Mucor* sp.	18	1.49	+	–	+
3.	*Rhizopus arrhizus*	102	8.49	+ + +	+ +	+ + +
4.	*Rhizopus mocrosporus*	35	2.91	+ +	–	+
5.	*Rhizopus schipperae*	37	3.08	+ +	+	+
6.	*Syncephalastrum* sp.	7	0.58	+	–	–
Ascomycetes						
7.	*Chaetomium* sp.	17	1.41	+ +	–	–
Deuteromycetes						
8.	*Alternaria alternate*	95	7.91	+	+ + +	+ +
9.	*Arthrinium* sp.	7	0.58	–	+	–
10.	*Aspergillus candidus*	10	0.83	+	–	–
11.	*Aspergillus flavus*	100	8.32	+ + +	+	+ + +
12.	*Aspergillus glaucus*	15	1.24	–	–	+
13.	*Aspergillus niger*	144	11.99	+ + +	+	+ + +
14.	*Aspergillus restrictus*	16	1.33	–	+	–
15.	*Aspergillus ustus*	19	1.58	+ +	+	–
16.	*Candida* sp.	9	0.74	–	–	+
17.	*Chrysosporium* sp.	2	0.16	–	–	+
18.	*Cladosporium herbarum*	70	5.82	+	+ + +	–
19.	*Colletotrichum* sp.	16	1.33	–	+	–
20.	*Curvularia lunata*	85	7.07	+ +	+ +	+
21.	*Drechslera* sp.	17	1.41	+	+	+
22.	*Epicoccum purpurascens*	105	8.74	+ +	+ + +	+
23.	*Fusarium oxysporum*	10	0.83	+	–	+
24.	*Fusarium* sp.	12	0.99	+	–	+
25.	*Geotrichum* sp.	26	2.16	+	–	+
26.	*Helminthosporium* sp.	31	2.58	+	+ +	–
27.	*Nigrospora* sp.	32	2.66	+	+ +	+
28.	*Ochroconis* sp.	6	0.49	–	+	–
29.	*Penicillium citrinum*	55	4.57	+ +	+ +	+
30.	*Pestalotiosis* sp.	15	1.24	+	–	–
31.	*Stemphylium* sp.	4	0.33	–	+	–
32.	*Trichoderma* sp.	22	1.83	+	–	+
33.	*Trichothecium roseum*	35	2.91	+	–	+
	Unidentified colonies	17	1.41	–	–	–
	Total	1201	99.84			

– absent, + present, + + moderately present, + + + abundantly present.
*Average data of two years.

Table 3.8: Number, types, percentage contribution and seasonal frequency of colonies trapped on exposed Petriplates at Motilal Nehru Park during the investigation period between January, 2000 to December, 2001.*

S.No.	Fungal types	No. of colonies	Contribution (%)	Seasonal Frequency		
				Rain	Winter	Summer
Phycomycetes %						
1.	*Rhizopus arrhizus*	25	4.10	+ +	–	+ +
2.	*Syncephalastrum* sp.	3	0.49	–	–	+
Ascomycetes						
3.	*Chaetomium* sp.	15	2.46	+	+	–
Deuteromycetes						
4.	*Alternaria alternata*	75	12.31	+ +	+ + +	+ +
5.	*Aspergillus candidus*	15	2.46	+	–	+
6.	*Aspergillus flavus*	51	8.37	+ +	+	+ + +
7.	*Aspergillus niger*	108	17.73	+ + +	+ +	+ + +
8.	*Aspergillus terreus*	12	1.97	+	–	+
9.	*Aspergillus ustus*	7	1.14	+	–	+
10.	*Aureobasidium pollulans*	6	0.98	–	+	–
11.	*Cladosporium herbarum*	62	10.18	+ +	+ + +	+
12.	*Curvularia lunata*	22	3.61	+	+ +	+
13.	*Drechslera* sp.	9	1.47	–	+	+
14.	*Epicoccum purpurascens*	40	6.56	+ +	+ +	+
15.	*Fusarium moniliforme*	27	4.43	+	–	+ +
16.	*Fusarium oxysporum*	12	1.97	+	–	+
17.	*Geotrichum* sp.	4	0.65	–	+	–
18.	*Helminthosporium* sp.	26	4.26	+ +	+	–
19.	*Nigrospora* sp.	9	1.47	–	+	–
20.	*Penicillium citrinum*	14	2.29	+	+	–
21.	*Penicillium purpurogenum*	19	3.11	+	+ +	–
22.	*Pestalotiopsis* sp.	15	2.46	+	+	–
23.	*Scolecobasidium* sp.	4	0.65	–	+	–
24.	*Trichothecium roseum*	14	2.29	–	+	–
25.	*Ulocladium* sp.	3	0.49	+	–	–
	Unidentified colonies	12	1.97			
	Total	609	99.87			

– absent, + present, + + moderately, + + + abundantly present.
*Average data of two years.

contributed 45.36% to the total air spora. *A. niger* was the major type and contributed 27.19%, followed by *Alternaria alternata* (12.08%), *Aspergillus flavus* (10.57%), *Cladosporium herbarum* (9.06%), *Curvularia lunata* (6.64%), *Epicoccum purpurascens* (6.64%), *Aspergillus glaucus* (5.43%) and *Nigrospora* sp. (5.13%), while *Trichothecium* (0.60%) was trapped in minimum concentration.

Table 3.9: Number, types, percentage contribution and seasonal frequency of colonies trapped on exposed Petriplates at Paliwal Park during the investigation period between January, 2000 to December, 2001.*

S.No.	Fungal Types	No.of Colonies	Contribution (%)	Seasonal Frequency		
				Rain	Winter	Summer
Phycomycets						
1.	*Mucor* sp.	8	0.99	–	–	+
2.	*Rhizopus arrhizus*	26	3.23	+	–	+ +
3.	*Rhizopus microsporus*	9	1.11	–	–	+
Ascomycetes						
4.	*Choetomium* sp.	30	3.73	+ +	+ +	–
Deuteromycetes						
5.	*Alteraria alternata*	44	5.47	+ +	+ + +	+
6.	*Alternaria* sp.	12	1.49	+	+	–
7.	*Aspergillus niger*	180	22.38	+ +	+ + + +	
8.	*Aspergillus flavus*	40	4.97	+ +	+ + +	
9.	*Aspergillus glaucus*	24	2.98	+ +	–	–
10.	*Aspergillus restrictus*	30	3.73	+ + +	–	
11.	*aspergillus terreus*	45	5.59	+ +	+ +	+
12.	*Cladosporium herbarum*	72	8.95	+ + +	+ +	+
13.	*Curvularia lunata*	43	5.34	+	+ +	–
14.	*Chrysosporium* sp.	20	2.48	–	–	+ +
15.	*Drechslera* sp	14	1.74	–	+	+
16.	*Epicoccum purpurascens*	25	3.10	+	+ +	–
17.	*Fusarium oxysporum*	30	3.73	+	–	+ +
18.	*Fusarium moniliforme*	20	2.48	+	–	+ +
19.	*Geotrichum* sp.	7	0.87	–	+	–
20.	*Helminthosporium* sp.	15	1.86	+ +	–	–
21.	*Nigrospora* sp.	14	1.74	+	+	–
22.	*Penicillium purpurogenum*	12	1.49	–	+	+
23.	*Penicillium citrinum*	34	4.22	+ +	+	–
24.	*Pestalotiopsis* sp.	18	2.23	+ +	+	–
25.	*Trichoderma* sp.	6	0.74	+	–	–
26.	*Trichothecium roseum*	5	0.62	–	+	–
27.	*Ulocladium* sp.	2	0.24	+	–	–
	Unidentified colonies	19	2.36			
	Total	602	99.89			

– absent, + present, + + moderately present, + + + abundantly present
* Average data of two years.

Trichoderma sp. (0.90%), were recorded in the rainy season, *Trichothecium roseum* (0.60%) were recorded in winter and *Bipolaris* sp. (1.20%) were recorded only in the summer season (Table 3.10).

Table 3.10: Number, types, percentage contribution and seasonal frequency of colonies trapped on exposed Petriplates at Ravali Temple during the investigation period between January, 2000 to December, 2001. *

S.No.	Fungal types	No. of colonies	Contribution (%)	Seasonal Frequency		
				Rain	Winter	Summer
Phycomycetes						
1.	*Rhizopus* sp.	15	4.53	+	–	+
Deuteromycetes						
2.	*Alternaria alternata*	40	12.08	+	+ + +	+ +
3.	*Aspergillus flavus*	35	10.57	+	+ +	+ +
4.	*Aspergillus glaucas*	18	5.43	+	–	+ +
5.	*Aspergillus niger*	90	27.19	+ +	+ +	+ + +
6.	*Aspergillus ustus*	9	2.71	+	–	+
7.	*Bispolaris* sp.	4	1.20	–	–	+
8.	*Cladosporium herbarum*	30	9.06	+	+ +	–
9.	*Curvularia lunata*	22	6.64	+	+ +	–
10.	*Epicoccum purpurascens*	22	6.64	–	+ +	+
11.	*Fusarium oxysporum*	9	2.71	+	–	+
12.	*Nigrospora* sp.	17	5.13	+	+ +	–
13.	*Penicillium citrinum*	15	4.53	+	+	–
14.	*Trichoderma* sp.	3	0.90	+	–	–
15.	*Trichothecium roseum*	2	0.60	–	+	–
	Total	331	100.02			

– absent, + present, + + moderately present, + + + abundantly present.
* Average data of two years.

Mankameshwar Temple

It is evident from the Table 3.11 that 20 fungal types belonging to 15 genera were recorded from Mankameshwar temple. Four species of *Aspergillus* were recorded and contributed 37.4% to the total air spora, of which, *A. niger* (18.05%) was the dominant type. After *Aspergillus, Alternaria alternata* (10.45%) was the dominant type, followed by *Cladosporium herbarum* (6.65%), *Curvularia lunata* (4.51%), *Alternaria* sp. (4.27%), *Penicillium citrinum* (4.27%) and *Trichoderma* (4.27%).

Members of Phycomycetes i.e. *Rhizopus arrhizus* and *Mucor* sp. were recorded only in the rainy season, while *Aureobasidium pollulans* and *Nigrospora* sp. were restricted to the winters. On the other hand, *Aspergillus fumigatus, Bipolaris* sp. and *Fusarium* sp. were recorded only in the summer season (Table 3.11).

Indoor Environment

District Hospital

It is clear from Table 3.12 that 602 colonies belongimg to 17 genera were trapped from the District hospital. Three genera were recorded from Phycomycetes, of

Table 3.11: Number, types, percentage contribution and seasonal frequency of colonies trapped on exposed Petriplates at Mankameshwar during the investigation period between January, 2000 to December, 2001.*

S.No.	Fungal Types	No. of colonies	Contribution (%)	Seasonal Frequency		
				Rain	Winter	Summer
Phycomycets						
1.	*Mucor* sp.	2	0.47	+	–	–
2.	*Rhizopus arrhizus*	14	3.32	+ +	–	–
Deuteromycetes						
3.	*Allernaria alternata*	44	10.45	+ +	+ + +	+
4.	*Alternaria* sp.	18	4.27	+	+ +	–
5.	*Aspergillus flavus*	67	15.91	+ +	+	+ + +
6.	*Aspergillus glaucus*	13	3.08	+	+	–
7.	*Aspergillus niger*	76	18.05	+ +	+	+ + +
8.	*Aspergillus fumigatus*	7	1.66	–	–	+
9.	*Aureobasidium pollulans*	8	1.90	–	+	–
10.	*Bipolaris* sp.	5	1.18	–	–	+
11.	*Cladosporium herbarum*	28	6.65	+ +	+ +	–
12.	*Curvularia lunata*	19	4.51	+	+	+
13.	*Epicoccum purpurascens*	9	2.13	+	+	–
14.	*Fusarium moniliforme*	17	4.03	+	–	+ +
15.	*Fusarium* sp. 7	1.66	–	–	+	
16.	*Nigrospora* sp.	13	3.08	–	+ +	–
17.	*Penicillium citrinum*	18	4.27	+	+ +	–
18.	*Pestalotia* sp.	10	2.37	+	+	–
19.	*Trichoderma* sp.	18	4.27	+	–	+ +
20.	*Trichothecium roseum*	15	3.56	+	+ +	–
	Unidentified colonies	13	3.08			
	Total	421	99.90			

– absent, + present, + + moderately present, + + + abundantly present.
* Average data of two years.

which *Rhizopus arrhizus* was dominant and it contributed 4.15% to the total air spora. The colonies of Phycomycets were restricted to rainy season. Fourteen genera were recorded from Deuteromycetes. Among them, genus *Aspergillus* was found to be the dominant type and contributed 33.54%, followed by *Alternaria alternata* (14.11%), *Cladosporium herbarum* (10.29%), *Curvularia lunata* (5.14%), *Penicillium citrinum* (4.15%) and *Fusarium moniliforme* (3.65%), while *Aureobasidium pollulans* (1.16%) was found in least concentration.

The maximum number of fungal genera were recorded in the rainy season followed by winter and summer seasons. *Aspergillus candidus* (1.99%) was recorded only in summers, while *Aureobasidium pollulans* (1.16%), *Drechslera* (1.99%) and *Geotrichum* sp. (1.49%) were restricted to the winter (Table 3.12).

Table 3.12: Number, types, percentage contribution and seasonal frequency of colonies trapped on exposed Petriplates at District Hospital during the investigation period between January to December, 2000

S.No	Fungal Types	No.of Colonies	Contribution (%)	Seasonal Frequency		
				Rain	Winter	Summer
Phycomycets						
1.	*Mucor* sp.	4	0.66	+	–	–
2.	*Rhizopus arrhizus*	25	4.15	+ + +	–	–
3.	*Syncephalastrum* sp.	3	0.49	+	–	–
Deuteromycetes						
4.	*Alternaria alternata*	85	14.11	+ +	+ + + +	
5.	*Aspergillus candidus*	12	1.99	–	–	+
6.	*Aspergillus flavus*	80	13.28	+ + +	+	+ + +
7.	*Aspergillus glaucus*	18	2.99	+	+	–
8.	*Aspergillus niger*	92	15.28	+ + +	+	+ + +
9.	*Aureobasidium pollulans*	7	1.16	–	+	–
10.	*Cladosporium herbarum*	62	10.29	+ +	+ + +	–
11.	*Curoularia lunata*	31	5.14	+	+ +	+
12.	*Drechslera* sp.	12	1.99	–	+	–
13.	*Fusarium moniliforme*	22	3.65	+ +	–	+
14.	*Fusarium* sp.	18	2.99	+	–	+
15.	*Geotrichum* sp.	9	1.49	–	+	–
16.	*Helminthosporium* sp.	17	2.82	+	–	+
17.	*Penicillium citrinum*	25	4.15	+	+ +	–
18.	*Pestalotiopsis* sp.	9	1.49	+	–	–
19.	*Pithomyces* sp.	13	2.15	+	+	–
20.	*Trichoderma* sp.	21	3.48	+	–	+ +
21.	*Trichothecium* sp.	18	2.99	+	+	–
	Unidentified colonies	19	3.15			
	Total	602	99.89			

– absent, + present, + + moderately present, + + + abundantly present

S.N. Medical College

It is evident from Table 3.13 that 30 fungal types belonging to 23 genera were recorded from S.N. Medical College. A total of 820 number of fungal colonies were trapped. In this site also, *Aspergillus* was recorded as the dominant type. Among all species of *Aspergillus, Aspergillus niger* (13.41%) was found to be a major type. The other dominant genera were *Alternaria alternata* (11.58%), *Aspergillus flavus* (10.48%), *Cladosporium herbarum* (8.04%), *Penicillium citrinum* (5.12%) and *Curvularia lunata* (5.0%).

The maximum number of fungal genera were recorded in the rainy and winter seasons. Most of the species of *Aspergillus* were abundantly recorded in

Table 3.13: Number, types, percentage contribution and seasonal frequency of colonies trapped on exposed Petriplates at S.N. Medical College during the investigation period between January to December, 2000

S.No	Fungal Types	No.of Colonies	Contribution (%)	Seasonal Frequency		
				Rain	Winter	Summer
Phycomycets						
1.	*Mucor* sp.	6	0.73	–	–	+
2.	*Rhizopus arrhizus*	40	4.87	+ +	–	+ +
3.	*Rhizopus schipperae*	9	1.09	+	–	–
4.	*Syncephalastrum* sp.	3	0.36	+	–	–
Ascomycetes						
5.	*Chaetomium* sp.	14	1.70	+	+	–
Deuteromycetes						
6.	*Alternaria alternata*	95	11.58	+ +	+ + +	+ +
7.	*Arthrinium* sp.	6	0.73	+	+	–
8.	*Aspergillus Candidus*	24	2.92	–	–	+ +
9.	*Aspergillus flavus*	86	10.48	+ + +	+	+ +
10.	*Aspergillus glaucus*	25	3.04	–	+ +	–
11.	*Aspergillus niger*	110	13.41	+ + +	+	+ + +
12.	*Aspergillus restrictus*	10	1.21	–	+	–
13.	*Aureobasidium pollulans*	12	1.46	–	+	–
14.	*Bipolaris* sp.	3	0.36	–	–	+
15.	*Candida sp*	4	0.48	–	–	+
16.	*Cladosporium herbarum*	66	8.04	+ +	+ + +	+
17.	*Curvularia lunata*	41	5.00	+	+ +	+
18.	*Drechslera* sp.	9	1.09	+	–	+
19.	*Fusarium Moniliforme*	28	3.41	+	+	+
20.	*Fusarium oxysporum*	24	2.92	+ +	–	+
21.	*Geotrichum* sp.	15	1.82	–	+	–
22.	*Helminthosporium* sp.	32	3.90	+ +	–	+ +
23.	*Nigrospora* sp.	24	2.92	+	+ +	–
24.	*Penicillium citrinum*	42	5.12	+ +	+ +	–
25.	*Pencillum purpurogenum*	14	1.70	+	–	+
26.	*Pestalotiopsis* sp.	8	0.97	+	–	–
27.	*Pithomyces* sp.	18	2.19	+	+ +	–
28.	*Scolecobasidium* sp.	4	0.48	–	+	–
29.	*Trichoderma* sp.	25	3.04	+	–	+ +
30.	*Ulocladium* sp.	3	3.36	+	–	–
	Unidentified colonies	20	2.43			
	Total	820	99.81			

– absent, + present, + + moderately present, + + + abundantly present the summer season. *Rhizopus arrhizus* was also frequently observed in summer and rainy seasons. On the other hand, *Aspergillus glaucas, Aspergillus restrictus, Aureobasidium pollulans, Geotrichum* sp. and *Scolecobasidium* sp. were restricted to winter, whereas *Rhizopus schippearae, Syncephalastrum* and *Ulocladium* were recorded only in the rainy season (Table 3.13).

The results of the present investigation showed that moderate temperature, high RH and mild rain fall favoured fungal population. An average of two years data indicates that the spore population of fungi follows a definite periodicity in the weather conditions prevailing over Agra. In general, the fungal spores were observed in considerable concentrations from September to March, with peak incidence during November and February. The results of the present study are also supported by observations by Agarwal and Shivpuri (1974) who have reported that extreme heat and cold are unfavourable for the sporulation of fungal spores.

The results of the present observation are also in accordance with those of Bhatt and Rajasab (1991) at Gulburga, Bhandari and Mishra (1992) at Jabalpur, Kalkar and Patil (1994) at Nagpur, Tiwari and Sahu (1994) at Raipur, Reddy and Reddy (1996) at Vikarabad, Appana and Atluri (1999) at Vijaywada, Atluri and Murthy (1999) at Srikakulam (A.P.), Sahney and Purwar (1999) at Allahabad and Tiwari *et al.* (1999) at Raipur.

In the indoor survey maximum concentration was obtained during the rainy months (more in July and August) followed by winters and summers. The peak month expressing highest concentration was July with an average temperature of 30.15°C, RH 91% and 145.1 mm rainfall being recorded. The incidence of fungal spores in the indoor environment was high during the rainy season, which could be due to rainfall, high humidity and availability of dead and decaying matter in the hospital wards (Babu, 1983). The present study also reveals that local sources particularly crop patterns and vegetables around the sampling sites (living and dead) chiefly influence the concentration of air spora of a particular sampling site.

A total number of 102 fungal types were trapped during present investigation. The reason for the large variety of fungal spores obtained in the present study might be attributed to the continuous periodical sampling carried out at different sites having varied topography and vegetation and to the seasonal variations in meterological parameters.

The spores of *Sclerospora* were also trapped in the month of October at Dayal Bagh, Agra. Only a few workers have reported the spores of *Sclerospora* from the air (Srivastava and Shukla, 1990 at Balrampur and chiley, 1993 at Jabalpur). The appearance of these spores in the atmosphere can be assigned to the onset of rains.

Thirty fungal types were encountered from Ascomycetes. Maximum ascospores showing both quality and quantity were observed during the rains. It is clearly evident from the present observations that the effect of rainfall is an essential factor for the release of ascospores. In fact, many workers (Vaidhya, 1990 and Agashe and Sudha, 1995) have established a direct relationship between rainfall and the release of ascospores. In many ascomycetes, free water is required for the expansion of ascospores. Rain splash plays an important role in dispersal of ascospores.

The highest amount of rust spores were observed during the study period. This might have been due to the presence of wheat crops in the surrounding localities. Rust infects ferns, gymnosperms and angiosperms but the cereals suffer the most. Rust spores in highest amounts were also reported by Chile (1993) at Jabalpurs and Srivastava and Shukla (1990) at Balrampur.

Among all Basidiomycetes, smut spores were found to be dominant. Some important diseases like loose smut of wheat, loose smut of barley, smut of maize, whip smut of sugarcane are caused by smut fungi. Tilak *et al.* (1981) have also recorded the highest incidence of smut spores among Basidiomycetes. According to Reddy and Reddy (1996), smut spores occupied second position and contributed 16.37% to the total air spora at Vikarabad.

The spores of smuts were recorded mainly in clumps. Although smut spores were trapped throughout the year but their maximum occurrence was mainly in the dry period between October to February. Bhatt and Rajasab (1991) have observed its to be maximum in the months of January / March.

During the course of the present investigation, *Alternaria* was found to be the most dominant spore type among Deuteromycotes in the atmosphere of Agra. Its contribution to the total air spora was maximum in both the years at all sites except vegetable markets and hospitals. This might be because of its occurence as a saprophyte, as well as leaf spot parasite on various crops. The present findings were supported by Majumdar and Bhattacharya (2000) at Cooch Behar town (W.B).

In the present findings, the spores of *Aspergilli* occupied second position during the investigation at most of the sites, but in the vegetable markets and hospitals they were found to be a major type. Similar observations were made by Bhatt and Rajasab (1988) at Gulbarga and Appana and Atluri (1999) at Vijaywada. *Aspergilli* were trapped in higher amounts at each site, because of their ability to grow in various substrates.

Rhizopus is a very common saprophyte and facultative parasite. It is clear from the investigations that the concentration of *Rhizopus* was closely related with moderately high temperature, and high relative humidity favoured spore release in *Rhizopus*, as has been encountered by Mishra, 1985. The source of *Rhizopus* spores is dead and decaying plant parts around sampling sites.

In Basidiomycetes, only one genera ie, *Schizophyllum commune* was reported for the first time from Agra. It is ubiquitous in the environment as a plant and human pathogen attacking a wide range of host trees. Its a thermophillic fungus, growing saprophytically or parasitically on woody substrate (Sutton *et al.* 1988).

In the present investigation the maximum concentration of fungal population was encountered from Dayalbagh and adjacent areas, followed by Taj Mahal, Sikandara and Red Fort respectively. The maximum number of fungal spores from Dayalbagh and Taj Mahal were due to high vegetation. The cultivation of crops Taj Mahal has directly affected the increase in fungal concentration. Similarly, Dayalbagh area also has a luxuriant vegetation with crops the Wheat, Bajra, Maize and vegetables being cultivated by the people of the nearby villages.

Fungal biopollutants are responsible for spoiling the walls of the monuments during otherwise favourable meterological factors the excreta of birds and bats inside the walls of monuments serve as a substrate (Tilak and Kulkarni, 1972).

As compared to Motilal Nehru Park, Paliwal Park has shown a higher occurence of fungal spores. Paliwal Park has a luxuriant vegetation which directly affects the atmospheric fungal concentration. It is also responsible for the availability and prevalence of dead and decaying matter, which provides sites for the growth of fungal population.

The present findings also exhibited a low concentration of *air spora* in Ravali and Mankameshwar temple during the year 2000 and 2001. Since both the temples are situated in thickly populated areas high, population density at the city area drastically curtail the available sites for the growth of herbaceous flora, and consequently vegetable matter for the growth of fungi, thus reducing the quantity of airborne fungal spores.

The hospital is an indoor environment especially conducive to the spread of airborne pathogens. It contains a reservior of individuals infected with pathogenic microorganisms that are potentially transmissible to other persons, including patients, hospital personnel and patient visitors. The environment is contaminated up by sneezing, coughing, spitting and the presence of rotting bandages, cotton, etc. which provided substrate for the multiplication of various types of pathogenic and allergic fungi.

District hospitals contained lower concentration of spores as compared to S.N. Medical College which may be due to more vehicular traffic in this area. The chemical pollutants also affect the viability of the air spora.

REFERENCES

Agrawal, M.K. and Shivpuri, D.N. 1974. Fungal spores, their role in respiratory allergy. Adv. in Pollen Spores Res., 1: 78–128.

Agashe, S.N. and Sudha, P. 1995. Studies of circadian periodicity of fungal spores in Bangalore city. Ind. J. Aerobial., 8(1&2): 17–23.

Appanna, N. and Atluri, J.B. 1999. Fungal air spora at Vijaywada (A.P.). 10th National Conference on Aerobiology and its application P. 38 (Abstract).

Arya, A., Shah, A.R. and Sadasivan, S. 2001. Indoor aeromycoflora of Baroda museum and deterioration of Egyptian mummy. Current Science, 81 (7) : 793–798.

Atluri, J.B. and Murthy, D.V.V.S. 1999. The fungal air spora of Srikakulam (A.P.). 10th National Conference on Aerobiology and its application P–35 (Abstract).

Babu, M.1983. Indoor aeromycoflora and its relevance to human allergy Part II. Ph.D. Thesis, Marathwada University, Aurangabad.

Bhandari, S. and Mishra, R.P.1992. Monthly variation in the occurrence of air borne fungal and pollen flora. Proc. 81st Indian Sci. Cong. Part III, (Abstracts).

Bhatt, M.M. and Rajasab, A.H. 1988. Incidence of viable fungal air spora in Gulbarga during 1984–86. Ind. J. Aerobiol., 1 (1) : 10–16.

Bhatt, M.M. and Rajasab, A.H. 1991. Incidence of airborne fungal spores at two different sites in Gulbarga. Ind. J. Aerobiol., 4 (1& 2) : 1–6.

Burnett, H.L. and Hunter, B.B. 1972. Illustrated genera of Imperfect Fungi. 3rd ed. Burgess Publishing Company. Minneapolis Minnesota.

Chakraborty, S.; Sen, S.K. and Bhattacharya, K. 2000. Indoor and outdoor aeromycological survey in Burdwan, West Bengal. Aerobiologia, 16 (2) : 211–219.

Chile, S. 1993. Study of airborne fungal spores of Jabalpur with special reference to allergenicity Ph.D. Thesis, Rani Durgavati University, Jabalpur (M.P.).

Ellis, M.P. and Ellis, J.P. 1985.Microfungi on land plants. An identification hand book, International books and Periodicals supply service, New Delhi, India.

George, A.M. and Verma, K.S.2000.Seaonal and diurnal variations of air borne fungal spores in Jabalpur. Vsundhara, 1 & 2 (2 & 1) : 23–26.

Gilman, J.C. 1959. A manual of soil fungi 2nd ed. Lowa U.S.A. lowa state University press.

Kakde, V.B., Kakde, H.U. and Saoji, A.. 2001. Seasonal Variation of fungal propagules in a fruit market environment, Nagpur. Aerobiologia, 17(2) : 177–182.

Kalkar, S.A. and Patil, G.V. 1994. Airborne biocomponents in the air of Nagpur. Indian. J. Aerobiol. 7, (1 & 2) : 1–7.

Khandelwal, A. 2001. Survey of air spora by Rotorod Sampler in Lucknow, India : quantitative and qualitative assessment. Aerobiologia, 17 (1) : 77–83.

Majumdar, M.R. and Bhattacharya, K. 2000. Aeromycoflora at the foot hills of Eastern Himalaya. Ind. J. Aerobiol., 13 (1 & 2) : 14–19.

Mishra, R.P. 1985. Aerobiological studies with special reference to pollen grains and fungal spores of Jabalpur (M.P.). Ph.D. Thesis, R.D. University, Jabalpur.

Reddy, R.R. and Reddy P.R.1996. Aeromycological survey of Vikarabad. Indian J. Aerobiol., 9 (1 & 2) : 9–13.

Sahney, M. and Purwar, A. 1999. Aeromycological survey of two different localities of Allahabad. 10th National Conference on Aerobiology and its application. P. 32 (Abstract).

Srivastava, S.K. and Shukla, D.S. 1990. A volumetric investigation of air spora of Balrampur city. Ind. J. Aerobiol., 3 (1 & 2) : 26–31.

Sutton, D.A., Fothergill, A.W. and Rinaldi, M.G. 1998. Guide to "Clinically significant fungi".Williams and Wilkins Company, London.

Tilak, S.T. and Srinivasu, B.V. 1967. Air spora of Aurangabad. Ind. J. Microbiol., 7 : 167–170.

Tilak, S.T. and Kulkarni, R.L. 1972. Microbial content of air inside and outside the caves at Aurangabad. Curr. Sci. 23 : 850–851.

Tilak, S.T., Saibaba, M. and Pillai, S.G. 1981. Air spora inside the poultry shed. Proc. Nat. Conf. Env. Biol. India, PP. 221–225.

Tilak, S.T. 1989. Air borne pollen and fungal spores. Vaijayanti Prakashan, Aurangabad.

Tiwari, K.L. and Sahu, S.K. 1994. Survey of air borne fungi at Raipur. Ind. J. Aerobiol., 7 (1 & 7): 20–27.

Tiwari, P., Jadhav, S.K. and Tiwari, K.L. 1999. Fungal air spora of college campus at three different heights 10th National Conference on Aerobiology and its application. P. 30 (Abstract).

Vaidya, K.K. 1990. Studies in air spora at Aurangabad and its relevance to environmental parameters. Ph.D. Thesis, Marathwada University, Aurangabad.

4

Perspectives of Microbial Diversity in Forests of Central India

JAMALUDDIN

Tropical Forest Research Institute, Jabalpur, India

SUMMARY

Bacteria and fungi are two major groups of microbes significant in the tropical forests of Central India. A large number of bacterial species are nitrogen-fixing. *Fluorescent pseudomonas* are helpful in the solubilisation of complex phosphates. Some of the bacterial species like *Bacillus subtilis* is helpful in biocontrol of root pathogens. *Bacillus thuringiensis* is used for the control of teak skeletonisers and defoliators. Some bacteria are potential pathogens, causing diseases. *Xanthomonas* sp. causes leaf curl and kills the young leaves of teak. Certain bacteria are dominant in calcium, iron and bauxite mine spoil over burdens.

Fungal diversity in the tropical forests is very high. A large number of fungi are thermophilic and are responsible for breaking down of complex salts. A vast majority of plant species in tropical forests are associated with vesicular arbuscular mycorrhizal fungi while *Shorea robusta* and *Eucalyptus* spp. are ectomycorrhizal. Diverse groups of fungal species belonging to different fungal taxonomic groups cause diseases of sal, teak, bamboo, legumes, tree species, casuarinas and miscellaneous tree species. The entomogenous fungi control insect larvae, whereas species of *Trichoderma* and *Scytalidum* are helpful in the control of many root pathogens like *Macrophomina* and *Ganoderma* sp. A large number of fungi have been found to possess cellulolytic and lignolytic properties. The variety of microbes occurring in sal, teak and miscellaneous forests as well as in mine-disturbed ecosystems need systematic study and documentation of effective beneficial microbial populations for enhancement of forest productivity.

Keywords: Microbial diversity, Forest fungi

The tropical forests of Central India are very rich in biological diversity. Microbes constitute a major part of this biodiversity and are also a potential source of useful bioactive compounds. Therefore, the need for their *in situ* conservation cannot be overemphasised. Because of the uncertainty of long-term security of this strategy and due to the fact that isolation of many microorganisms from nature is problematic, *ex situ* collection alone can ensure conservation of microbial diversity. While many microbes are potentially useful sources for bioremediation and biotransformation, others produce oxygen, thereby reducing the CO_2 levels in the atmosphere. Microbial diversity of the world is widely covered at an IUBS / IUMS Workshop in 1991 (Hawksworth, 1991a, b; Hawksworth and Colwell, 1992).

A Variety of microbes occur in the global resources of microorganisms. As habitat destruction and species extinction rates are accelerating, biotechnology is identifying new ways to use information and products from biota. Management of microbial diversity is a serious task as there is little baseline information about the amount and distribution of microbes. In the majority of cases, microorganisms remain either on the fringes of biodiversity programmes or are left out entirely. Microbial biodiversity provides the foundations for biotechnology, ie the basis for new products discovery, bioremediation and genetic manipulation of organisms for new commercial products and processes. Thus, retention and conservation of microbial genomic variations for future applications is not only an obligation of the highest order, but also of great importance for global economy.

SIGNIFICANCE OF MICROBIAL DIVERSITY

(i) Microbial diversity must be treated as a global resource, to be indexed, used and preserved.

(ii) As habitat destruction and species extinction rates accelerate, biotechnology is identifying new ways to use information and products from the microbiota.
- Basis for new product discovery.
- Bioremediation.
- Genetic manipulation of organisms for new commercial products and processes processes.

(iii) Retention and conservation of microbial genomic variations for future applications.

(iv) Inventory of microbial species.

SPECIES DIVERSITY

Comparisons of species numbers between microorganisms is reflected into several variations in the species concept. This difficulty arises because sexual processes are either absent or difficult to detect in many microorganism groups. In practice, microbiologists tend to be pragmatic and recognise species or strains on the basis of morphological, biochemical or molecular similarity. The process of assimilation, substrate utilisation and cultural attributes are all extensively used as a taxonomic criteria, more in bacteria and yeasts than filamentous fungi and protozoa.

GENETIC DIVERSITY

The genetic diversity exhibited by the microorganism group is very vast. The extent of genetic diversity now demonstrated between the higher ranks of microorganisms is also reflected at the species level.

ECOSYSTEM HABITAT DIVERSITY

The variety of ecological niches exploited by the major groups of macro and microorganisms is directly related to their geological age. The greatest niche breadth is seen in bacteria, with a declining sequence in algae, protozoa and fungi. There is every reason to believe that regions and habitats with a maximum diversity of macroorganisms will also be rich in the genetic base of microorganisms. A sequence of a larger number of host-specific parasites, mutualists and saprobes is considered for the conservation of a diversity of macroorganisms. Extreme environment also continue to be a rich source of previously unknown microorganisms, belonging to diverse groups. Studies on the number of species of particular and general families of fungi in different geographic regions can particularly lead to the recognition of centres of diversity. Various edaphic and biotic factors also contribute towards changes in species and genera.

BIODIVERSITY MAINTENANCE

Biodiversity has played a major role in the evolution and diversification of macroorganisms. As mutualists, mycorrhizae are either involved in nutrient supply or in performance by other biochemical processes on which they depend. Bacteria, fungi and protozoa in the guts of insects and herbivores perform a crucial role in their digestive processes, particularly in the breakdown of cellulose and lignin.

About 85 % of the earth's vascular plants form mycorrhizas. This association in the life cycle of symbionts is often obligate in nature, the mycorrhizas being crucial to the absorption of growth limiting nutrients. The very existence of many macroorganisms is consequently dependent on the continued availability of the requisite mutualistic microorganisms. In the absence of mutualistic microorganisms (crustess cerallinelgae), coral reef ecosystems simply cannot exist. The formation of soil biomass through biodegradation of residue, breakdown of rock and nutrient cycling is performed mainly through fungi. In monitoring microorganisms with reference to conservation of biodiversity in macroorganisms, the maintenance of functional groups rather than individual species can be presumed to be limiting, except where a particular microorganism is a keystone species. Microorganisms also contribute to the maintenance of ecosystem structures through natural biocontrol. Plant pathogenic microorganisms can limit the plant population. Similarly, entomophagus fungi (microorganisms) can limit population of insects that would otherwise become major pests.

MICROORGANISMS IN BIOSPHERE FUNCTION

Microorganisms play a major role in the maintenance of the global ecology through various biogeochemical cycles. Some of the bacteria produce oxygen, thereby reducing CO_2. Supply of other elements like carbon, nitrogen, sulphur and minerals are also maintained in the biosphere by microorganisms.

Microorganisms are beneficial in sustaining development, particularly by increasing productivity of forest and agricultural lands through: (i) efficacious nitrogen-fixing strains into legume crops, (ii) enhancement of natural nitrogen fixation by application of cyanobacterial inocula, (iii) role of mutalists (mycorrhizas) in nutrient cycling in higher plants, (iv) the use of bacteria and fungi as biocontrol agents, (v) role of microbes in genetic manipulation, (vi) biotechnological aspects and (vii) CO_2 fixation, etc.

MICROBIAL DIVERSITY IN CENTRAL INDIA

Biodiversity in Madhya Pradesh is very rich. The forest area comprising sal forest, teak forest and miscellaneous forests, covers a vast area of the state. Due to the rich diversity macro-organisms, microbial diversity is equally very high. The microbes belonging to different groups particularly fungi, bacteria, actinomycetes etc. live closely with the macroorganisms in diverse habitats. Some of these organisms cause diseases, particularly fungal and bacterial diseases and belong to different taxonomic groups. These microbes are capable of damaging leaves, stem roots, stored wood, flowers and seeds etc.

These microbes are also capable of decomposing forest litter. The major group of microbes belongs to thermophilic or thermotolerant bacteria and fungi. The decomposition of forest flora is fastened by microbial activity and ultimately nutrient release results in thus maintaining the nutrient dynamics for the plant community. Many of the higher fungi belonging to different taxonomic groups are responsible for decaying wood, some of which spoil the timber beyond the economic threshold level and are thus categorised as harmful microbes. Some of the higher fungi are responsible for decomposing the woody plant material of the forest floor and converting it into humus. They help in maintaining the C and N ratio in the soil. Some of the highly cellulytic organisms amongst the thermophill are helpful in preparation of compost from forest waste material (Table 4.1).

There is another functional group of symbiotic fungi belonging to a different taxonomic position, which are capable of harming ecto and endo-mycorrhizae. The mycorrhiza is specific to different plant communities and thus it is helpful in maintaining biodiversity in the natural forest or in the degraded ecosystem. Many of the fungi and bacteria help in nitrogen fixation, phosphate solubilisation and fixation of atmospheric nitrogen.

The microbial control of diseases of the tree species is also one of the important functions microbial diversity. They antagonize the pathogens and protect the plants from diseases. Some of these (are entomophagus) which parasitise the insect larvae like Baueria, Numuria and *Aspergillus flavus* etc. parasitise *Bacillus thringensis* is also effective in killing insect larvae. Many species of fungi like *T. harzianum, T. viride, T. psdukongii, Gliocladium virens,* etc. act as biocontrol agents for a number of diseases. Besides these actinomyces like *Streptomyces* sp. fluorescent *Pseudomonas* and *Bacillus substitutes* are also widely used for the microbial control of diseases.

Termitomyces sp. is also well recognised for developing an association with termite combs. A large number of fungi and bacteria belonging to thermophillic and thermo tolerant groups have been isolated and identified from the rhizosphere of tree species and soil of M.P. forests. The major groups of fungi among thermophillis are species of *Scytallidium, Thielavia, Chaetomium, Thermomyces* and *Talaromyces* etc.

The thermotolerant group of fungi are sp. of *Aspergillus, Paecilomyces, Chaetomium, Mucor, Rhizopus, Scopulariopsis* etc. From a biotechnological point of view, thermophillic fungi are potentially important as they posseses battery of extra-cellular enzymes able to hydrolise polymers such as starch, protein,

Table 4.1: Diversity of Fungi in the Forests of Central India.

S. No.	Forests	No. of Fungi
1.	Teak forests	65
2.	Sal forest	29
3.	Miscellaneous forest	96
4.	Tree legumes	68
5.	Bamboos	49
6.	Wood decay	55
7.	Bacteria on forest tree species	48

pectin, hemi-cellulose and cellulose etc. They are also known to produce a number of antibacterial and antifungal antibiotics, extra cellular phenolic compounds and organic acids etc. (Satyanarayana *et al.*, 1992). The enzyme produced by this group of fungi are thermostable.

SCOPE OF MICROBIAL DIVERSITY

The studies on microbial diversity are very scant. Thus, there is a need to investigate and list out the fungi and bacteria and actinomycetes from the different soils and habitat in the forest ecosystem. There is a need to identify the major functional groups of microorganisms responsible for the biological diversity of macro-organisms, particularly the mycorrhizae, nitrogen-fixers, phosphate solubilising bacteria and many other biocontrol agents.

The microbes specifically involved in biochemical processes like bio-remediation, bio-indication, as well as synthesis of important chemicals, need identification and documentation.

Ex-situ conservation of the microbes by making a life culture herbaria is very essential for exploiting their potential. *In situ* maintenance and marking of the potentially important microbes is equally important.

Since there is no specific listing of the microbial diversity of M.P., it is necessary to prepare a project to document the relevant literature on microbial diversity and also list out the significant microbes in different habitats.

REFERENCES

Hawksworth, D.L. (1991a). The fungal dimension of biodiversity magnitude, significance and conservation. *Mycological Research*, **95:** 641–655.

Hawksworth, D.L. (1991b). The biodiversity of microorganisms and Invertebrates : Its Role in sustainable Agriculture, Wallingford : CAB International.

Hawksworth, D.L. and Colwell, R.R. (1992). Microbial diversity 21 : biodiversity amongst microorganisms and its relevance. *Biodiversity and Conservation*, 1 : 221–226.

Jamaluddin and Chandra, K.K. (1999).Application of VAM technology in bamboo cultivation in mine soils. *In:* Modern Approaches and Innovation in Soil Management. Editors : Bagyaraj, D.J., Verma, Ajit, Khanna, K.K. and Kehri, H.K., Rastogi Publications, Meerut.

Satyanarayana, T.; Johri, B.N. and Klein, J. (1992). Biotechnological potential of thermophillic fungi. In: Arora, D.K.; Elander, R.P. and Mukerji, K.G. (eds.). Handbook of Applied Mycology Vol. 4 Marcel Dekker, New York, pp. 729–761.

5

Observations on Waterborne Fungi of Uttara Kannada Region

C. RAMESH AND S. VIJAYKUMAR

Mycology Laboratory, Department of Botany, Karnataka University Dharwad, Karnataka, India

SUMMARY

The four streams, Kali, Panda, Bidaralla and Magod flowing in the Western Ghat forest region of Karnataka state, India, were selected and surveyed during 1991–1993 for the occurrence of waterborne fungi. Leaf litter incubation, Foam analysis, Boiled Seed and Test wood block methods were employed in the study. Twenty-four waterborne fungi were recorded from Kali stream. Twenty-eight species of waterborne fungi were recorded from Panda streams, while nine species each of waterborne fungi were recorded from Bidaralla and Magod streams. Greater number of fungal species and conidia were encountered during the monsoon and post-monsoon periods than in the summer. Seasonal periodicity of waterborne fungi in relation to rainfall, availability of leaf litter, water temperature pH and dissolved oxygen has also been discussed.

Keywords: Waterborne fungi, Seasonal periodicity, Stream mycoflora

Uttara Kannada is one of the 19 districts of Karnataka, India. The Western Ghats divide the district into two parts. While five of its taluks are in the coastal plains six taluks are above the Ghats with parts of their territory stretching to the fringe of the Deccan Plateau. The district has three distinctive regions, the Coastal lands, the Sahyadrian interior and the Eastern margin, where the table-land begins. There are four leading rivers, the Kalinadi in the north, the Bedthi or Gangavali in the south, the Aghanashini Tadri, far south and the Sharavati or Gersoppa to south of Tadri. In the slopes of the Western Ghats, from north to south of the Uttara Kannada district, there is a beautiful cover of dense forests of magnificient timber. The Uttara Kannada region, especially the Ghat sections are very rich in plant material of varied types and are thus of interest to taxonomists, plant geographers, foresters etc.

The most significant contribution on freshwater fungi was the pioneer study of Prof. C.T. Ingold in 1942, which extended to over three decades. Aquatic fungi have been reported from a variety of environments ranging from marine to fresh water (Jones 1976). In freshwater habitats they are represented by two major groups, namely Saprolegniales (water moulds) and Aquatic Fungi Imperfecti (Hyphomycetes). A number of Ascomycetes, other Mastigomycetes (Zoosporic fungi) and a few Basidiomycetes also occur in freshwater habitat, though in the dynamic of freshwater stream ecosystems Hyphomycetes fungi are the most important (Barlocher 1982, 1985, 1982a; Barlocher and Kendrick 1973a, 1973b, 1974, 1976; Kaushik and Hynes 1968, 1971, Graca 1993, Graca et al., 1993a, 1993b).

Research on freshwater Hyphomycetes in the Indian subcontinent commenced with a record of 4 aquatic Hyphomycetes species from streams near Madras by Ingold and Webster (1973). Subramaniam and Bhatt (1981) gave a general floristic picture of aquatic Hyphomycetous fungi in several streams and rivers of the Western Ghats in Southern India. The ecology and biology of freshwater Hyphomycetes from the streams and rivers of the Western Ghats in Karnataka have been extensively studied by Kaveriappa and his students in Mangalore University since 1982 (Sridhar et al., 1992). Those of the Himalayan mountain streams have been investigated by Sati and his students since 1988 in Nainital (Sati and Tiwari 1989, 1992, 1997).

The aquatic fungal population in fresh water consists of representatives of the Phycomycetes, fungi imperfecti, Ascomycetes and Basidiomycetes (Sparrow 1968, Johnson, 1968). These fungi occur on a wide variety of substrata in fresh water but also in marine waters, although to a lesser extent. Fungi belonging to Mastigomycotina form the prevalent group of fungi along with conidial fungi. Among the aquatic fungi zoosporic fungi are comprising of Chytridiomycetes and Oomycetes. This division of fungi comprise 204 genera and 1160 species (Manoharachary, 1991).

Aquatic Hyphomycetes occur on decaying plant materials in fast flowing well aerated streams and rivers (Ingold 1975). Their occurrence has been reported from many freshwater streams and rivers in temperate regions (Gonczol 1971, 1975, lqbal and Webster 1973, 1977; Lindsey and Clover 1976; Lamose and Goos 1978; Muller-Haeckel and Marvanova 1979; Barlocher and

Rosset 1981; Del Frate and Caretta 1983;Akridge and Kochn 1987). A few studies their occurrence in tropical streams are also available (Padgett 1976; Subramaniam and Bhatt 1981; Sridhar and Kaveriappa 1984) therefore, due to the paucity of information on aquatic fungi in tropical streams the present study was undertaken. Experiments were designed to relate the aquatic Hyphomycetes flora and seasonal fluctuations to other experimental factors in the streams of evergreen forests of the Western Ghats region of the Uttara Kannada District of Karnataka.

MATERIALS AND METHODS

Frequent visits were made to different areas in the locality (Uttara Kannada) during different seasons. The various groups of waterborne fungi were isolated by different methods of isolation.

Phycomycetes

Collection of Water Samples and Baiting in the Laboratory

Phycomycetes are found in nearly all kinds of freshwater habitats, as well as in seawater. They were isolated as saprophytes or facultative parasites on dead insects, algae, twigs, seeds, dead leaves and fruits, using them as baits. Baiting techniques have been used successfully after Crouch (1939) and Sparrow (1960). A few millimetres of the collected water is added to each petridish. Various baits are then added and the cultures incubated and examined at daily intervals from the 4th to 14th day. The following have been used as baits-autoclaved houses, Drosophila flies, termite wings, boiled hemp seeds, boiled grass leaves, autoclaved wettable cellophane, pollen grains of various conifers and white human hair. Fungi, which grow on these baits, were isolated onto pure cultures.

Baiting in the Field

The above method was very effective for isolating a variety of fungi, but some escaped. Most of them were isolated by using still baiting techniques in the field. The materials used as baits were placed in a small quarter inch galvanised wire mesh basket, plastic covered wire baskets. These baskets were suspended in the water by nylon ropes for a period of 1–5 weeks and determined by withdrawing samples at certain intervals.

Ascomycetes

Freshwater Ascomycetes are to be found growing on reed swamp plants (Ingold and Chapman 1952; Ingold 1968b) on driftwood (Ingold 1954, 1955) and on submerged wood (Jones and Oliver 1964). Jones and Eaton (1969) have described Ascomycetes on wood test-blocks which are similar to marine Ascomycetes.

Wood Baiting

Two sampling methods were employed. One consisted of collecting naturally occurring submerged wood debris. The second consisted of baiting for wood, inhabiting fungi by submerging autoclaved pieces of *Bursera delpechianum, Pongamia glabra, Pinus roxburghii* and *Bambusa arundinacea*. These baits prepared from dead decorticated branches which were 3–4 cm in diameter and in 7–12 cm in length.

The incubated baits were examined directly for fungi, before they were cut into smaller pieces and placed in a petridish and incubated. All wood samples were examined at weekly intervals under a dissecting microscope for the presence of fungi.

Fungi Imperfectii

Analysis of Natural Foam Sample

On studying the aquatic Hyphomycetes of particular streams, ponds or rivers, an analysis of naturally formed foam is of great interest. Fresh foam is generally white in colour and contains recently liberated and accumulated conidia. This is ideal for sampling. Often one can notice oily or brown coloured foam cakes. Such samples are generally old or flood flown and may contain either germinated or distorted conidia.

When the foam is collected in a jar and briefly agitated, it quickly breaks down into a bubble-free liquid without any treatment. If a drop of this is mounted on a slide and examined microscopically, it contains several conidia of aquatic Hyphomycetes. A wide mouthed jar with a few drops of FAA fixative is used and the foam is scooped into it. The four fingers are kept against the edge of the jar mouth and excess water is drained off.

Isolation of Spores from Submerged Leaves

The submerged decaying leaves are collected individually from the stream in a polythene bag. They are rinsed thoroughly in stream water before being placed in the bags. These samples are placed in an ice pack container for transporting them to the laboratory and then processed, following Ingold (1942, 1975).

Quantification of Conidia in Water

In order to estimate the number of conidia present in a unit volume of stream or river water and to study the quantitative ecology of fresh water fungi Iqbal and Webster (1973b) were followed.

The four streams Kali, Panda, Bidaralla and Magod streams flowing in the Western Ghat forest region of Karnataka state, India, were selected for the study. These streams are lined by thick vegetation including cultivated crops (Table 5.1), and the stream beds are mostly rocky and muddy. They carry a lot of organic materials including leaf litter and account for heavy rains during

Table 5.1: Fungi Isolated on Different Substrates

Fungi	Substrate
Pseudoeurotia platisporous (Satio & Minouralstolk)	Pongamia glabra, Vent wood
Thielaviopsis paradoxa (Dade) C. Moreau.	Cannabis sativa, seeds
Hunicola fuscodatra Treeam	Marselia sp. Leaf litter
Bactrodesuniella aasonii (Hughes) M.B. Ellis	On wood
Helicomyces hyderabadensis Rao and Dev Rao	Cocos nucifera Litter
Honacrosporium phyaatophagus (Dreschler) Subram	Typha angustaata. Bory
Fusariella hughesii Chabelksa-Frydamn	Cannabis sativa Wild
Aspergillus fumigatus Freseniu	Brusera delpechianua Poiss. wood
Phoma capitulum-Pawar, Mathur and Thirus	Marselia sp. Leaf litter
Phoma capitumlum Pawar Mathurand Thirus	Polygonum sp. Wood

the southwest monsoon. Three sampling sites at a distance of 1–3 km along each stream were selected and submerged leaves, foam and water samples were collected from three spots in each site, at monthly intervals, over a period of two years during 1991–1993.

RESULTS

Kali streams: Twenty-four species of waterborne fungi were recorded, out of which six species belong to Phycomycetes, eight species to Ascomycetes and ten species to fungi imperfectii. All these forms were collected by three methods, namely baiting boiled seeds, test-wood blocks and leaf litter. Phycomycetes were isolated only from boiled seeds in laboratory, whereas Ascomycetes were isolated on test wood blocks. Fungi imperfectii were isolated on both test wood blocks and leaf litter.

Out of these *Dictyuchus monosporus, Zopfiella latipes, Savoryella lignicola, Didymosphaeria dandeliensis, Helicosporium panachaeum* and *Phaeoisaria sparsa* were recorded abundantly. All these organisms were collected through all the methods. *Jahnula australiensis, Gliocladiopsis tenuis, Sporoschisna mirabile* and *Stagnospora vitensis* were less abundant. The remaining sixteen species occurred rarely.

In Phycomycetes, *Dictyuchus monosporus* was the most predominant fungus and the frequency of occurrence of the rest of the fungi was low. In case of Ascomycetes, *Savoreylla lignicola* was the predominant fungus, whereas *Zopfiella lundquistii* and *Didymosphaeria dandeliensis* were less abundant. The frequency of occurrence of the other organisms was low.

In case of fungi imperfectii, *Phaeoisaria sparsa* was the predominant fungus. *Helicosporium panachaeum* was less abundant and the frequency of occurrence of the other organisms was low (Table 5.2). Thirty-two species of foam spora were recorded, out of which *Triscelophorus accuminatus, Campyllospora filicladia*

Table 5.2: Comparison of Occurrence of waterborne Fungi

S. No.	Fungus	Kali Stream	Panda Stream	Magod Stream	Bidralla Stream	OWB
Phycomycetes						
1.	*Polychytrium aqqregatum* Ajello	–	+	–	–	+
2.	*Howakowskiella ramosa* Thaxter	–	+	–	–	+
3.	*Balstocladia ramosa* Thaxter	+	–	–	–	–
4.	*Allomyces macrogynus* (Emerson) Emer. Wills	–	+	–	–	–
5.	*Allomyces moniliformis* Coker and Braxton	–	–	–	–	+
6.	*Gonapodya polymorpha* Thaxter	+	+	+	+	–
7.	*Gonapodya prolifera* (Cornu) Fischer	+	+	–	+	+
8.	*Saprolegnia eccentrica* (Coker) Seymour	–	–	–	–	+
9.	*Isoachlya rodriqueziana* F.T.Wolf	–	–	–	–	+
10.	*Protoachlya oryzae* Khulbe	+	–	–	–	–
11.	*Achlya racemosa* Hildebrand	–	+	–	–	–
12.	*Aphanomyces stellatus* de Bary	–	+	–	–	–
13.	*Dictyuchus monosporous* Leitgeb.	+	+	–	–	+
14.	*Honblepharis macrandra* (Lagerheim) Woronin	+	–	+	+	–
Ascomycetes						
15.	*Pseudoeurotium multisporum* (Saito and Mina) Stok	+	+	–	–	–
16.	*Aniptodera chesapeakensis* Shearer et Miller	–	+	–	–	+
17.	*Halosaroheia lotica* Shearer	–	–	–	–	+
18.	*Chaetomium medusarum* Meyer and Lanneau	–	–	–	–	+
19.	*Zopfiella laptipes* (Lundqvist) Malloch and Cain	+	+	–	–	+
20.	*Zopfiella lundqvistii* Shearer and Crane	+	–	–	–	–
21.	*Savoryella aquatica* Hyde.	+	–	–	–	–
22.	*Savoryella lignicola* Jones and Eaton.	+	+	+	–	–
23.	*Nectria dentifera* Samules	+	–	+	–	–
24.	*Frondicola tunitricuspis* Hyde.	–	+	–	–	+
25.	*Didymosphaeria dandeliensis* sp. Nov	+	–	–	–	–
26.	*Jahnula australiensis* Hyde	+	+	–	–	–
27.	*Letosphaeria peruviana* Spegazzini	+	–	–	–	–
28.	*Molisia pinosa* sp nov	–	–	–	–	+
Fungi Imperfectii						
29.	*Scytalidium lignicola* Pasante	–	–	–	–	+
30.	*Thielaviopsis paradoxa* (Dade) C. Moreau.	–	–	–	–	+
31.	*Zalerion varium* Anastasiou	–	–	–	–	+
32.	*Acremoniula sarcinellae* (Pat and Har) Arn. Ex Deiq	–	+	+	–	–
33.	*Humicola fuscoatra* Traen	+	–	–	–	–
34.	*Tripospermum myriti* (Lind) Hughes	–	+	–	–	–
35.	*Septonema Sacendens* Corda	–	–	–	–	+

Table 5.2: (Contd.)

S. No.	Fungus	Kali Stream	Panda Stream	Magod Stream	Bidralla Stream	OWB
36.	*Bactrodesmiella masonii* (Hughes) M.B. Ellis	–	–	–	–	+
37.	*Endophragnia elliptica* (Berk and Br.) M. B. Ellis	–	+	–	–	–
38.	*Helicosporium indicum* Rao and Dev Rao	–	+	–	–	–
39.	*Helicosporium panacheum* Moore	+	–	–	–	–
40.	*Helicomyces hyderbadensis* Rao and Dev Rao	–	–	–	+	–
41.	*Helicomyces bellius* Morgan	–	–	–	–	+
42.	*Drepanospora pannosa* Barkely and Curtis	+	–	–	–	–
43.	*Phaeoisaria sparsa* Sutton	+	–	–	–	–
44.	*Phaedactylium alpiniae* (Sawada) M.B. Ellis	–	+	–	–	–
45.	*Monacrosporium phymatophagum* (Dreschs) Subras	–	–	–	–	+
46.	*Dematophora necatrix* Ha rtig	+	–	–	–	–
47.	*Phaeoisariopsis griseola* (Sacc.) Ferraris	–	+	–	+	–
48.	*Phialosporostilbe setosa* Bhat and Kendrick	–	–	–	+	–
49.	*Beltraniella odinae* Subram.	–	–	+	–	–
50.	*Verrucispora proteacearum* Shaw and Alcon	–	+	+	+	–
51.	*Higrospora panici* Zimm	–	–	+	–	–
52.	*Pariconia Sarawatipurensis* Bilgrami	–	–	–	–	+
53.	*Gliocladiopsis tenuis* (Bugnicourt) Corous and Wing	+	+	–	–	–
54.	*Drechslera spicifer* Nelson	–	–	–	+	–
55.	*Sporoschisna wirabile* Berk and Br.	+	+	–	–	–
56.	*Fusariella hughesii* Chabelska Frydman	–	–	–	–	+
57.	*Acrophialophora fusispora* (Saksena) M.B. Ellis	+	+	–	–	+

and *Flabellospora verticillata* were the most abundant, while the rest of the fungi were low (Table 5.3) in frequency of occurence.

The data on the number of species encountered in the three methods and the average number of conidia present in the foam at quarterly intervals were created. In all the methods the total number of species encountered was the highest during the July–September months and it was lowest in March–May months. The number of species encountered in the three methods is less as compared to the foam method. The maximum number of species occurred during the months July–August and the minimum during the March–May.

Panda stream: Twenty-eight species of waterborne fungi were recorded out of which six species each were of Phycomycetes and Ascomycetes, while fifteen species were of fungi imperfectii. All these forms were encountered in three methods, namely baiting, boiled seeds, test-wood blocks and leaf litter.

Table 5.3: Comparison of Occurrence of Waterborne Fungi in the Foam Spora of Different Streams

S. No.	Fungus	Kali Stream	Panda Stream	Magod Stream	Bidralla Stream
1.	Actinospora megalospora Ingold	+	+	–	–
2.	Anguillospora curvula Iqbal	+	+	+	+
3.	Anguillospora filliformis Greathead	+	–	–	+
4.	Anguillospora gigantea Ranzoni	–	+	+	–
5.	Anguillospora longissima Ingold	+	+	+	–
6.	Articulospora moniliforma Ranzoni	+	–	–	–
7.	Articulospora tetracladia Ingold	–	+	–	–
8.	Beltrania rhombica Penzig	+	+	+	–
9.	Campylospora chaetocladia Ranzoni	+	–	+	–
10.	Campylospora filicladia Nawawi	+	+	+	–
11.	Clavariana aquatica Nawawi	+	+	+	–
12.	Clavatospora tentacula (Umphlett) Nilsson	+	–	–	–
13.	Diplocladia scalaroides Arnaud	+	+	+	–
14.	Flabellospora multiradiata Nawawi	+	+	+	–
15.	Flabellospora verticilata Alasoadura	+	+	+	+
16.	Helicomyces colligatus Moore	+	+	+	+
17.	Helicomyces troquarous Lane and Sharer	–	+	+	+
18.	Hydrometerospora symmetrica Gonczol and Revay	+	–	–	–
19.	Isthmotricladia gombakiensis Nawawi	+	+	+	+
20.	Laridospora appendiculata Nawawi	–	–	–	–
21.	Lateriramulosa uni-inflata Matsuishima	+	+	+	+
22.	Lunulospora curvula Ingold	+	+	+	–
23.	Mycoleptodiscus lateralis Alcorn and Sutton	+	+	+	+
24.	Pestalotia sp.	+	+	+	–
25.	Phalengispora constricta Nawawi & Webster	+	+	+	+
26.	Speiropsis hyalospora Subram & Lodha	+	+	+	–
27.	Speiropsis pedatospora Tubaki	+	+	+	–
28.	Tetracladium setigerum (Groove) Ingold	+	+	–	–
29.	Tetraploa aristata Berk and Br.	+	+	+	+
30.	Tricladium angulatum Ingold	+	+	+	–
31.	Tricladium splendens Ingold	+	+	+	–
32.	Tripospermum cameleopardus Ingold	+	+	–	+
33.	Triscelophorus acuminatus Nawawi	+	+	+	+
34.	Triscelophorus konajensis Sridhar and Kaveri	+	+	+	+
35.	Triscelophorus monosporus Ingold	+	+	+	–
36.	Wiesneriomyces javanicus Koorders	+	+	+	+

Out of these, *Nowakowskiella profusa, Frondicola tunitricuspis, Savoreylla lignicola, Sporoschisna mirabile* and *Septoria bambusae* were recorded abundantly. *Pseudoeurotium multisporum, Coniothyrium obiones, Phaeoisariopsis grisecola* and *Verrucispora proteacearum* were less abundant. The remaining 19 species occurred rarely.

All the species belonging to Phycomycetes, *Nowakowskiella profuse* and *Frondicola tunitricuspis* were very predominat fungus. *Savoryella lignicola* was less abundant, while the frequency of other organisms was low.

All fifteen species of fungi imperfectii were collected either on test wood blocks or leaf litter. *Sporoschisna mirabile* was the most predominant fungus, whereas, *Septoria bambusae, Phaeoisariopsis griseola, Verrucispora proteacearum* and *Coniothyrium obiones* were less abundant. The frequency of occurrence of other organisms was low. Thirty species of foam spora were recorded out of which *Triscelophorus accuminatus* was the most predominant, *Triscelophorus konajensis* and *Lunulospora curuvla* were less predominant, while the occurrence of other fungi was low (Table 5.2 and 5.3).

In all the methods, the total number of species encountered was the highest during the July–September and it was lowest during the month of March–May. The number of species encountered in these three methods is less when compared to the foam method. The maximum number was July–September and it was the lowest during the month of March–May.

Bidaralla stream: Nine species of waterborne fungi were recorded, out of which three species were of Phycomycetes and six species were of fungi imperfectii. No isolate of Ascomycetes could be recorded due to natural disturbances as the test wood blocks were washed away. The data on the number of species encountered in the two methods, the average number of conidia present in the foam at quarterly intervals and the monthly data on water temperature, rainfall and pH were recorded.

All the three species belonging to Phycomycetes were isolated by the baiting technique. The frequency of the occurrence of *Gonapodya polymorpha, G. prolifera* and *Monoblepharis macrandra* was same and all three occured commonly.

All the six species of fungi imperfectii were collected on leaf litter *Phaeoisariopsis griseola* as the most predominant fungus. *Helicomyces hyderabădensis* and *Verrucispora proteacearum* were less abundant. The frequency of occurrence of other organisms was low (Table 5.2). Fourteen species of foam spora were recorded. Out of which *Triscelophorus accuminatus, Helicomyces torquatus* and *Tripospermum camelopardus* were less dominant. The occurrence of other fungi was quite low (Table 5.3).

The number of species encountered in the three methods is less when compared to the foam analysis. The maximum number of species occurred in foam during the months of July–August in 1991–1992, 1992–1993 and they were the lowest in December–February.

Magod streams: Nine species of waterborne fungi were recorded, out of which two species each belonged to Phycomycetes and Ascomycetes, while the rest of the five species belonged to fungi imperfectii.

Both the species belonging to Ascomycetes were isolated by test wood blocks; *Savoryella lignicola* was the predominant fungus. The frequency of occurrence of *Nectria dentifera* is less abundant when compared to *Savoryella lignicola.* All the species of fungi imperfectii were recorded on leaf litter and the test wood blocks. *Beltraniella odinae* was the most predominant fungus. The frequency of occurrence of the rest of the organisms was low (Table 5.2). In foam thirty species were recorded from this stream, out of which *Triscelophorus accuminatus* was predominant, while *Isthmotricladia gombakiensis* and *Triscelophorus konajensis* were dominant, and the occurrence of other fungi was low (Table 5.3).

The average number of conidia present in the foam at quarterly interval and the monthly data on water temperature, rain fall and pH were recorded. The number of species encountered in all the three methods was less when compared to the foam method. The maximum number of species were encountered during the month of July–August and the lowest during months of December–February.

Other water bodies: Twenty-six species of waterborne fungi were recorded, out of which seven species belonged to Phycomycetes, six species to Ascomycetes and thirteen species of fungi imperfectii. All these organisms were isolated by three methods.

Seven species of Phycomycetes were encountered on 'Seed Baiting' technique. *Dictyuchus monosporus* was the most predominant fungus and very close to it was *Saprolegnia eccentrica.* The frequency of occurrence of other organisms was low.

Six species of Ascomycetes were encountered on test wood blocks. *Chaetomium medusarum* was the most predominant fungus and the frequency of occurrence of other organisms was low. Thirteen species of fungi imperfectii were encountered on leaf litter and test wood blocks. *Acremoniula sarcinellae* and *Phoma capitulum* were the most predominant. The frequency of occurrence of other organisms was low (Table 5.2). The data on water temperature, rain fall, pH was not represented as the collections were randomly taken and hence they could not be recorded.

Comparison of the Occurrence of Fungi in Four Streams

A total number of hundred and three species belonging to 86 genera were recorded from the four streams during 1991–1993. Twelve species, namely *Gonapodya polymorpha, Anguillospora curvula, Flabellospora verticillata, Helicomyces colligatus, Isthmotricladia gomabarkiensis, Lateriramulosa uniinflata, Mycoleptodiscus lateralis, Phalangispora constricta, Tetrapola aristata, Triseelophorus accuminatus, T. konajensis* and *Wiesneriomyces javanicus* were common to all the streams.

There were fifty-eight species in Kali stream, sixty in Panda stream, thirty-eight species in Magod stream and twenty-three species in Bidaralla stream, while twenty-three species were recorded from other bodies (Table 5.2 and 5.3).

DISCUSSION

Various mycologists have studied waterborne fungi by isolating them by baiting methods, leaf litter, test wood blocks, foam and water (Iqbal and Webster 1973b; Willoughby and Archer 1973; Barlocher and Kendrick 1974; Ingold 1975a; Sridhar et al., 1992; Sati and Tiwai 1997). In the present study, it was possible to record hundred and three species belonging to eighty-six genera in the four streams with four methods of analysis over a period of two years. Twelve species were common to all the four streams. Further, all the mycoflora recorded in the streams were not found throughout the collections; while some species were recorded in most of the collections, almost throughout the year, others were restricted to certain parts of the year. These observations indicate that for a correct assessment of waterborne fungi in a stream, not only different methods of isolation will have to be adapted, but samplings also need to be done during different parts of the year.

Of all the other organisms recorded in the present study, *Gonapodya polymorpha, Anguillospora curvula, Flabellospora verticillata, Helicomyces colligatus, Isthmotricladia gomabankiensis, Lateriramulosa uniin. flata, Mycoleptodiscus lateralis, Phalangispora constricta, Tetrapola aristata, Triscelophorus accuminatus, T. konajensis* and *Wiesneriomyces javanicus* were not only common to the four streams but were also found in all the seasons throughout the two years of investigation. Hence, these can be considered as the most common waterborne fungi of this region.

The occurrence of waterborne fungi was also found to be affected by rainfall, water temperature and availability of leaf litter. The influence of rainfall and availability of leaf litter on the occurrence on waterborne fungi have been reported earlier from temperate regions. Iqbal and Webster 1973b; Muller–Haeckel and Marvanova 1976; 1979a; Lamore and Goos 1978; Delfrate and Caretta 1983; Willoughby and Archer 1973, reported the occurrence of maximum number of species during wet periods rather than dry periods. Conway (1970) observed a considerable increase in the number of aquatic fungi two weeks after the beginning of leaf fall. The conidia of waterborne Hyphomycetes in water was reported to be more after the deposition of leaves in the streams (Iqbal and Webster 1973b; Muller-Haeckel and Marvanova 1976, 1979b Del frata and Caretta 1983). While Lamore and Goos (1978) observed more number of species, Willoughby and Archer (1973) recorded maximum number of conidia in foam samples after heavy rainfall; lqbal and Webster (1973b) correlated the abundance of conidia of *Clavariopsis aquatica* and *Flagellospora curvula* to the period of leaf deposition. Gonczol (1975) found the sporulation peak of *Flagellospora curvula* coinciding with intensive leaf deposition. Similar results on the influence of leaf deposition and rainfall on the occurrence of waterborne Hyphomycetes were also obtained in the present investigation. In the four streams, the maximum number of species were recorded mostly from the June–July (monsoon) and November (Post-monsoon) collections and the minimum in the month of May (Summer). The rainfall data of Dandeli, Londa, Yellapur and Sirsi regions during the two years of investigations shows that in this region, the south west monsoon starts during April–May, reaches maximum level during in the months of July–August and declines thereafter.

The collection of waterborne Hyphomycetes in the Bidaralla stream was also the lowest during the summer months, where there were no rains. Leaf fall usually takes place during March–June in this region. The heavy rainfall during July–August carries the fallen leaves into the streams along with the terrestrial fungi thereby enriching the organic substrata and mycoflora of the streams.

Wood-Eggenschwiler and Barlocher (1983) studied the waterborne fungi of sixteen streams in France, Germany and Switzerland and correlated the species maxima with that of riparian vegetation living in the streams. In the present investigation, the number of species of waterborne fungi was found to be more in the Kali and Panda streams which could have been due to heavy litter in flow, pH, temperature, rainfall and vegetation on the stream banks.

The population of aquatic Hyphomycetes was reported to decline with an increase in, pH (Ranzoni 1953; Conway 1970; Wood-Eggenschwiler and Barlocher 1983). Also, the occurrence and distribution of aquatic Hyphomyeetes was found to decline with a decrease in pH (Suzuki and Nimura 1960a, 1960b, 1961; Nilsson 1964). Barlocher and Rosset (1981), in a comparative study of their own investigation with those of Conway (1970), Willoughby and Archer (1973) Barlocher and Kendrick (1974) lqbal and Webster (1977) have found that the maximum number of fungal species occurring in the streams as a function of its pH value formed a unimodel curve with a maximum at a pH 6.7.

The rate of aeration has also been reported to influence the sporulation of aquatic fungi. The increased rates of aeration enhance the sporulation due to higher levels of oxygen (Webster and Towfik 1972; Webster 1975). However, according to Nilsson (1964) the requirement of oxygen for growth and sporulation varies from species to species.

Suberkropp and Klug (1981) reviewed the work on the growth of aquatic Hyphomycetes at different temperatures and concluded that the growth of aquatic Hyphomycetes mostly depended on the availability of suitable leaf litter at lower temperature. They have also reported that optimum growth of most of the species occurred at a range of 15–25°C. Growth and sporulation of some aquatic Hyphomycetes have been restricted to colder temperatures (Nilsson 1964; Conway 1970; lqbal and Webster 1973b; Koske and Duncan 1974; Gonczol 1975; Delfrate and Caretta (1983). Aquatic fungi were found to be lesser during summer than in winter in some of the streams of England (Iqbal and Webster 1973b) and Italy (Delfrate and Caretta 1983). Ingold (1975b) also observed more conidia in autumn than in summer in some of the English streams.

In the present investigation, the number of species recorded during monsoon and winter seasons was more than in the summer season. However, temperature may not be the sole reason for this variation. As stated earlier, the availability of substrate during monsoon and winter appears to be a more plausible reason.

Among the fungi, which were reported to occur in temperate regions (Nilsson 1964; Ingold 1975a) *Gonapodya polymorpha, G. prolifera, Monoblepharis macrandra, Savoryella lignicola, Anguillospora longissima, Lunulospora curvula Triscelophorus monosporus* were observed in the present study. All were found in the Kali stream, where the temperature ranged from 19–30°C. These fungi were also classified as very common species in temperate and colder regions (Nilsson 1964). *Lunulospora curvula* and *Triscelophorus monosporus* were reported to have a wide distribution in tropical countries (Ingold 1956, 1958, 1959, 1960a; Dixon 1959; Lejohn 1965; Great head 1961). The high temperature optimum for growth, sporulation and spore germination of these fungi may be responsible for their wide tropical distribution. Apart from the present investigation, no other detailed investigation on waterborne fungi of tropical regions in India was carried out, but Kaveriappa and his students worked in details on waterborne Hyphomycetes only and that too from outside India. It is from Puerto Rico (Padgett 1976) that *Campylospora chaetocladia, Lunulospora curvula* and *Triseclophorus monosporus* were reported to be the predominant fungi colonising leaf discs. In the present study, *Gonapodya polymorpha* on boiled seeds, *Gonapodya proliphora, Dictyuchus monosporus, Zopfiella latipes* and *Savoryella* were found to be predominant colonisers.

REFERENCES

Ainsworth, G.C. and Bisby, G.R. 1963. Dictionary of fungi (5th edt.) C.M.I Kew, Surrey, 447 pp.

Akridge, R.E. and Koehn, R.D. 1987. Amphibious Hyphomycetes from the San Marcos river in Texas. *Mycologia* **79:** 228–233.

Barlocher, F. 1982. The contribution of fungal enzymes to the digestion of leaves by *Gammarus fossarum* koch. (Ampphipods) *Oecologia* **52:** 1–4

Barlocher, F. 1985. The role of fungi in the nutrition of stream invertebrates. *Bot. J Linn. Soc.* **91:** 83–94

Barlocher, F. 1982a. The ecology of aquatic Hyphomyeetes, ecological studies **94:** 225 pp, Springer-Verlag, Heidelbergs.

Barlocher, F. and Kendrick, B. 1973a. Fungi and food preference in Gammarus pseudolinnaceous. *Arch. Hydrobiol.* **72:** 501–516

Barlocher, F. and Kendrick, B. 1973b. Fungi in the diet of Gammarus pseciolim nacreous (Ampipoda) *Oikos,* **24:** 295–300

Barlocher, F. and Kendrick, B. 1976. Hyphomycetes intermediates of energy flow in streams. In Recent Advances in Aquatic Mycology, Jones, E.B.G (ed) Elck, London. 435–446 pp

Barlocher, F. and Kendrick, B. 1974. Dynamics of fungal population on leaves in a stream. *J. Ecol.* **62:** 761–791 pp.

Barlocher, F. and Rosset, J. 1981. Aquatic Hyphomyeetes of two black forests and two swiss jura streams.*Trans.Br.Mycol.sco.* **76:** 479–483.

Chandrsekhar, K.R., Sridhar, K.R. And Kaveriappa, K. 1986. Aquatic Hyphomyc'etes of Kempu hole in the Western Ghat forests of Kamataka. *Ind. Phytopathology.* **39:** 368–372

Conway, K.E. 1970. Aquatic Hyphomycetes of Central New York. *Mycologia* **62:** 516–530

Crouch, J.N. 1939. Technique for the collection, isolation and culture of Chytrid, *J. Elisha Mitchall Scient. Soc.* **55:** 208–214

Del frate, G and Caretta, G Dix. 1983. Aquatic Hyphomycetes of a mountain stream in Valsesia. *Hydrobiologia:* **102:** 69–71

Dixon, P.A. 1959. Stream spora in Ghana. *Trans. Br. Mycol. Soc.* **42:** 174–176

Gonczol, J. 1971. Aquatic Hyphomycetes from the mountains of Borzsony. *Ann. Hist. Nat. Mus. Natl. Hung.* **63:** 57–75.

Gonczol, J. 1975. Ecological observations on the Aquatic Hyphomycetes of Hungary. *Acta. Bot. Acad. Sci. Hung* **21:** 243–264

Graca, M.A.S 1993. Patterns and Process, in detritus based stream systems. *Limnology.* **23:** 107–114.

Graca, M.A'.S., Maltby, L. and Calow, P. 1993a. Importance of the fungi in the diet of Gammaarus Pulex Asellus aquaticus–1 feeding Strategies. *Ocology* **93:** 139–144.

Graca, M.A.S. Maltby, L. and Calow, P. 1993b. Importance of the fungi in the diet of Gammaarus Pulex Asellus aquaticus-11, Effect on Growth, Reproduction and Physiology. **96:** 304–309.

Greathead, S.K. 1961. Some Aquatic Hyphomycetes in South Africa. *J. South African Bot.* **27:** 95–228.

Ingold, C.T. 1942. Stream spora in Nigeria of decaying of leaves. *Trans. Br. Mycol. Soc.* **76:** 339–417

Ingold, C.T. 1954. Aquatic Ascomycetes : Discomycetes from lakes. *Trans. Br. Mycol. Soc.* **37:** 1–18.

Ingold, C.T. 1955. Aquatic Ascomycetes : Further species from the English lake District. *Trans. Br. Mycol. Soc.* **38:** 157–168.

Ingold, C.T. 1956. Stream spora in Nigeria. *Trans. Br. Mycol. Soc.* **39:** 108–110

Ingold, C.T. 1958. Aquatic Hyphomyeetes from Uganda and Rhodesia. *Trans. Br. Mycol. Soc.* **41:** 109–114.

Ingold, C.T. 1959. Aquatic spora of Omo forest, Nigeria. Trans. Br. Mycol. Soc., **42:** 479–485.

Ingold, C.T. 1960a. Aquatic Hyphomycetes in Southern Rhodesia. Proc. *Trans. Rhod Sci. Assoc.* **48:** 49–53.

Ingold, C.T. 1960b. Aquatic Hyphomycetes forrn Canada. *Can. J. Bot.,* **38:** 803–806.

Ingold, C.T. 1975a. An illustrated Guide to aquatic and waterborne Hyphomycetes (Fungi imperfectii) with notes on their biology. Fresh water Biological Association, Sci. publ. No.30, ferry House, Ambleside, cumbria England 96 pp.

Ingold, C.T. 1975b. Conidia in the foam of two English streams. *Trans. Br. Mycol. Soc.,* **65:** 522–527.

Ingold, C.T. and Chapman, B. 1952. Aquatic Ascomycetes: *Loramyces junicola* Weston and *L. macrospora* n, sp. *Trans. Br. Mycol. Soc.,* **35:** 268–272.

Ingold, C.T. and Webster, J. 1973. Some Aquatic Hyphomycetes from India. *Kavaka,* **1:** 5–9.

Iqbal, S.H. and Webster, J. 1973. The trapping of aquatic Hyphomycetes spore by air bubbles. *Trans. Br. Mycol. Soc.,* **60:** 37–48.

Iqbal, S.H. and Webster, J. 1973b. Aquatic Hyphomycetes spora of the river Exe and its tributaries. *Trans. Br. Mycol. Soc.,* **61:** 331–346

Iqbal, S. H. and Webster, J. 1977. Aquatic Hyphomycetes spora of some Dartmoor streams. *Trans. Br. Mycol. Soc.,* **69:** 233–241.

Johnson, T. W. 1968. Saprobic Marine Fungi in the fungi (G.C. Ainsworth, and Sussman). Academic Press New York **vol. 3**, 95–104pp.

Jones, E.B.G. and Oliver, A. C. 1964. Occurrence of Aquatic Hyphomycetes on wood submerged in fresh brackish water. *Trans. Br. Mycol. Soc.,* **47:** 45–48.

Jones, E. B. G and Eaton, R, A. 1961. *Savoryella lignicola* gen. et.sp. from water cooling towers. *Trans. Br. Mycol. Soc.,* **52:** 161–164.

Jones, E.B. and Eaton, R, A 1976. Lignicolous and Algicolous fungi. In: Recent Advances in Aquatic Fungi: Jones, E.B.6 (Ed.) Elak Series, London, 1 –51 pp

Kaushik, N.B. and Hynes, H.B.N. 1968. Experimental study on the role of autumn shed leaves in aquatic environments. *J. Ecol.* **56:** 229–243

Kaushik, N.B. and Hynes, H.B.N. 1971. The fate of dead leaves that fall into the streams. *Arch. Hydrobiol.,* **68:** 1465–15 1 5.

Koske, R.E. and Duncan, I.W. 1974. Temperate Effects on Growth, Sporulation and Germination of some Aquatic Hyphomycetes. *Can. J. Bot.,* **52:** 1387–1391.

Lamore, B.J. and Goos, R.D. 1978. Wood inhabiting fungi of freshwater streams in Rhode Island. *Mycologia.,* **70:** 1025–1034.

Lejohn, H.B. 1965. Sierra Leone freshwater Hyphomycetes. *Trans. Br. Mycol. Soc.,* **48:** 261–264.

Lindsey, B.T. and Glover, B.J. 1976. Ecological Studies of spores of Aquatic Hyphomycetes in the Cringle Brook, Lines. *Hydrobiologia.,* **51:** 201–208.

Manoharachary, C. 1991. Aquatic myco-ecology from India, an over view. In Current Trends in Limnology **vol.1** Narendra Publishing House, Delhi 79–90 pp.

Muller-Haeckel, A. and Marvanova, L. 1976. Konidien Produktion und Kolonisation Von Susswasser Hyphoyzeten im Kaltisjokk (Lappland) Bot. Not. **129:** 405–409.

Muller-Haeckel, A. and Maryanova, L. 1979. Freshwater Hyphomycetes in brackish seawater, Botanica Marina, **22:** 421–424.

Nilsson, S. 1964. Freshwater Hyphomycetes., *Symb. Bot. up sal.,* **18:** 1–130

Padgett, D.E. 1976. Leaf decomposition by fungi in tropical rain forest streams. *Biotropica,* **8:** 166–178.

Ranzoni. F.V 1953. The Aquatic Hyphomycetes of California Farlowia, **4:** 353–398.

Sati, S.C. and Tiwari, N. 1989. Freshwater Hyphomycetes from Jageshwar Stream, Kumaun Himalaya, *India. Nat. acad. sci. Letters* **13:** 7–9

Sati, S.C. and Tiwari, N. 1990. Some aquatic hyphomycetes of Kumaun Himalaya, India. Mycotaxon, 39: 407–414.

Sati, S.C. and Tiwari, N. 1992. Colonisation, Species composition conidial production of aquatic Hyphomycetes on Chirpine needle litter in a freshwater Kumaun Himalayan stream. *Int. Rev. Ges. Hydrobiol.,* 77: 445–453.

Sati, S.C. and Tiwari, N. 1992. A new species of *Tricladium* from Kumaun Himalaya, India. Mycol. Research, 96: 229–232.

Sati, S.C. and Tiwari, N. 1997. Glimpses of conidial aquatic fungi in Kumaun Himalaya. In *"Himalayan Microbial Diversity"*. Eds. S. C. Sati, J. Saxena and R.C. Dubey. *Today and Tomorrow Printers & Publishers,* New Delhi. 17–37.

Sparrow, F.K. Jr. 1960. Aquatic Phycomycetes 2nd (Ed) Michigan Univ. Press, Ann.Arber. 1187pp

Sparrow, F.K. Jr. 1968. Ecology of freshwater fungi. In The Fungi (G.C.Ainsworth and A.S. Sussman eds.) Vol. **3**, 41–93pp, Academic Press, London, New York.

Sridhar, K.R. and Kaveriappa, K.M. 1984. Aquatic Hyphomycetes of Western Ghat Forests in Karnataka. *Indian Phytopahology.* **37:** 546–548.

Sridhar, K.R., Chandrasekhar, K.R. and Kaveriappa, K.M. 1992. Research on Indian subcontinent in the ecology of Aquatic Hyphomycetes, ed.F Barlocher, Ecol. Stud. Vol. 94, Springr verlag Heidelberg, 182–211 pp.

Subramanian, C.V. and Bhatt, D.J. 1981. Conidia from freshwater foam samples from the Western Ghats. Southern India. *Kavaka,* **9:** 45–62.

Suberkroppa, K. and Klug, M.J. 1981. Degradation of Leaf Litter by Aquatic Hyphomycetes, in Wicklow, D. T; Carrol, G.C; (eds). The fungal community. Marcel Decker, New York, 761–776 pp.

Suzuki, S. and Nimura, H. 1960a. Aquatic Hyphomycetes in the lake of Mt. Hakkoda. *J. Jap. Bot.,* **35:** 265–268.

Suzuki, S. and Nimura, H. 1960b. The Microbiological Studies of the lakes of Volcano Bandai–11 Ecological Study in Goshikinuma and Akanuma lake group. *Bot.Mag.Tokyo.,* **73:** 360–364.

Suzuki, S. and Nimura, H. 1961. Relation between the distrubtion of Aquatic Hyphomycetes in the Japanese lake types. *Bot. Mag. Tokyo.,* **74:** 51–55.

Webster, J. 1975 and Towfik, F.W. 1972. Further Studies of Sporulation of Aquatic Hyphomycetes in Relation to Aeration. *Trans. Br. Mycol. Soc.,* **59:** 353–364.

Webster, J. 1975. Further Studies of Sporulation of Aquatic Hyphomycetes in Relation to Aeration. *Trans. Br. Mycol. Soc.,* **64:** 119–127.

Willoughby, L.G. and Archer, J.F. 1973. The fungal spora of freshwater stream and its colonization pattern on wood. *Fresh water biol.,* **3:** 219–239.

Wood-Eggenschwiler, S. and Barlocher, F.1983. Aquatic Hyphomycetes in sixteen Streams in France, Germany and Switzerland. *Trans. Br. Mycol. Soc.,* **81:** 371–379.

6

A New Species of *Periconiella* from Kathmandu Valley, Nepal

USHA BUDATHOKI

Central Department of Botany, Tribhuvan University, Kirtipur, Kathmandu, Nepal

SUMMARY

Periconiella litseae sp.nov. collected on *Litsea* from the forest of Kathmandu Valley, Nepal, is described, illustrated and compared with allied species.

Keywords: Hyphomycetes, *Periconiella*, Taxonomy

During the survey of parasitic fungi in and around the forests of Kathmandu valley, Nepal, the author came across many interesting fungi, specially Hyphomycetes. The genus *Periconiella* Sacc, was at first described by Saccardo (1885), Apud Sacc and Berlese in Atti. Till now twenty-three species of *Periconiella* have been described (Ellis, 1971). The literature survey on Nepalese fungi (Adhikari 1990, 1991; Pawsey 1989) records the presence of a few Hyphomycetes from Kathmandu valley (Deighton, 1967; Khadka *et al.*, 1967–68; Adhikari and Manandhar, 1984). However, at present no *Periconiella* species have been reported from Nepal.

DESCRIPTION OF SPECIES

***Periconiella litseae* sp. nov.** [Fig. 6.1]

Colonies hypogenous, brown, effuse; Mycelium of hyphae superficial, hylaine, septate, branched, thin and smooth walled, associated with trichome;

Conidiophores macronematous, mononematous, straight to slightly flexous, thick smooth-walled, divided into stripes and marked head, bearing small fertile branches; striped septate, stout, smooth, dark brown, up to 19.5–137.5 × 4.5–6.5 μm branches of the head-pale coloured, simple branched into primary and secondary branches, measuring upto 22.0–32.5 × 3.5–5 μm (at the apex); Conidiogenous cells-polyblastic, integrated, terminal on stipe and branches, sympodical cylindrical, cicatrized, scars prominent, obclavate, Conidia solitary to catenate, acropluerogenous, dry, simple, hyaline, finely verrucose, rounded at the apex and truncated at the base, single transversely septate, obvoid, measuring 5.5–9.5 × 1.5–4.5 μm.

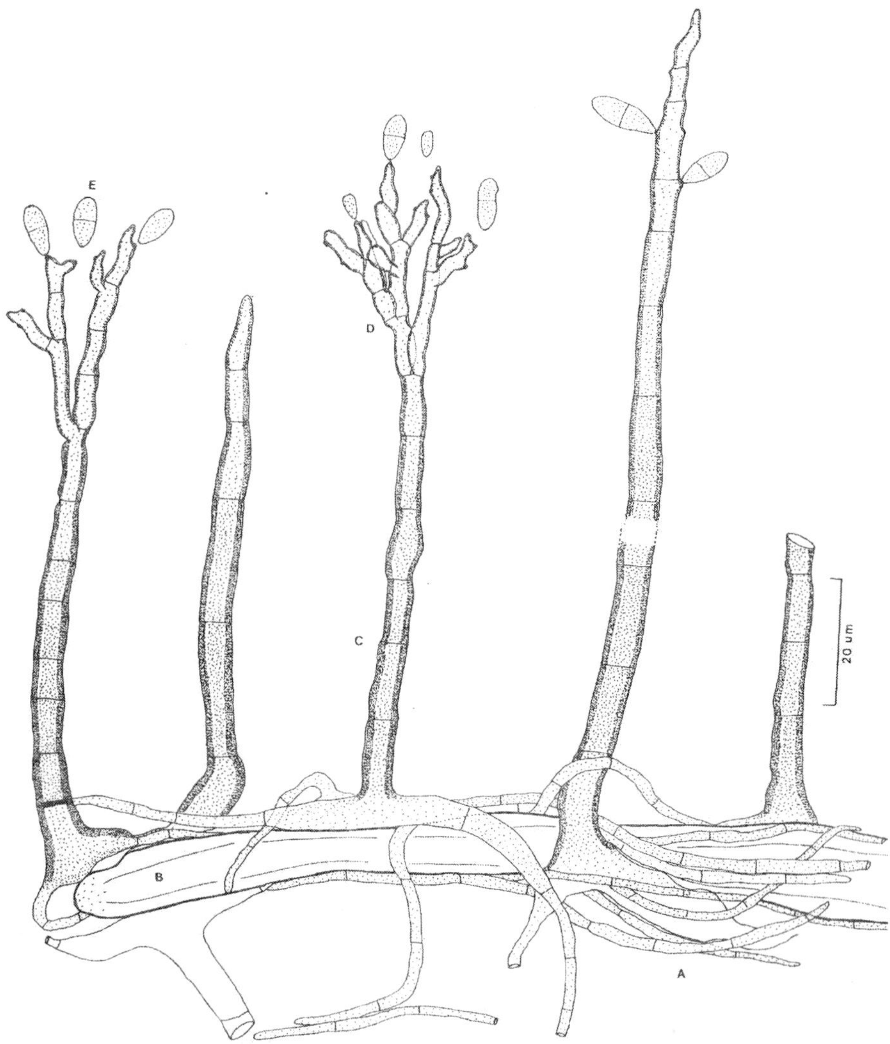

Fig. 6.1: A—Vegetative Hyphae on trichome; B—Trichome; C—Conidiophore; D—Head of branches; E—Conidia.

On living leaves of *Litsea* sp. (Lauraceae), Hatiban, Kathmandu Valley, Nepal, Dec. 1987, leg. Usha Budathoki, KU 336 holotype, IMI 303946 isotype.

Maculae plerumque hypogenae, effusae, brunneaevel acrofuscae; Mycelium ex hyphis superficials hyalinae, septatae, ramosae; Conidiophora plerumque, solitariae, recta vel, sub flexuosa, crassitunicata, brunnea, in stipem et capitem bene definitum ramose parvos fertiles frenetem divisa, stipe usque ad 19.5–137.5 × 4.5–6.5 μm, septatis glaberis, ramosis, pallide olivaceis, plerumque 22–32.5 × 3.5–5.0 μm (as apides), saepe in conidiophora secunderia et deinole tertiaria, capita fertile ferentia proliferans; Cellulae conidiogenae integratae, terminalis, polyblasticae, saepe cum raims, capitis, conjugentia, sympodialis, cylindricaevel obclavatae et irregulares, cicatricibus, bebe evolutis donata; Conidia solitaria, sicca, simplicia, acropleurogene, hyaline, usque 5.5–9.5 × 1.5–4.5 μ.

In folis vivis *Litseae* sp. (Lautaceae), Hatiban. Kathmandu valley, Nepal December, 1987, leg. Usha Budathoki. KU 336 holotypus, IMI 303946 isotypus.

In Nepal the host *Litsea* has been recorded parasitised by *Karnella lauricola* (rust) only. No Hyphomycetes on this host have been reported so far. In Indian literature review Bilgrami *et al.*, (1979) and Kamal *et al.*, (1979) have recorded this genus from India. But the present species differs from *P. longiospora* in a measurement of stipe (800 × 3.5–8.5 μm), branches of head (40–175 × 4.5–11.5 μm) and conidia 36–130 × 44.5–9.5 μm). Therefore, the present isolate has been proposed as a new species and described here.

REFERENCES

Adhikari, M.K. 1990. History of mycological exploration in Nepal Cryptog. Mycol. 11(2): 111.

Adhikari, M.K. 1991. Research Trends on Nepal Mycoflora, J. Ins. Agri, Anim. Sci. Vol. 12: 17–27.

Adhikari, M.K. and Manandhar, V. 1984. Parasitic fungi on some Rosacdeous medicinal plants from Kathmandu valley, Nepal. J. Nep. Pharm. Aasso. 11(1–2): 19–28.

Bilgrami, K.S., Jamaluddin and Rehawi, M.A. 1979. Fungi Fungi of India. Part I list and references Univ. Bhagalpur, India.

Deighton, F.C. 1967. Studies on *Cercospora* and allied genera 2. Mycol. Pep 112: 80; Common Wealth and Mycological Institute (CMI), Surrey UK.

Ellis, M.B. 1971. Dematiaceous Hyphomycetes, CMI, Kew, England.

Ellis, M.B. 1976. More Dematiaceous Hyphomycetes, CMI, Kew, England.

Kamal, Khan, M.K. and Budathoki; U. 1989. A peculiar foliicolous hyphomycetes from Kathmandu valley, Ana Net 15: 5.

Khadaka, B.B. and Shah S.M. 1967. Preliminary list of plant diseases recorded in Nepal Nepalese J. Agri. 2:447–76.

Khadaka, B.B., Shah, S.M. and Rawat, K. 1968. Plant diseases in Nepal. A supplementary list. Tech. Doc. 66 FAO plant prot. Con. South East Asia and pacific Region Bangkok, Thailand.

Pawsey, B.D. 1989. Survey A check-reference list of plant pathogens in Nepal. FRIC Occasional paper No. 1/89 Kathmandu.

7

New Fungal Record from Kathmandu Valley, Nepal

N. Pandey, M.K. Adhikari*, P. Maharjan and Usha Budathoki

Central Department of Botany, TU, Kirtipur, Kathmandu.
**National Herbarium and Plant Laboratories, Godavari, Nepal*

SUMMARY

Thelephora fuscella Lloyd., *Microporous flabiliformis* Klotzschi, wild inedible mushrooms of Basidiomycetes class, have recently been collected from the pine cultivated area of Kirtipur. The mushroom was spotted in swampy places just near the Tribhuvan University, along with some other herbaceous plants. Both are being recorded as new species to Nepal.

Keywords: Basidiomycets, *Thelephora*, *Microporous*, Fleshy fungi

The diverse phytogeographical situation of Nepal has resulted in its having rich fungal diversity. Most of the new species are collected and described from the temperate, sub-alpine and alpine belts, as well as from the subtropical region (Adhikari, 2000). Phytogeographically, the area under the present study is the one surrounding the Tribhuwan University Kirtipur. The area of collection includes coniferous and other similar trees, all of which are cultivated. The area lies at an altitude of 1300–1350 metres towards the south of the Kathmandu valley.

Polypores are tough or woody fungi, which are generally wood-inhabiting. They are commonly known as 'bracket' or shelf fungi. The basidiocarp may be of various shapes and sizes, the hymenium composed of tubes or pores with openings facing towards the ground. They are called 'takhte chayu' in Nepali. Berkeley (1854a) reported seventeen species from Hooker's collection, which included eight new species to science. Since then, several authors have added more species to the list. Recently, Adhikari (1988a) reported fifty-four genera and one hundred and fifty species of polypores from Nepal.

METHODOLOGY

During the ethnomycological study of mushrooms of central Nepal and the survey of Kirtipur (7th July, 2002), an interesting mushroom was collected on marshy land. The specimen was studied for its morpohological characters (habit, habitat, colour, texture), all of which were noted. The specimen was sun-dried and brought to the laboratory for a detailed study. Slides were prepared and measurements of basidia and basidiospores were taken. The specimens were presented at the National Herbarium, Godavari and Herbarium of Central Department of Botany, Tribhuwan University, Kirtipur, Nepal.

DESCRIPTION OF SPECIES

***Thelephora fuscella* Lloyd.** Myc. Notes (1923) 1288, f. 2544,

Corner, 1968 A Monograph of Thelephora (basidiomycets) 60 pg. 31

Fruit bodies: 7–12 cm high, pluropodial, palmately divided from a common trunk.

Stipe: 0.5–5 cm × 2 –5 mm, consisting a compact dark core, 0.8 –1.3 mm thick and a loose tomentous layer, 0.5–3 mm thick; branches flattened, 10 mm broad or narrow (2–5 mm), becoming spiculose or digitate, acute or obtuse, often fimbriate, the upper side more or less strigose or appressedly spiculose fibrillose pale dirty yellowish, gray to pale yellow livid blister; hymenium inferior, smooth, fuscous purple, inclining to umber; flesh dry, floccose-coriaceous, tough, smell none or slightly rotten.

Hymenium thickening 200 μm, subhymenium 30 μm deep, composed of loosely interwoven hyphae, no cystidiae. Hyphae 2.5 –5 × 6 μm.

Basidia: 3.5–4.5 × 6–7 μm, with fucous brownish cytoplasm, sterigmata, 2–4, 5 μm long.

Spore: 6–8 × 4.8–6 μm, fuscous, umber, irregularly ellipsiod to subglobose, angular, nodolus, not echinulate.

Thelephora is a large genus. It contains most of the coriaceous and resupinate basidiomycetes that are without gills, pores and spines (Burt, 1966). E.J.H. Corner's book "A Monograph of *Thelephora*" has described fifty-one species. They are mostly terricolous or humicolous, and a few are lignicolous and cosmopolitan.

The specimen was found in cultivated pine forests, especially in swampy areas along with other herbacious plants. They are generally found from June to October at an altitude of 1300–1350 m –N. 22204 –TUCH, Pandey, N. 7/7/2002. Until now, three species viz., *Thelephora caryophylla, Thelephora palmata* and *Thelephora terrestris,* have been recorded from Nepal (Adhikari, 2000). The present species are recorded as new to Nepal.

Distribution : North America, Europe, Japan, India and Nepal.

Microporus flabelliformis. Klotzsch Imazeki, Otani and Hongo 1988, Fungi of Japan, 454. –Lee, J.Y., 1993, *Coloured Korean mushrooms,* 363.

Synonym : *Polyporous flabelliformis* K1, -Bakshi, 1971, *Indian Polyporaceae,* 102. –Teng, 1986, *A contribution to our knowledge of the higher fungi of China,* 395.

Basidiocarp annual, stipitate, growing in groups, adjacent pilei, leathery, drying hard, flabelliform to semicircular and some are reniform, 2.5–4 × 3–6.5 × 0.1–0.3 cm; stalk lateral, expanded at base into flat globular disc; brownish yellow to reddish brown to almost black, 0.5 cm long and 0.3 to 0.4 cm diameter; upper surface yellowish brown, pinkish brown, reddish brown to almost black with yellow margin, velvety tomentose, becoming glabrous with age, narrowly concentrically zonate, sulcate, margin entire; context pale yellow, floccose, 0.05–0.2 cm thick; hymenial surface cream coloured, margin sterile up to 1.5 mm, pores globose to subglobose, 7–8 per mm, pore tube 0.05 to 0.1 cm long; basidia clavate, 4–5 μm broad. Basidiospore light brown, oval, 2–2.5–3.3–5 μm.

The present species was found on the stump of *Castonopsis indica* (Katush). Raniban, Pumdibhumi VDC, 1000m 30.04.057/14th July 2000, Mahajan, P. No. 006.

Distribution: Worldwide.

ACKNOWLEDGEMENT

We are very thankful to Mrs Vidya Manandhar (National Herbarium and Plant Laboratories, Godavari) for her immense help in the identification of mushrooms. We are also thankful to Bhaskar Adhikari (Assistant lecturer), and Geeta Timsina, a former student of the Central Department of Botany, T.U., for their assistance in the collection phase.

REFERENCES

Adhikari, M.K. 1988.. Poplypores (Wood rotting fungi) of Nepal. *Banko Janakari.* 2(1): 9–20.

Adhikari, M.K. 2000. *Mushrooms of Nepal.* Edited by G. Durrieu, P.U. Printers, Battishputali, Kathmandu, Nepal.

Bakshi, B.K. 1971. *Indian Polyporaceae,* ICAR, New Delhi.

Berkeley, M. J. 1854a Decades XLI–XLII Indian fungi. *Hooker's Jour. Bot.* Kew Gard. Misc. Jour. 6; 129–143. b-Decades XLIV–XLVI. Indian fungi; Ibid. 6; 225–235 (reprint 1969). Decade of fungi. Ducas 1–62. Ashe & Co., Amsterdam.

Burt A.E. ,1966. *The Thelephoraceae of North America I–XV,* Hafner Publishing Company, Inc.

Corner, E.J.H., 1968. *A Monograph of Thelephoraceae (Basidiomycetes). Nova Hedwigia. 27 Verlag von J. Cramer.*

Imazeki, R., Kobayashi, Y. and Aoshima, K. 1966. *Fungi.* The Flora of Eastern Himalaya. Ed. H. Hara. University of Tokyo, Japan. 611–629.

Imazeki, R., Otani, Y. and Hongo, T. 1988. *Fungi of Japan.* Yama—Kei Pub. Co. Ltd, London.

Lee, Ji-Yul., 1993. *Coloured Korean Mushrooms.* Vol. 1. Korea.

Teng, S.C. 1986 *A contribution to our knowledge of higher fungi of China.* National Ins. of Zoology and Botany, India.

8

Pythium Species in the Water Bodies of Kumaun Himalaya, India: A Review

ARCHANA TIWARI AND R.D. KHULBE

Department of Botany, D.S.B. Campus, Kumaun University, Nainital, India

SUMMARY

Twenty-one species of *Pythium* have been reported from thirteen different water bodies of Kumaun Himalaya. Of them, *P. undulatum* and *P. vexans* showed a wide occurrence under different ecological conditions. Nanaksagar reservoir and Kosi river yielded the maximum number of species. A few species, namely, *P. aphanidermatum, P. afertile, P. echinulatum* and *P. undulatum* were associated with fish species.

Keywords: Watermoulds, *Pythium*, Kumaun Himalaya, Water bodies

Kumaun Himalaya is famous for its large number of aquatic bodies. Several fungal species have been reported from different habitats time to time by many mycologists including Sati (1981, 1997), Khulbe (1989, 1992), Khulbe and Bhargava (1981, 1983), Mer *et al.* (1984), Khulbe *et al.* (1995) and Bisht *et al.* (1996).

Out of the different fungal species, *Pythium*, a member of Pythiaceae (Peronosporales) was first reported from Kumaun region by Khulbe and Bhargava (1977). Later, several species have been reported as saprophytes (Khulbe, 1980; Khulbe and Bhargava, 1983; Mer *et. al* 1980, 1984; Mer, 1992; Mer and Khulbe, 1983; Sati and Khulbe, 1980, 1983) fish parasites (Sati, 1981, 1983; Khulbe, 1989; Khulbe and Sati, 1980) and parasites of seedlings and roots of different cultivated plants (Verma, 1987a,b, 1985, 1986; Sati and Tiwari, 1992; Sati *et al.*, 1992). All the previous reports are found scattered in different journals.

The present paper deals with a review on the occurrence, seasonal periodicity and parasitic nature of *Pythium* species as reported by different investigators from different water bodies of Kumaun Himalaya, India so far.

Eight lakes, Nainital, Khurpatal, Bhimtal, Naukuchiyatal, Ramtal, Naldamyantital, Garurtal and Nanak Sagar and five rivers, namely, Ramganga (west), Kosi, Saryu, Gomati and Gaula have been explored by different investigators for the study of watermoulds. Six fish species, namely, *Puntius ticto, P. conchonius, Carassius auratus, Nemacheilus rupicola, Gudusia chapra* and *Tor tor* were also studied for mycotic infection.

Baiting technique of Robertson (1980) was followed by most of the investigators. Hemp seeds, mustard seeds, snake skin, ants, houseflies and green grass have been used as baits. CMA and Czapax agar have been found best for culturing. Pure culture and identification were made with the help of techniques defined by Middleton (1943), Waterhouse (1967, 1968), Robertson (1980), Dick (1990), and several taxonomic papers, including Mer and Khulbe (1984).

Occurrence of *Pythium* species has also been studied in relation to physico-chemical factors of water, following A.P.H.A. (1989).

SPECIES DIVERSITY

A total of twenty-one species of *Pythium* have been isolated from different habitats in Kumaun Himalaya. Of them, tweleve species were found to be distributed in lentic and thirteen species in lotic water bodies of Kumaun Himalaya (Table 8.1). In lentic water, *P. monospermum, P. afertile, P. proliferum* and *P. undulatum* showed a wider adaptability (Bisht *et al.*, 1996; Khulbe *et al.*, 1995; Khulbe, 1980b; Mer *et al.*, 1980). The other species had limited range of occurrence viz, *P. gracile* and *P. middletonii* occurred in Ramtal (Mer *et al.*, 1980); *P. aphanidermatum, P. echinulatum, P. hypogynum, P. inflatum, P. torulosum* and *P. vexans* in Nanak Sagar (Khulbe *et al.*, 1993b). These species flourished under moderate range of temperature (15–24°C), pH (7.2–9.6) and dissolved oxygen 6.5–15.0 mg/l.

River aquatic system also showed a considerable number of *Pythium* species. *P. debaryanum, P. undulatum, P. vexans* and *P. aphanidermatum* were isolated from different rivers of Kumaun region and they showed their adaptability under different ecological conditions (Khulbe *et al.*, 1993b, 1994; Mer *et al.*, 1980). The other species had limited range of occurrence. *P. butleri, P. periplocum, P. pulchrum* and *P. ultimum var. sporangiiferum* flourished in Gaula river only. *P. aquatile, P. echinulatum, P. multisporum, P. pluroticum* and *P. dissotocum* were isolated from Kosi river (Khulbe *et al.*, 1993a, 1994). These species require special features like a narrow range of temperature between 14.5–25°C, pH from

Table 8.1: Occurrence of *Pythium* species in Lentic and Lotic water bodies of Kumaun Himalaya (after Khulbe et al, 1980b, 1988, 1993b; Khulbe and Durgapal, 1997).

	Lentic water (Lakes)									Lotic water (Rivers)					
Species	1	2	3	4	5	6	7	8	Total	A	B	C	D	E	Total
P. afertile	+	+	+	+	+	+	+	–	7	–	–	–	–	–	0
P. aphanidermatum	–	–	–	–	–	–	–	+	1	–	+	–	–	+	2
P. aquatile	–	–	–	–	–	–	–	–	0	–	+	–	–	–	1
P. butleri	–	–	–	–	–	–	–	–	0	–	–	–	–	+	1
P. debaryanum	–	–	–	–	–	–	–	–	0	–	+	–	+	–	2
P. dissotocum	–	–	–	–	–	–	–	–	0	–	–	+	–	–	1
P. echinulatum	–	–	–	–	–	–	–	+	1	–	+	–	–	–	1
P. gracile	–	–	–	–	+	–	–	–	1	–	–	–	–	–	0
P. hypogynum	–	–	–	–	–	–	–	+	1	–	–	–	–	–	0
P. inflatum	–	–	–	–	–	–	–	+	1	–	–	–	–	–	0
P. middletonii	–	–	–	–	+	–	–	–	1	–	–	–	–	–	0
P. monospermum	+	+	+	+	–	–	–	–	4	–	–	–	–	–	0
P. multisporum	–	–	–	–	–	–	–	–	0	–	+	–	–	–	1
P. periplocum	–	–	–	–	–	–	–	–	0	–	–	–	–	+	1
P. pluroticum	–	–	–	–	–	–	–	–	0	–	–	–	–	+	1
P. proliferum	+	+	+	+	–	–	–	+	5	–	–	–	–	–	0
P. pulchrum	–	–	–	–	–	–	–	–	0	–	–	–	–	+	1
P. torulosum	–	–	–	–	–	–	–	+	1	–	–	–	–	–	0
P. ultimum var. *sporangiiferum*	–	–	–	–	–	–	–	–	0	–	–	–	–	+	1
P. undulatum	+	+	+	+	+	+	+	+	8	+	+	+	+	–	4
P. vexans	–	–	–	–	–	–	–	+	1	+	+	–	+	+	4
Total	4	4	4	4	4	2	2	8		2	8	2	3	6	

1–Nainital, 2-Khurpatal, 3-Bhimtal, 4-Naukuchiyatal, 5-Ramtal, 6-Naldamyantital, 7-Garurtal, 8-Nanaksagar, A-Ramganga, B-Kosi, C-Saryu, D-Gomati, E-Gaula

7.7–8.8, dissolved oxygen from 7.0–10.6 mg/l and a wide range of organic matter between 10–170 ppm.

SEASONAL PERIODICITY

Pythium species showed a marked seasonal periodicity in lentic and lotic water systems of the Himalayan region (Mer *et al.*, 1980). In lentic waters, four species, namely *P. afertile, P. undulatum, P. gracile* and *P. proliferum* flourished in all seasons and showed their wide adaptability from winter to summer except for the last species which was absent in spring. *P. middletonii* and *P. aphanidermatum* were isolated from rainy to winter seasons (Table 8.2).

Table 8.2: Seasonal periodicity of *Pythium* species in Lentic (LE) and Lotic (LO) Water Systems (after Khulbe *et al*, 1993a; Mer *et al*, 1980)

Species	Spring		Summer		Rainy		Autumn		Winter	
	LE	LO	LE	LO	LE	LO	LE	LO	LE	LO
P. afertile	+	–	+	–	+	–	+	–	+	–
P. aphanidermatum	–	–	–	–	–	+	–	+	+	–
P. aquatile	–	–	–	–	–	+	–	–	–	–
P. butleri	–	–	–	+	–	–	–	–	–	–
P. debaryanum	–	+	–	+	–	+	–	+	–	+
P. dissotocum	–	+	–	–	–	+	–	–	–	–
P. echinulatum	–	–	–	–	–	–	+	+	+	–
P. gracile	+	–	+	–	+	–	+	–	+	–
P. hypogynum	+	–	–	–	–	–	+	–	+	–
P. inflatum	–	–	–	–	+	–	+	–	+	–
P. middletonii	–	–	–	–	+	–	+	–	+	–
P. monospermum	+	–	+	–	–	–	–	–	+	–
P. multisporum	–	–	–	–	–	–	–	–	–	+
P. periplocum	–	–	–	–	–	–	–	+	–	–
P. pluroticum	–	–	–	–	–	–	–	+	–	–
P. proliferum	–	–	+	–	+	–	+	–	+	–
P. pulchrum	–	+	–	–	–	–	–	–	–	+
P. torulosum	–	–	–	–	+	–	–	–	–	–
P. ultimum var. sporangiiferum	–	–	–	–	–	–	–	+	–	+
P. undulatum	+	+	+	+	+	+	+	+	+	+
P. vexans	+	+	–	–	–	–	–	+	+	–
Total	6	5	5	3	7	5	8	8	11	5

Some species occurred in one to three different seasons, namely *P. aphanidermatum* in winter, *P. torulosum* in rainy, *P. echinulatum* in autumn and winter and *P. vexans* in spring and winter. Similarly, *P. monospermum* and *P. hypogynum* were found during spring and winter seasons. Besides, these species were also isolated during summer and autumn seasons, respectively.

In lotic water system, two species *P. debaryanum* and *P. undulatum* were found to be common in all the seasons. The other species were reported in one or more seasons. For example, *P. aphanidermatum, P. periplocum, P. echinulatum* and *P. pluroticum* appeared only in autumn, *P. multisporum* in winter, *P. aquatile* in rainy season and *P. butleri* in summer; *P. vexans* and *P. dissotocum* showed their presence during spring and rainy seasons. The former species was also collected in autumn. Some species were isolated during two seasons viz, *P. pulchrum* in spring and winter, and *P. ultimum var. sporangiiferum* in autumn to winter seasons.

FISH PATHOGENICITY

Species of *Pythium* have also been found to be associated with living fish under natural conditions (Sati, 1981; Sati and Khulbe, 1983). Eight species were found to be associated with six fish species of the lentic and lotic water (Table 8.3). *P. aphanidermatum*, *P. undulatum* and *P. inflatum* exhibited pathogenic association with three fish species, namely *Puntius ticto*, *P. conchonius* and *Carassius aurantus* (Khulbe, 1983a). The other five species were associated with only two fish species (Khulbe, 1983a; Sidhu and Khulbe, 1998). However, *Tor tor* and *Nemacheilus rupicola* were found free from *Pythium* infection, while *Puntius ticto* and *P. conchonius* were parasitised by maximum number of *Pythium* species (Khulbe, 1983b).

Table 8.3: *Pythium* Species associated with Fish in Kumaun Himalaya (after Khulbe, 1983a,b; Sidhu and Khulbe, 1998).

Species	*Puntius ticto*	*Puntius conchonius*	*Carassius aurantus*	*Gudusia chapra*	Total
P. aphanidermatum	+	+	+	–	3
P. debaryanum	+	+	–	–	2
P. echinulatum	+	–	–	+	2
P. inflatum	+	+	+	–	3
P. pulcharum	+	+	–	–	2
P. spinosum	+	+	–	–	2
P. undulatum	+	+	+	–	3
P. vexans	+	+	–	–	2
Total	8	7	3	1	

PYTHIUM IN RELATION TO PHYSICO-CHEMICAL FACTORS

Pythium species showed their occurrence at a wide range of temperature (12.7–30°C), pH (7.2–10), dissolved oxygen (6.5–21.5 mg/l) and organic matter (10–300 ppm). However, most of the species were isolated from waters having lower temperature (15°C), low alkaline (pH below 9), less dissolved oxygen (below 10 mg/l) and higher organic matter (Table 8.4).

Being facultative in nature, *Pythium* species flourished in waters having more organic matter in dissolved or undissolved states. Various point and non-point (diffused) sources of pollution (in the catchment areas of the water bodies of Kumaun region), such as litter falls, domestic wastes, kitchen and market products, dung and fertilisers have been found as good sources of organic matter (Khulbe and Durgapal, 1997; Khulbe *et al.*, 1988).

Table 8.4: Occurrence of *Pythium* species in relation to temperature, pH, dissolved oxygen (D.O.) and organic matter (after Khulbe and Durgapal, 1997; Khulbe et al, 1988).

Species	Temperature (°C)	pH	D.O. (mg/l)	Organic Matter (ppm)
P. aphanidermatum	15–24	8.2–9.3	8.2–12.0	30–122
P. aquatile	24–25	7.7–8.0	9.0–8.0	350–360
P. butleri	21–22	8.8–8.5	8.6–8.7	80–122
P. debaryanum	11–30	7.4–10.0	6.5–14.0	40–350
P. dissotocum	18–23	7.7–8.8	7.0–10.6	80–254
P. echinulatum	15–24	8.2–9.0	8.0–21.5	10–124
P. hypogynum	15–24	8.2–9.0	16.5–21.5	10–60
P. inflatum	15–28	7.2–9.4	8.5–16.0	30–110
P. multisporum	18–22	8.5–9.0	8.0–9.5	85–124
P. periplocum	19–22	9.5–9.3	8.6–8.9	80–122
P. pleroticum	21–22	9.0–8.6	9.2–8.0	96–124
P. proliferum	15–30	7.2–9.6	9.0–16.5	10–90
P. pulcharum	15–14.5	8.9–9.2	8.9–9.0	64–80
P. torulosum	15–28	7.2–9.4	12.0–16.0	30–90
P. ultimum var. sporangi ferum	12.7–22	8.4–8.8	7.5–8.7	40–122
P. undulatum	15–30	7.2–9.6	6.5–15.0	50–170
P. vexans	15–25	7.2–10.0	7.4–21.0	10–300

REFERENCES

A.P.H.A. 1989. Standard method for the examination of water and wastewater. (17th edition) American Public Health Association, Washington DC.

Bisht, G.S.; Bisht, D.; Joshi, C. and Khulbe, R.D. 1996. Potential threat to reservoir fishery by fungi in Kumaun Himalaya, India. Current Science, 71: 720–723.

Dick, M.W. 1990. Keys to *Pythium*. Department of Botany, University of Reading, UK. 64p.

Johnson, T.W. Jr. 1956. The Genus *Achlya*: Morphology and Taxonomy. University of Michigan Press, Amn. Arbor. 180 p.

Khulbe, R.D. 1980. Occurrence of watermoulds in some lakes of Nainital, Kumaun Hills, India. Hydrobiologia, 74: 77–80.

Khulbe, R.D. 1983a. Pathogenicity of some species of *Pythium* Pringsheim on certain freshwater temperate fishes. Mykosen, 26: 273–275.

Khulbe, R.D. 1983b. Studies on aquatic fungi in relation to fish disease in Kumaun Himalaya. Plant and Nature, 1: 65–67.

Khulbe, R.D. 1989. Infection ability of watermoulds in some temperate fishes of Himalayas, India Mykosen, 32: 84–86.

Khulbe, R.D. 1992. Watermoulds and their activity in Kumaun Himalaya, India, Water Sci. Tech., 26: 2595–2598.

Khulbe, R.D. and Bhargava, K.S. 1977. Distribution and seasonal periodicity of watermoulds in some lakes of Nainital Hills, India. Hydrobiologia, 54: 67–72.

Khulbe, R.D. and Bhargava, K.S. 1981. Studies in parasitic watermoulds of Kumaun Himalayas: The host range of five species of *Olpidiopsis*. Hydrobiologia, 76: 45–47.

Khulbe, R.D. and Bhargava, K.S. 1983. Frequency of watermoulds in relation to nitrates, sulphate and phosphate in some lakes of Nainital. Tropical Ecology, 24: 180–187.

Khulbe, R.D. and Durgapal A. 1997. Evaluation, impact and remedies of water pollution in Kumaun Himalaya with reference to microorganisms. In: S. C. Sati, *et al.* (Eds.) Recent Researches in Ecology, Environment and Pollution, Vol. XI. Today and Tomorrow's Print. and Publ., New Delhi, pp. 361–370.

Khulbe, R.D. and Sati, S.C. 1980. Studies of parasitic watermoulds of Kumaun Himalaya; Host range of *Achlya americana* Humphrey on certain temperate fish. Mykosen, 24: 177–180.

Khulbe, R.D., Joshi, C. and Bisht, G.S. 1993a. Factors affecting seasonal occurrence of watermolds in Kosi river, Kumaun, India. Ind. J. Microbiol. Ecol., 3: 37–45.

Khulbe, R.D., Sati, M.C. and Dhyani, A.P. 1988. Water pollution in Nainital lake: A survey. In: Khulbe, R. D. (Ed.) Perespectives in Aquatic Biology. Papyrus Publication, New Delhi. Pp 269–279.

Khulbe, R.D., Joshi, C. and Bisht, G.S. 1995. Fungal diseases of fish in Nanak Sagar, Nainital, India. Mycopathologia, 130: 71–74.

Mer, G.S. 1992. Species composition and habit preference of watermoulds in the district of Kumaun Himalaya, India. Int. Revul. Ges. Hydrobiol., 77: 315–324.

Mer, G. S. and Khulbe, R. D. 1983. *Pythium vexans var. minuta* Var. Nov. from Kumaun Himalaya, India. Current Science, 52: 735–736.

Mer, G.S., Sati, S.C. and Khulbe, R.D. 1980. Occurrence, Distribution and Seasonal Periodicity of Some Aquatic Fungi in Sattal (Nainital) India. Hydrobiologia, 76: 201–205.

Mer, G.S., Verma, B.L. and Khulbe, R.D. 1984. Aquatic fungi of Kumaun Himalaya, India: *Pythium* Pringshim. Sydowia, 37: 208–221.

Middleton, J.T. 1943. The taxonomy, host range and geographic distribution of the genus *Pythium* Mem. Torrey Bot. Club., 20: 1–171.

Robertson, G.J. 1980. The genus *Pythium* in New Zealand, New Zealand J. of Botany, 18:73–102.

Sati, S.C. 1981. Aquatic fungi of Kumaun in relation to fish infection. Ph.D. Thesis, Kumaun University, Nainital. 111 pp.

Sati, S.C. 1983. Use of certain chemicals in controlling fungal parasites of fish. Bibliotheca Mycologia, 91: 563–573.

Sati, S.C. 1997. Diversity of aquatic fungi in Kumaun Himalaya: Zoosporic fungi. In: Sati, S.C. *et al.* (Eds.) Recent Researches in Ecology, Environment and Pollution, Vol. XI. Today and Tomorrow's Printers and Publishers, New Delhi. pp. 1–16.

Sati, S.C. and Khulbe, R.D. 1980. Occurrence of aquatic fungi in some soils of Nainital hills. Geobios, 7: 42–43.

Sati, S.C. and Khulbe, R.D. 1983. *Pythium gracile* Schenk as parasite of fish gills. Indian Phytopath., 587–588.

Sati, S.C. and Tiwari, N. 1992. Studies on five species of *Pythium* parasitic on mustard and Cabbage in India. Mycopathologia, 119: 97–100.

Sati, S.C., Tiwari, N. and Adhikari, T. 1992. Zoosporic fungi associated with vegetable growing soils of Khurpatal (Nainital) Kumaun Himalaya. Proc. Nat. Acad. Sci., 62:433–435.

Sidhu, R.K. and Khulbe, R.D. 1998. A survey of impact and evaluation of water pollution on reservoir fisheries in Kumaun Himalaya, India with special reference to microorganisms. Current Science, 75: 1303–1308.

Verma, B.L. 1987a. A serious root disease of tomato caused by *Pythium inflatum* Matthews. Current Science, 56: 616–617.

Verma, B.L. 1987b. Zoosporic fungi in wheat field of Kumaun Himalaya. Madras Agric. J., 74: 1–5.

Verma, B.L. and Khulbe, R.D. 1985. *Pythium polytylum* Drechsler and *Pythium rostratum* Butler in some crop field of Tarai, Nainital. Science and Culture, 51: 156–157.

Verma, B.L. and Khulbe, R.D. 1986. *Pythium inflatum* Mathews, a new record for India. Current Science, 55: 47–48.

Waterhouse, G.M. 1967. Key to *Pythium* Pringsheim. Mycological Papers. Common Wealth Mycological Institute, 109: 15.

Waterhouse, G.M. 1968. The genus *Pythium* Pringsheim. Mycological Papers. Common Wealth Mycological Institue, 110: 1–171.

Section B

Fungal Diseases/Management

9

Ecofriendly Disease Management Approach in Horticulture

R.C. Sharma, Sanjeev Sharma and N.P. Dohroo

Department of Mycology and Plant Pathology, Dr Y.S. Parmar University of Horticulture and Forestry, Nauni, Solan – 173 230 (H.P.), India

SUMMARY

The inappropriate use of fungicides and other agrochemicals (in our Horti-ecosystem) in the last two decades has resulted in the pollution of the environment resulting in diseases of a complex nature vis-à-vis new pest problems. With more and more reports of resistance of pathogens to fungicides, the number of diseases has increased and many weak parasites have attained the status of serious pathogens. The disease outbreaks have been maximum in areas where crops were subjected to various fungicide applications. For example, the indiscriminate use of pesticides in vegetables is of such high magnitude that a re-appraisal of our strategies for pest control has become necessary. In addition, quality aspects such as pest-free and residue-free agricultural products are becoming increasingly important in the light of market liberalisation and increasing importance of agricultural exports. Biocontrol agents may thus, substitute chemicals by protecting the environment or these may be used in conjunction with minimal amounts of chemicals in an integrated disease management system. Biological control still remains an area, which needs to be further exploited. Genetic engineering has made it possible to select mutants with enhanced activities or to combine efficacy of different mutants through protoplast fusion. After obtaining active new strains, further selection could be carried out for strains tolerant to fungicides which are used regularly in the crop chain, allowing the use of integrated control. There is a need to optimise production, formulation, storage, application procedures and registration of their use.

Keywords: Disease management, Biocontrol agents, Biological control

Of the various strategies available for disease management, chemical based strategies have been so far dominating. Overemphasis on chemical control of plant diseases has caused serious imbalances in the agro ecosystems. Strangely, about 70 per cent of the chemicals sprayed on field crops does not stick to the plants. Enormous quantities of chemicals adversely affect microbial life that fall onto the earth get mixed up with soil. Some problems like non-target effects of chemicals as well as chemically induced diseases/malformations are being experienced. A shift towards non-chemical strategies is likely to correct the imbalance in our approach and the most logical and ecofriendly approach in this direction is biological control.

Biological control employs the introduction of antagonists into cropping systems as well as manipulation of the eco-environment to favour resident beneficial organisms through crop rotation and a wide range of cultural practices. People around the world are now aware of possible deleterious effects of pesticides in the biosphere and with the growing interest in chemical-free agricultural food products—the biological control of plant pathogens is receiving considerable attention these days.

Biological control is the reduction in the amount of inoculum or disease-producing activities of a pathogen in its active or dormant state by one or more organisms, accomplished naturally or through manipulation of the environment, host, or antagonist, or by mass introduction of one or more antagonists.

The antagonists in biological control of plant pathogens include bacteria, actinomycetes, fungi, yeasts and predatory microfauna such as protozoa. Antagonists of fungal or bacterial origin have long been studied in the management of plant diseases. Among the fungi, *Trichoderma* species have been fully exploited for their use in biocontrol agents (Lisansky, 1993). Among these, *T. viride* and *T. harzianum* have shown much promise in the biocontrol of plant pathogens (Singh *et al.*, 2000).

KINDS OF ANTAGONISTS

Bacteria

Bacteria are extremely useful in the biological control of plant pathogens. These rank first in order of importance in maintaining an eco-balance in the soil, largely because of their rapid growth and ability to utilise different forms of nutrients. The species of *Agrobacterium, Arthrobacter, Bacillus, Erwinia,*

Enterobacter, Pseudomonas and *Xanthomonas* act as antagonists (Pal and Jalali, 1998). Out of these, the species of *Pseudomonas* and *Bacillus* have been most extensively studied. The premier example of a bacterial biocontrol agent is *A. radiobacter* strain 84, which controls crown gall caused by *A. tumefaciens*. Strain 84 is the first bacterium used commercially in biocontrol and has been successful worldwide. Among the *Pseudomonas* species, fluorescent pseudomonads make up a dominant population in the rhizosphere and possess several properties that have made them into biocontrol agents of great value and choice (Johri *et al.*, 1997). *Bacillus* represents the most utilised biocontrol agent because of its inherent superiority in the biological control of plant pathogens. The resistant endospores that *Bacillus* spp. produce, provide tolerance to heat and cold as well as to pH extremes, pesticides, fertilisers and storage.

Actinomycetes

Actinomycetes are second only to bacteria in use, but are important in maintaining a satisfactory eco-balance in the soil, largely because of their widespread ability to produce antibiotics. It is known that three-fourth of the species of *Streptomyces* produce antibiotics. Among these, a most are effective against pathogenic bacteria, but many are effective against fungi too. Some of these have established ecto enzyme systems required to break down proteins, cellulose and chitin, which thus have great potential in the biocontrol of microorganisms, of pathogenic nature, affecting horticultural plants.

Fungi

Fungi have received a lot of attention as antagonists, possibly because they are easier to handle and identify than bacteria and actinomycetes. However, fungi probably rank third as potential antagonists in biological control. The most common fungi of ecofriendly nature are the species of *Trichoderma, Gliocladium, Pythium, Chaetomium, Coniothyrium, Talaromyces* and *Paecilomyces*. There is a growing interest among researchers to deploy mycorrhizal fungi like *Glomus, Gigaspora, Acaulospora* and *Scutellospora* etc. in the microenvironment of horticultural plants affected with diseases.

Yeasts

The species of *Cryptococcus* and *Rhodotorula* have been reported to be antagonistic against *Botrytis cinerea* (Elad *et al.*, 1994). *Pseudozyma flocculosa* is a commonly used microbial antagonist against powdery mildew.

Protozoa

Very little importance has been placed on soil amoeba in comparison to bacteria, fungi and actinomycetes. There are many examples where these amoebae feed on the conidia, chlamydospores, sporangia, teleutospores, basidiospores and hyphae of various pathogens (Old and Patrick, 1976). Amoebae cause perforation and lysis of spores and hyphae. The important parasitic amoebae in use are *Arachnula impatiens* and *Thecamoeba granifera*.

FORMS OF ANTAGONISM/MECHANISMS OF BIOLOGICAL DISEASE CONTROL

An understanding of plant-microbe ecology and the mechanism(s) of action of biocontrol agents / products is a pre-requisite for their effective use in the control strategy of diseases in horticultural plants. In general, parasitism or predation of one organism by another; antibiosis through the production of antibiotics, siderophores and other metabolites and; competition for nutrients and niche have been suggested as the major mechanisms of action of microorganisms in the biological control of various plant diseases. Besides, protection against pathogens can also be obtained through cross protection or induced resistance. Fluorescent *Pseudomonas* spp. produce 2,4 –DAPG which have biocontrol ability.

Parasitism and Predation

In the ecofriendly biocontrol context, these usually refer to the destruction of plant pathogens by biocontrol agents. Parasitism and predation are common mechanisms of fungal biocontrol agents viz, *Trichoderma* sp., *Gliocladium virens, Coniothyrium miniforms, Laetisaria arvalis, Pythium nunn* and *Talaromyces flavus* (Adams, 1990). However, certain rhizosphere bacteria have also been reported to parasitise plant pathogenic bacteria and fungi. Amino sugars and other cell wall constituents of the pathogen cells are dissolved by the lytic enzymes, glucanase and peptidase (Goto, 1990). In addition to peptidase and glucanase, the role of chitinase and other hydrolysing enzymes has also been established in the lysis of some plant pathogens.

Antibiosis

Antibiosis perhaps, is the most investigated mechanism of biocontrol agents. The term, antibiosis applies to the production and release of certain inhibitory metabolites by microorganisms. These toxic or antimicrobial metabolites mainly include antibiotics, bacteriocins and several other extracellular metabolites. The phenomenon of antibiosis in biocontrol was first reported by Weindling (1934). Trichodermin, a sesquisterpene, is an antibiotic produced by *Trichoderma,* while gliotoxin and virdin are produced by *Gliocladium*. Among bacterial biocontrol agents, fluorescent *Pseudomonas* is considered to be the most promising, as it produces a wide spectrum of inhibitory compounds like DAPG.

Competition

Competition for nutrients or ecological niche is another mechanism by which biocontrol agents reduce diseases of horticultural plants. Competition for the nutrients supplied by root and seed exudates occurs between rhizosphere microorganisms and plant pathogens. It seems logical that fast-growing rhizosphere microorganisms would effectively compete with slow-growing plant pathogens for the available nutrients in root exudates in the rhizosphere. This mechanism was proposed by Elad and Chet (1987) who found a correlation between the ability of various strains of *Arthrobacter* sp. to compete for nutrients with germinating oospores of *Pythium aphanidermatum* causing nursery diseases of horticultural plants.

The infected sites have also been suggested as an important ecological niche in the biocontrol of plant pathogens (Suslow and Schroth, 1982). Usually, the points of emergence of lateral roots and cell junctions of roots appear to be favoured for colonisation by many kinds of rhizosphere microorganisms due to abundant availability of nutrients in root exudates. The application of ecofriendly biocontrol agents prevents the establishment of most of the pathogens at these sites.

METHODS OF APPLICATION OF BIOCONTROL AGENTS

Colonised natural substrates are one of the methods of *Trichoderma* delivery in soil (Elad *et al.*, 1980, 1982; Strashnow *et al.*, 1985). Chet (1990) demonstrated that four different techniques ie broad cast, furrow, root zone, and seed coating methods for the application of *Trichoderma* sp. as biocontrol agents. Each method has been shown to be effective in this field.

Seed Inoculation

The treatment of seeds with biocontrol agents, such as fungi and bacteria offers protection against a number of soil-borne pathogens, especially the damping-off of fungi (Paulitz, 1991; Harman, 1991). It is done either by seed priming, seed coating or seed pelleting. The basic strategy of this method is to establish a large population of the antagonist on a spermosphere as a protectant, which colonises the seed coat by protecting the young germinating seedlings against attack by species of *Pythium, Phytophthora, Rhizoctonia* etc. Compared to other strategies of introducing biocontrol agents, seed treatment requires relatively small amounts of inoculum and is an economical and commercially feasible method of applying a biocontrol agent, especially for high value hybrid seeds.

Soil Inoculation

Considerable interest in the biocontrol of plant pathogens has been developed among growers during recent years, especially in soil-borne pathogens, by inoculation of soil with *Trichoderma* and *Gliocladium* preparations. Biocontrol agents can be augmented in soil through broadcast, furrow and root zone application methods.

Biocontrol by adding large amounts of *T. harzianum* to soil with its food base is exemplified by the work of Wells *et al.* (1972). *Trichoderma* species, especially *T. harzianum* grown on solid substrates have been tested with varying degrees of success against white root rot of onion (*Sclerotium cepivorum*) in Egypt (Abd-El Moity *et al.*, 1982; Papavizas *et al.*, 1982), cucumber diseases (*Verticillium dahliae*) in erstwhile USSR (Fedorinchik *et al.*, 1975), Rhizoctonia damping-off and blight (*Rhizoctonia solani*) of several crops in Israel (Elad *et al.*, 1980) and Rhizoctonia fruit rot of cucumber (Lewis and Papavizas, 1980).

Foliage Application

Trichoderma can be applied as a foliar spray to control diseases. *T. harzianum* when sprayed on cucumber and tomato plants in greenhouse and fields achieved effective control of grey mold (Elad and Ziamand, 1992; Elad *et al.*, 1993). Application of a spore suspension of the fungus *Colletotrichum lagenarium* to the cotyledons or first leaves of cucumber made the same plants subsequently resistant not only to anthracnose caused by *C. lagenarium* but also to a range of other pathogens (Dean and Kuc, 1985). Direct application of bacteria to leaves and stems has been shown to regularly decrease diseases caused by a range of necrotrophs and biotrophs. Application of *Bacillus subtilis* onto leaves of bean, decreased the incidence of bean rust (*Uromyces phaseoli)* up to 75 per cent and when applied three times a week it gave control equivalent to weekly treatments with the fungicide, mancozeb (Baker *et al.*, 1983). Long-term control can also be achieved for diseases of perennials. *Penicillium frequentans* has been used to control twig blight on peaches caused by *Monilinia laxa* and to control equivalent to that, obtained with captan (De Cal *et al.*, 1990).

TIMING OF DELIVERY

The time of the application of biocontrol agents should be considered with respect to the stage of crop development, the effect of the environment upon the survival and activity of the agent and the target pathogen, and other management practices like cultivation, pruning, pesticide application etc. Decisions on biocontrol applications should be based not on our convenience, but on understanding the effects of the above factors especially these that impinge upon the performance of the agents. Which development stages of the plant are susceptible, how long it remains susceptible, and the duration of exposure to the pathogen need to be considered to determine application rates and frequencies. The effects of temperature and moisture upon the survival and activity of the agent should be also considered (Spurr and Knudsen, 1985). The number of times the agent needs to be applied will depend not only upon plant parameters (eg age, duration of susceptibility, root growth and morphology, canopy architecture), but also upon agent parameters (eg survival,

Table 9.1: Modes of Action Exhibited by Potential Biocontrol Agents

Antagonist	Pathogen	Reference
Parasitism or predation		
Fauna		
Amoebae	*Thielaviopsis basicola*	Old and Patrick, 1976
Bacteria and actinomycetes		
Arthrobacter spp.	*Pythium debaryanum*	Mitchell and Hurwitz, 1965
Fungi		
Acremonium alternatum	*Sphaerotheca fuliginea*	Malathrakis, 1985
Ampelomyces quisqualis	*S. fuliginea*	Phillip, 1985
Gliocladium catenulatum	*Fusarium* spp., *Sclerotinia sclerotiorum*	Huang, 1978
Pythium nunn	Numerous	Deacon, 1976
P. oligandrum	Numerous	Laing and Deacon, 1990
Trichoderma hamatum	*Pythium* spp., *Rhizoctonia solani*	Chet *et al*, 1981
Trichoderma spp.	*R. solani* and *Sclerotium rolfsii*	Elad *et al*, 1983
Antibiosis		
Bacteria and actinomycetes		
Agrobacterium radiobacter	*A. tumefaciens*	Kerr and Tate, 1984
Bacillus subtilis	Numerous	Swinburne *et al*, 1975
Enterobacter aerogenes	*Phytophthora cactorum*	Gupta and Utkhede, 1987
Streptomyces griseovirdis	*Alternaria brassicicola, R. solani, F. oxysporum* f.sp. *dianthi*	Lahdenpera *et al*, 1991; Rattink, 1989
Fungi		
Gliocladium virens	*Pythium ultimum, R. solani*	Howell and Stipanovic, 1983; Roberts and Lumsden, 1990
Talaromyces flavus	*Verticillium dahliae*	Fravel *et al*, 1987
Penicillium oxalicum	*F. oxysporum* f.sp. *conglutinans*	Reyes, 1984
Trichoderma spp.	Numerous	Papavizas, 1985
Competition		
Bacteria and actinomycetes		
Enterobacter spp.	*Pythium* spp.	Nelson *et al*, 1986
Pseudomonas putida	*F. oxysporum*	Scher and Baker, 1985
Pseudomonas spp.	*F. oxysporum, F. oxysporum* f.sp. *cucumerinum, F. solani, P. aphanidermatum*	Elad and Baker, 1985; Sneh *et al*, 1984; Elad and Baker, 1985; Elad and Chet, 1987
Fungi		
Pythium nunn	*P. ultimum*	Paulitz and Baker, 1988
P. oligandrum	*P. ultimum*	Martin and Hancock, 1986
Trichoderma harzianum	*F. oxysporum* f.sp. *melonis, F. oxysporum* f.sp. *vasinfectum*	Sivan and Chet, 1989
Cross protection		
Viruses		
Tobacco necrosis virus	*Colletotrichum lagenarium*	Jenns and Kuc, 1977

Tomato mosaic virus race MII-16	Tomato mosaic virus	Fulton, 1986
Mild tristeza virus	Tristeza virus	Fulton, 1986
Bacteria and actinomycetes		
Bacillus subtilis	*F. roseum* f.sp. *dianthi*	Aldrich and Baker, 1970
Pseudomonas syringae	*Colletotrichum lagenarium*	Caruso and Kuc, 1979
Pseudomonas spp.	*F. oxysporum* f.sp. *dianthi*	Van Peer and Schippers, 1991
Fungi		
Colletotrichum lagenarium	*C. lagenarium*	Dean and Kuc, 1985
Didymella bryoniae	*D. bryoniae*	Descalzo *et al*, 1990
Phytophthora parasitica	*P. cinnamomi, P. citricola*	Dolan *et al*, 1986

Table 9.2: Commercial Biocontrol Products

Biocontrol Agent	Target Pathogen	Disease/Host	Product	Company
Bacteria and actinomycetes				
Agrobacterium radiobacter strain K84	A. tumefaciens	Crown gall of several crops	Diegall	Fruit Growers Chemical Co., NZ
			Galltrol	Ag Biochem. Inc. CA, USA
			Norbac 84c	New Bio Products Inc. CA, USA
A. radiobacter strain K1026			No Gall	Root Nodule Pvt. Ltd., Bio Cave, Australia
Bacillus subtilis strain GB 03	*Pythium ultimum, Rhizoctonia solani*	Damping-off of cotton, legumes	Kodaik	Gustafson Inc. Tx, USA
B. subtilis strain GB 07			Epic	-do-
B. subtilis strain A-13/GB03			Quantum	-do-
B. cepacia	*Pythium* sp.	Damping-off of several hosts	Deny	CCT Crop Carlsbad, USA
Pseudomonas fluorescens	*P. tolaasii*	Bacterial blotch of mushrooms	Conqueror	Mauri Foods, Australia
P. fluorescens WCS 374r	*Fusarium oxysporum* f.sp. *raphani, F. oxysporum* f.sp. *dianthi*	Fusarium wilt, Carnation wilt	Biocoat	S and G seeds, BV, Netherlands
P. fluorescens strain A 506	*Erwinia amylovora*	Fire blight	Blightban	Plant Health, Boise, ID, USA

Streptomyces griseovirdis	Alternaria brassicicola, *F. oxysporum* f.sp. *dianthi*	Damping-off of crucifers, Fusarium wilt	Mycostop	Kemira Agro Oy, Finland
Fungi				
Coniothyrium minitans	*Sclerotinia sclerotiorum*	Sunflower disease	Coniothyrin	Russian Govt.
Pythium oligandrum	*P. ultimum*	Damping-off of suger beet	Polygandron	Vyzkummyustov, Czechoslovakia
Trichoderma harzianum	*Pythium* spp.	Damping-off	F-Stop	Environment Protection Agency and Eastman Kodak Co.
T. harzianum strain 35	*Pythium, Rhizoctonia, Sclerotinia, Sclerotium, Fusarium*	Damping-off and root rot of vegetables and other plants	MTR-35 ST MTR-35 D MTR-35 WP MTR-35 G	Makhtashim Agam Koor Chem., Israel
T. viride			Biogaurd, Ecoderma, Fungil, Bio-Cure-F, Trieco	India
Trichoderma spp.	*Pythium, Fusarium, Rhizoctonia, Sclerotinia, Sclerotium, Verticillium* and other soil borne pathogens	Damping-off and root rot	BINAB –T Trichodermin Root Pro Trichoderma-2000, Trichodex Root Shield Bio-trek T-22G T-22 Planter Box Supresivit Trichopel Trichojeet Trichodowels Trichoseal	Sweden Bulgarian and Soviet Govt. Israel Israel Israel USA USA USA USA Newzealand Newzealand Newzealand Newzealand Newzealand

Source: Whipps, 1992; Pal and Jalali, 1998; Singh *et al*, 2000

dispersal). With this information, appropriate decision can be made for when and how to deliver the biocontrol agent to the environment or the plant.

CHARACTERISTICS OF AN IDEAL BIOCONTROL AGENT

Over the last 20 years, several investigations have been initiated to find biological control agents for the suppression and control of plant diseases. Though selection criteria for a potential biocontrol agent have been developed from different bases, the task of finding organisms that can function effectively under variable conditions prevailing during the period of plant growth, becomes very difficult. Wilson and Winiewski (1994) have presented nine criteria of ideal antagonists for the biological control of various plant diseases:

(i) Genetic stability
(ii) High consistent efficacy
(iii) Ability to survive under adverse environmental conditions
(iv) Effectiveness against a wide range of pathogens
(v) Amenability for growth on an inexpensive medium in fermentors
(vi) Stability of end products during storage
(vii) Non-production of secondary metabolites that might be toxic to humans
(viii) Strain resistant to standard fungicides
(ix) Compatibility with other chemical and physical treatments of the commodity.

Production and Formulation

During the last 50 years, the control of most of the plant diseases has fully relied on the use of resistant cultivars, cultural methods and the application of fungicides and other agrochemicals. Although, this has resulted in increased production of agricultural commodities, it is also led to the development of resistant strains to these agrochemicals, and increased pollution hazards. The increasing public awareness about these problems has stimulated research on the use of biocontrol agents and the development of commercial bioproducts. Jacobsen (1993) suggested some criteria for the development of ideal biocontrol products. These include:

(i) Biocontrol products must have a relatively wide spectrum of efficacy
(ii) The efficacy of the bioproducts must be high, consistent and reliable
(iii) Biocontrol products must also meet the acceptable standards for environmental and toxicological safety
(iv) The bioproducts must have an acceptable shelf life without special storage requirements
(v) The bioproducts must be cost effective
(vi) Thorough understanding of the mechanism(s) of action and ecological competence of the bioproducts must be made to assure an efficacious product
(vii) The application of the biocontrol products should be easy and possible with existing plant protection equipments
(viii) Biocontrol of bioproducts must also be highly compatible with control practices for other pests and with cultural and chemical control strategies.

Once a biocontrol agent is selected on the basis of its *in vitro* potentiality, production its biomass and suitable formulation becomes a major concern (Lumsden and Lewis, 1989). Formulation of antagonist biomass in a suitable form is another important step for the success of biological control. The formulation must be easy to handle and deliver at the site of action. The biomass production on semi-solid substrate is used as such for soil and seed treatment, while the inoculum grown on broth medium is formulated in powder, wettable form or pellets and so on.

REFERENCES

Abd-El Moity, T.H., Papavizas, G.C. and Shatla, M,N. 1982. Induction of new isolates of *Trichoderma harzianum* tolerant to fungicides and their experimental use for control of white root-rot of onion. *Phytopathology* 72: 396–400.

Adams, P.B. 1990. The potential of mycoparasites for biological control of plant diseases. *Annu. Rev. Phytopathol.* 28: 59–72.

Aldrich, J. and Baker, R. 1970. Biological control of *Fusarium roseum* f.sp. *dianthi* by *Bacillus subtilis*. *Plant Dis. Rep.* 54: 446–448.

Baker, C.J., Stavely, J.R., Thomas, C.A., Sasser, M. and Macfall, J.S. 1983. Inhibitory effect of *Bacillus subtilis* on *Uromyces phaseoli* and on development of rust pustules on bean leaves. *Phytopathology*. 73: 1148–1152.

Caruso, F.L. and Kuc, J. 1979. Induced resistance of cucumber to anthracnose and angular leaf spot of *Pseudomonas lachrymans* and *Colletotrichum lagenarium*. *Physiol. Plant Pathol.* 14: 191–201.

Chet, I. 1990. Biological control of soil-borne plant pathogens with fungal antagonists in combination with soil treatments. pp. 15–25. In: *Biological Control of Soil Borne Plant Pathogens*, (Chet I, Ed.). CAB, Internationals.

Chet, I., Harman, G.E. and Baker, R. 1981. *Trichoderma hamatum*: its hyphal interactions with *Rhizoctonia solani* and *Pythium* species. *Microbial Ecology*. 7: 29–38.

De Cal, A., Sagasta, E.M. and Melgarejo, P. 1990. Biological control of peach twig blight (*Monilinia laxa*) with *Penicillium frequentans*. *Plant Pathol*. 39: 612–618.

Deacon, J.W. 1976. Studies on *Pythium oligandrum*, an aggressive parasite of other fungi. *Trans. Br. Mycol. Soc.* 66: 383–391.

Dean, R.A. and Kuc, J. 1985. Induced systemic protection in plants. *Trends in Biocontrol* 3: 125–129.

Descalzo, R.C., Rahe, J.E. and Mauza, B. 1990. Comparative efficacy of induced resistance for selected diseases of greenhouse cucumber. *Can. J. Plant Pathol.* 12: 16–24.

Dolan, T.E., Cohen, Y. and Coffey, M.D. 1986. Protection of *Persea* species against *Phytophthora cinnamomi* and *P. citricola* by prior inoculation with a citrus isolate of *P. parasitica*. *Phytopathology* 76: 194–198.

Elad, Y. and Baker, R. 1985. The role of competition for iron and carbon in suppression of chlamydospore germination of *Fusarium* spp. by *Pseudomonas* spp. *Phytopathology* 75: 1053–1059.

Elad, Y. and Chet, I. 1987. Possible role of competition for nutrients in biocontrol of Pythium damping-off by bacteria. *Phytopathology* 77: 190–195.

Elad, Y. and Zimand, G. 1992. Integration of biological and chemical control for grey mould. In: *Recent Advances in Botrytis Research* (Verhoeff, K., Malathrakis, N.E. and Williamson, B., Eds.). Proceeding of 10th International *Botrytis* Symposium, Herakhin, Crete, Greece.

Elad, Y., Chet, I., Boyle, P. and Henis, Y. 1983. Parasitism of *Trichoderma* spp. on *Rhizoctonia solani* and *Sclerotium rolfsii*-scanning electron microscopy and fluorescens microscopy. *Phytopathology* 73: 85–88.

Elad, Y., Hadar, Y., Chet, I. and Henis, Y. 1982. Prevention with *Trichoderma harzianum* Rifaaggr. of reinfestation of *Sclerotium rolfsii* Sacc. and *Rhizoctonia solani* Kuhn of soil fumigated with methyl bromide and improvement of disease control in tomatoes and peanuts. *Crop Protection*. 1: 199–211.

Elad, Y., Katan, J. and Chet, I. 1980. Physical, biological and chemical control integrated for soil-borne diseases in potatoes. *Phytopathology*. 70: 418–422.

Elad, Y., Kohl, J. and Fokhema, N.J. 1994. Control of infection and sporulation of *Botrytis cinerea* on bean and tomato by saprophytic bacteria and fungi. *Eur. J. Plant Pathology*. 100: 315–336.

Elad, Y., Zimand, G., Zaqs, Y., Zuriel, S. and Chet, I. 1993. Use of *Trichoderma harzianum* in combination or alternation with fungicides to control cucumber grey mould (*Botrytis cinerea*) under commercial greenhouse conditions. *Plant Pathology*. 42: 324–332.

Fedorinchik, N.S., Tarunina, T.A., Tyunnikov, M.G. and Kudryavtseva, K.I. 1975. Trichodermin-4, a new biological preparation for plant disease control. *8th Int. Congr. Plant Prot.* 3: 67–72, Moscow: USSR Org. Comm.

Fravel, D.R., Kim, K.K. and Papavizas, G.C. 1987. Viability of microsclerotia of *Verticillium dahliae* reduced by a metabolite produced by *Talaromyces flavus*. *Phytopathology*. 77: 616–619.

Goto, M. 1990. *Fundamentals of Bacterial Plant Pathology*. Academic Press, NY.

Gupta, V.K. and Utkhede, R.S. 1987. Nutritional requirement for production of antifungal substance by *Enterobacter aerogenes* and *Bacillus subtilis* antagonists of *Phytophthora cactorum*. *J. Phytopathol.* 120: 143–153.

Harman, G.E. 1991. Seed treatments for biological control of plant diseases. *Crop Protection*. 10: 166–171.

Howell, C.R. and Stipanovic, R.D. 1983. Gliovirin, a new antibiotic from *Gliocladium virens*, and its role in the biological control of *Pythium ultimum*. *Can J. Microbiol*. 29: 321–324.

Huang, H.C. 1978. *Gliocladium catenulatum*: hyperparasite of *Sclerotinia sclerotiorum* and *Fusarium* species. *Can. J. Bot.* 56: 2243–2246.

Jacobsen, B.J. 1993. Biological and cultural plant disease controls: alternatives and supplements to chemicals in IPM systems. *Plant Dis.* 77: 311–315.

Jenns, A.E. and Kuc, J. 1977. Localised infection with tobacco necrosis for plant virus disease control. *Annu. Rev. Phytopathology*. 24: 67–81.

Johri, B.N., Rao Ch, V.S. and Goel, R. 1997. In: *Biotechnological Approaches in Soil Microorganisms for Sustainable Crop Production*, (Dadarwal K.R., ed.). Scientific Publishers, Jodhpur.

Kerr, A. and Tate, M.E. 1984. Agrocins and the biological control of crown gall. *Microbiology Sci*. 1: 1–4.

Lahdenpera, M.L., Simon, E. and Uoti, J. 1991. Mycostop –novel biofungicide based on *Streptomyces* bacteria. pp. 258–263. In: *Biotic Interactions and Soil–borne Diseases*, (Beemster, A.B.R., Bollen, G.J., Gerlagh, M., Ruissen, M.A., Schippers, B. and Temple, A., Eds.). Elsevier, Amsterdam.

Laing, S.A.K. and Deacon, J.W. 1990. Aggressiveness and fungal host ranges of mycoparasitic *Pythium* species. *Soil Biol. Biochem*. 22: 905–911.

Lewis, J.A. and Papavizas, G.C. 1980. Integrated control of *Rhizoctonia* fruit rot of cucumber. *Phytopathology*. 70: 85–89.

Lisansky, S.G. 1993. Crop protection without chemicals: the present and future of biopesticides. *World Agriculture*. 39–41.

Lumsden, R.D. and Lewis, J.A. 1989. Selection, production, formulation and commercial use of plant disease biocontrol fungi: problems and progress, pp. 171–190. In: *Biochemistry of Fungi for Improving Plant Growth* (Whipps, J.M. and Lumsden, R.D., Eds.). Cambridge University Press, Cambridge.

Malathrakis, N.E. 1985. The fungus *Acremonium alternatum*, a hyperparasite of the cucurbits powdery mildew pathogen *Sphaerotheca fuliginea*. *Z. PflKrankh. PflSchutz*. 92: 509–515.

Martin, F.N. and Hancock, J.G. 1986. Association of chemical and biological factors in soil suppressive to *Pythium ultimum*. *Phytopathology*. 76: 1221–1231.

Mitchell, R. and Hurwitz, E. 1965. Suppression of *Pythium debaryanum* by lytic rhizosphere bacteria. *Phytopathology*. 55: 156–158.

Nelson, E.B., Chao, W.L., Norton, J.M., Nash, G.T. and Harman, G.E. 1986. Attachment of *Enterobacter cloacae* to hyphae of *Pythium ultimum*: possible role in biological control of *Pythium* pre-emergence damping-off. *Phytopathology*. 76: 327–335.

Old, K.M. and Patrick, Z.A. 1976. Perforation and lysis of spores of *Cochliobolus sativus* and *Thielaviiopsis basicola* in natural soils. *Can J. Bot.*54: 2798–2809.

Pal, Vijai and Jalali, Indu. 1998. Rhizosphere bacteria for biocontrol of plant diseases. *Indian J. Microbiology*. 38: 187–204.

Papavizas, G.C. 1985. *Trichoderma* and *Gliocladium*: biology, ecology and potential for biocontrol. *Annu. Rev. Phytopathology*. 23: 23–54.

Papavizas, G.C., Lewis, J.A. and Abd-El Moity, T.H. 1982. Evaluation of new biotypes of *Trichoderma harzianum* for tolerance to benomyl and enhanced biocontrol capabilities. *Phytopathology*. 72: 126–132.

Paulitz, T.C. 1991. Effect of *Pseudomonas putida* on the stimulation of *Pythium ultimum* by seed volatiles of pea and soybean. *Phytopathology*. 81: 1282–1287.

Paulitz, T.C. and Baker, R. 1988. Interactions between *Pythium nunn* and *Pythium ultimum* on bean leaves. *Can. J. Microbiology*. 34: 947–951.

Phillip, D.W. 1985. Extracellular enzymes and nutritional physiology of *Ampelomyces quisqualis* Ces., hyperparasite of powdery mildew, *in vitro*. *Phytopathol. Zeit.* 107: 193–203.

Rattink, H. 1989. Possibilities of biological control of *Fusarium* wilt of carnation. *Mededlingen van de Faculteit Landbouwwetenschappen Rijksunivarsiteit Gent* 54: 517–524.

Reyes, A.A. 1984. Suppressive activity of different fungi against cabbage yellows in the soil and *Fusarium oxysporum* f.sp. *conglutinans in vitro*. *Phytopathology*. 80: 461–465.

Roberts, D.P. and Lumsden. R.D. 1990. Effect of extracellular metabolites from *Gliocladium virens* on germination of sporangia and mycelial growth of *Pythium ultimum*. *Phytopathology*. 80: 461–465.

Scher, P.M. and Baker, R. 1982. Effect of *Pseudomonas putida* and a synthetic iron chelator on induction of soil suppressive to *Fusarium* wilt pathogens. *Phytopathology*. 72: 1567–1573.

Singh, R.S., Singh, P. and Kaur, J. 2000. *Trichoderma* based biofungicides: progress and prospects. *Plant Dis. Res*. 15: 155–161.

Sivan, A. and Chet, I. 1989. The possible role of competition between *Trichoderma harzianum* and *Fusarium oxysporum* on rhizosphere colonisation. *Phytopathology*. 79: 198–203.

Sneh, B., Dupler, M., Elad, Y. and Baker, R. 1984. Chlamydospore germination of *Fusarium oxysporum* f.sp. *cucumerinum* as affected by fluorescent and lytic bacteria from *Fusarium* suppressive soil. *Phytopathology*. 74: 1115–1124.

Spurr, H.W. and Knudsen, G.R. 1985. Biological control of leaf diseases with bacteria. In: *Biological Control on the Phylloplane*, (Windles, C.E. and Lindow, S.E., Eds.). American Phytopathological Society, St. Paul, Minnesota.

Strashnow, Y., Elad, Y., Sivan, A. and Chet, I. 1985. Integrated control of *Rhizoctonia solani* by methyl bromide and *Trichoderma harzianum*. *Plant Pathology*. 34: 146–151.

Suslow, T.V. and Schroth, M.N. 1982. Role of deleterious rhizobacteria as minor pathogens in reducing crop growth. *Phytopathology* 72: 111–115.

Swinburne, T.R., Barr, J. and Brown, A.E. 1975. Production of antibiotics by *Bacillus subtilis* and their effect on fungal colonists of apple leaf scars. *Trans. Br. Mycol. Soc.* 65: 211–217.

Van Peer, R. and Schippers, B. 1991. Biocontrol of fusarium wilt by *Pseudomonas* sp. strain WCS 417r: induced resistance and phytoalexin accumulation. pp. 274–280. In: *Biotic*

Interactions and Soil-borne Diseases, (Beemster, A.B.R., Bollen, G.J., Gerlagh, M., Ruissen, M.A., Schippers, B. and Tempel, A., Eds.). Elsevier. Amsterdam.

Weindling, R. 1934. Studies on a lethal principle effective in the parasitic action of *Trichoderma lignorum* on *Rhizoctonia solani* and other soil fungi. *Phytopathology*. 24: 1153–1179.

Wells, H.D., Bell, D.K. and Jaworski, C.A. 1972. Efficacy of *Trichoderma harzianum* as a biocontrol for *Sclerotium rolfsii*. *Phytopathology*. 62: 442–447.

Whipps, J.M. 1992. Status of biological disease control in horticulture. *Biol. Sci. Technol*. 2: 2–24.

Wilson, C.L. and Wisniewski, M.E. 1994. In: *Biological Control of Post Harvest Diseases: Theory and Practices*. CRC Press, Boca Raton.

10

Seed Storage Fungi in Mustard and Changes in Protein, Oil, Starch and Sugars Under Storage

S.I. GHUGAL AND R.P. THAKRE

Department of Botany, Nagpur University Campus, Nagpur – 440033, India

SUMMARY

Twenty-eight seed storage fungi were isolated in mustard (*Brassica juncea* cv. Pusabold) seed samples by agar plate method and blotter method throughout the year on a monthly basis. Major predominant fungi were *Alternaria alternata, Aspergillus flavus, A. fumigatus, A. niger, Fusarium moniliforme, Penicillium multicolor, P. oxalicum* and *Curvularia lunata.* It was observed that germination percentage reduced from 90% to 60.50% at the end of storage. The crude protein, oil and starch decreased with an increase in storage period. It was observed that the percentage change over control in protein and starch was –30.00%, –11.09% and –61.137% respectively at the end of year. Total sugar, reducing and non-reducing sugars were found to be decreased till 9th month of storage period and subsequently increased at the end of storage period.

Keywords: Mustard, Storage fungi, Seed mycoflora

Brassica juncea (L.) Czern. and Coss. cv. Pusabold, also known as mustard is one of the most important and widely distributed oilseed crop. It is cultivated in the cooler agricultural regions at higher elevations, besides being a winter crop in the temperate zones. Large numbers of fungi are known to bring about several biochemical changes in mustard seeds and degrade seed constituents (Rai and Saxena, 1980). The mycoflora associated with the seed may be pathogenic, weak parasites or saprophytes and may be external or internal. The present investigation deals with storage fungi associated with mustard seeds and also changes in protein, oil, starch and sugars due to the association of seed fungi.

MATERIALS AND METHODS

Seed samples of *Brassica juncea* cv. Pusabold were selected for experimental study and collected from different oilseed Brassica growers of North India, and also from the Department of Botany, Nagpur. Seed samples were mixed together and selected randomly for further investigations. Germination percentage of the above seeds was calculated by roll-towel technique (Muskett, 1948). The seed germination count was taken on the sixth day of incubation.

Isolation of fungi was done by both blotter, as well as agar plate method, as recommended by ISTA (1966). The seed sample was stored in small cotton bags under normal room temperature condition for one year i.e. from Jan 1999 to Dec 1999. After every month 400 seeds were taken out randomly and the percentage incidence of fungi was recorded. Total percentage of fungal incidence was calculated by using the formula suggested by Sahai and Mehrotra (1982).

$$\text{Percentage frequency} = \frac{\text{Total no. of seeds in which particular fungus appeared}}{\text{Total no. of seeds studied}} \times 100$$

To study the effect of fungi on the seed constituents, the seeds were stored in separate cotton bags in a glass desiccator maintained at 96% relative humidity, prepared by a saturated salt solution of sulphuric acid (Johnston and Booth, 1983). The control seeds (surface sterilised and dried) were kept in separate desiccators at a relative humidity of 96%. At an interval of every 30 days, the seed samples were taken out randomly. The fungal mycelium was removed by washing the seeds gently in running tap water and then dried at 60°C for 48 hours. The surface sterilised, ie, control seeds were left as such. Both the infected as well as the control seeds were dried and powdered separately in a grinder which was followed by a quantitative estimation of nitrogen, protein, oil, starch and sugars (reducing, non-reducing and total). Nitrogen percentage and crude protein estimation was calculated by micro-kjeldahl method suggested by Davys and Pirie (1969). To calculate the crude protein percentage, N per cent per gram was multiplied by the factor, 6.25 (Sadasivam and Manikam, 1992).

Oil percentage was determined with the help of Oxford 4000 NMR (Nuclear Magnetic Resonance Spectroscopy Analyser) in a 2 ml assembly at the

Department of Botany, Nagpur, following the method of the European Economic Community (EEC Regulation 3519/84). The mode was introduced, tuned and calibrated with pure mustard seed oil.

For quantitative determination of sugars and starch, 0.5–1.0 g of dried powdered seeds were taken in a round bottom flask. 125 ml of 50% ethyl alcohol was added to it and the mixture was refluxed for 4 hours with an air condenser. The contents were kept overnight and centrifuged at 3000 rpm for 10 minutes. The supernatant was evaporated to a few ml by heating gently. This extract was diluted to 100 ml with distilled water. From this aqueous extract, 50 ml was taken for reducing sugar and 50 ml was kept for non-reducing sugar estimation. The residue after centrifugation was preserved in a deep freezer for starch analysis (Agrawal *et al.*, 1992). The amount of reducing sugar was determined by Phenol-Sulphuric Acid as suggested by Dubois *et al.* (1951).

RESULTS AND DISCUSSION

Table 10.1 shows the germination percentage of *Brassica juncea* cv. Pusabold seeds. The germination percentage of freshly collected seed samples was 90%, with 82.50% normal germination, and only 10% dead seeds at the beginning of storage. At the end of the year it reduced to 60.50% with a 39.50% mortality of seeds.

Table 10.1: Monthly Germination Percentage of *Brassica juncea* cv. Pusabold seeds.

S.No.	Month	Total Average Germination (%)	Normal Germination (%)	Abnormal Germination (%)	Dead Seeds (%)
1	Jan-99	90.00	82.50	7.50	10.00
2	Feb-99	85.00	72.00	13.00	15.00
3	Mar-99	86.00	72.50	13.50	14.00
4	Apr-99	80.50	69.00	11.50	19.50
5	May-99	82.00	70.00	12.00	18.00
6	Jun-99	80.00	60.00	14.00	20.00
7	Jul-99	80.50	68.50	12.00	19.50
8	Aug-99	76.50	68.50	8.00	23.50
9	Sep-99	76.00	64.00	12.00	24.00
10	Oct-99	70.00	60.00	10.00	30.00
11	Nov-99	62.00	50.00	12.00	38.00
12	Dec-99	60.50	42.50	18.00	39.50

The seed mycoflora obtained on Pusabold seeds from January 1999 to December 1999 is presented in Table 10.2. Altogether, 28 fungi were isolated, out of which some were recorded throughout the year. They were *Alternaria alternata, Aspergillus flavus, A. fumigatus, A. niger, Fusarium oxysprorum, Penicillium chrysogenum* and *P. oxalicum.*

Table 10.2: Monthly incidence of fungi during storage on *Brassica Juncea* cv Pusabold using blotter and agar plate method

S. No.	Name of Organism	Jan-99	Feb-99	Mar-99	Apr-99	May-99	Jun-99	Jul-99	Aug-99	Sep-99	Oct-99	Nov-99	Dec-99
1	*Alternaria alternata* a*	15.89	10.00	5.00	–	–	9.09	21.00	18.00	12.50	5.00	16.00	16.50
	b*	7.00	4.00	–	–	–	–	8.00	6.50	14.00	12.00	–	13.00
2	*Alternaria brassicae*	22.00	16.00	–	–	–	–	9.00	–	–	10.00	–	14.54
		8.00	–	–	–	–	–	–	5.50	–	14.54	2.00	5.05
3	*Aspergillus candidus*	7.62	–	–	5.72	–	3.00	–	–	–	–	–	10.00
		–	–	7.68	10.00	–	–	–	6.00	–	–	–	–
4	*Aspergillus flavus*	8.76	9.09	33.33	42.00	30.19	42.85	23.86	57.14	33.33	33.33	28.52	9.52
		5.05	–	–	40.00	–	–	–	10.00	14.28	61.90	38.09	–
5	*Aspergillus fumigatus*	–	37.50	47.00	20.20	47.61	9.52	19.04	19.04	12.08	–	42.85	13.27
		7.65	–	3.00	–	52.38	76.19	33.33	38.09	–	–	13.27	–
6	*Aspergillus glaucus*	–	–	–	–	–	–	8.76	10.00	19.52	14.28	–	16.70
		4.76	–	–	–	5.00	–	–	–	–	9.52	–	–
7	*Aspergillus nidulans*	16.00	8.33	5.00	10.00	9.00	–	–	–	–	–	–	15.50
		4.76	–	–	–	–	–	–	–	–	–	–	–
8	*Aspergillus niger*	4.54	–	–	47.61	52.38	33.33	57.14	66.66	61.90	19.52	–	13.60
		–	–	–	–	–	–	–	14.28	–	–	15.74	15.00
9	*Aspergillus versicolor*	–	–	9.09	5.54	–	8.00	–	–	–	–	–	–
		–	–	4.74	–	–	–	–	–	–	–	–	–
10	*Aureobasidium* sp.	–	8.96	–	10.62	12.00	–	–	–	–	–	–	–
		9.52	–	–	–	–	5.76	–	–	–	–	–	–
11	*Chaetomium bostrychodes*	–	–	10.00	–	–	–	–	15.16	15.60	9.05	–	–
		–	–	–	–	–	–	–	5.00	–	–	–	–
12	*Curvularia lunata*	22.72	30.17	15.42	–	–	–	4.76	15.16	–	–	–	9.52
		–	–	–	9.05	–	3.00	–	–	–	–	–	15.00
13	*Fusarium moniliforme*	4.54	25.00	31.00	–	4.76	–	–	–	15.00	27.00	19.05	–
		–	4.76	5.00	–	–	–	–	14.28	–	–	–	–
14	*Fusarium oxysporum*	18.18	45.58	61.38	65.05	20.00	18.18	30.19	–	23.80	–	39.12	–
		5.00	–	9.05	12.00	–	–	–	–	25.00	11.00	21.05	22.23

15	*Haplosporangium* sp.	–	–	–	–	–	4.76	–	–	–	–	–	–
		–	–	–	–	–	–	–	–	–	–	–	–
16	*Memnoniella* sp.	–	–	–	–	–	–	–	–	–	5.00	–	–
		–	–	–	–	–	–	–	–	–	–	–	–
17	*Penicillium chrysogenum*	–	28.57	12.00	–	–	–	38.50	–	–	28.57	19.05	9.52
		4.76	4.76	–	10.00	–	–	–	–	–	4.76	–	15.00
18	*Penicillium frequentans*	–	–	–	–	–	–	–	–	10.00	16.55	–	14.50
		–	–	–	–	–	–	–	–	–	–	5.00	–
19	*Penicillium multicolor*	–	30.00	–	–	30.54	–	19.05	11.76	–	13.49	28.09	–
		–	4.76	–	–	12.65	–	10.00	–	–	–	–	–
20	*Penicillium notatum*	18.18	20.00	21.51	19.09	–	4.76	26.32	33.33	9.00	–	–	–
		–	–	–	–	–	–	–	–	–	–	–	–
21	*Penicillium oxalicum*	21.47	–	9.54	30.00	10.00	–	47.61	18.18	23.00	30.00	23.80	33.33
		–	–	–	–	–	–	–	–	4.76	5.00	4.76	–
22	*Phoma lingum*	–	–	–	4.76	–	5.00	–	–	–	–	–	–
		–	–	–	–	–	–	–	–	–	–	5.54	–
23	*Phytophthora undulata*	–	–	–	4.54	–	–	–	–	–	–	–	–
		–	–	–	–	–	–	–	–	–	–	–	–
24	*pythium* sp.	–	–	5.00	–	–	–	–	–	–	–	–	–
		–	–	–	–	10.00	–	–	–	–	–	–	–
25	*Rhizopus nigricans*	–	–	20.00	–	–	–	–	–	–	6.90	43.25	10.00
		–	–	–	–	–	–	–	–	–	–	–	–
26	*Stachybotrys* sp.	–	–	–	4.76	–	–	–	–	–	–	–	–
		–	–	–	–	–	–	–	–	–	–	–	–
27	*Syncephalastrum* sp.	–	–	–	9.52	12.00	–	–	–	–	–	–	–
		–	–	–	–	–	–	–	–	–	–	–	–
28	*Trichoderma viridae*	–	–	–	–	7.71	–	–	–	–	–	–	–
		–	–	–	–	–	–	–	–	–	–	–	25.50

Fungal organisms recorded only in winter were *Alternaria brassicae, Aspergillus glaucus, Chaetomium bostrychodes, Curvularia lunata, Fusarium moniliforme, Penicillium frequentans,* and *Rhizopus nigricans*. The fungi, which were mainly confined to summer season, were *Aspergillus candidus, A. nidulans, A. versicolor, Aureobasidium* sp., *Penicillium multicolor, Penicillium notatum, Phoma lingum, Pythium* sp., and *Syncephalastrum racemosum*. The fungi occurring rarely during storage period were *Haplosporangium* sp., *Memnoniella* sp., *Phytophthora undulata, stachybotrys* sp. and *Trichoderma viride*.

In the beginning of the incubation of the storage period, the protein content was 30.63%. It gradually declined till the end of the storage period. After completion of storage, the percentage change in protein over control was – 6.82% in surfaced sterilised seeds, while in infested seeds, it was –30.00%. (Table 10.3)

In the beginning of the storage period the oil content was 36.89%. There was very small decline in oil content up to the 4th months of storage the oil content was 36.07% and percentage change over control was –2.22%. At the end of one year storage, oil content lowered to 32.80% with percentage change over control –11.09%. (Table 10.3)

Initially, at the beginning of storage, was 62.921% starch content as a control. Reduction in starch was observed after monthly intervals of storage in infested seeds. It showed 24.453% starch, with 61.137% decrease over control in infested seeds, and 61.633% starch content with –2.047% change over control in surface sterilised seeds. (Table 10.4)

Table 10.3: Changes in Protein and Oil Contents in *B. juncea* cv. Pusabold Seeds due to Fungal Association during Storage

S No.	Incubation Period (Months)	Protein Percentage		Percentage Change in Oil Protein over control		Oil Percentage		% Change in Oil over control	
		a*	b*			a*	b*		
1	0	30.63	30.63	0.00	0.00	36.89	36.89	0.00	0.00
2	1	30.63	30.10	0.00	–1.73	36.89	36.89	0.00	0.00
3	2	30.10	28.76	–1.73	–6.11	36.89	36.80	0.00	–0.24
4	3	30.10	27.50	–1.73	–10.22	36.84	36.42	–0.14	–1.27
5	4	30.10	25.69	–1.73	–16.13	36.80	36.07	–0.24	–2.22
6	5	29.84	25.42	–2.58	–17.01	36.72	35.40	–0.46	–4.04
7	6	29.58	24.56	–3.43	–19.82	36.64	34.82	–0.68	–5.61
8	7	29.58	22.56	–3.43	–26.35	36.50	34.26	–1.06	–7.13
9	8	29.32	22.56	–4.28	–26.35	36.32	33.87	–1.55	–8.19
10	9	29.06	22.00	–5.13	–28.17	36.20	33.21	–1.87	–9.98
11	10	28.80	21.70	–5.97	–29.15	36.00	33.10	–2.41	–10.27
12	11	28.54	21.44	–6.82	–30.00	35.84	32.80	–2.85	–11.09

a*-control seeds. b*-infested seeds.

4.350 mg/g total sugar was found at 0 month, of storage in both the seeds (infested as well as surface sterilised). Depletion in total sugar was observed up to 9th month of storage and that was 3.806 mg/g with –12.506% change over control. However, it was increased in 10th and 11th month of storage. In surface sterilised seeds, at the end of storage total sugar found to be 4.406 mg/g with a 1.287% increase over control and in infested seeds, it was 5.026 mg/g with a 15.540% increase over control.

Reduction in reducing sugar was observed in infested seeds till 9th month of storage i.e. 2.273 mg/g with 8.347% decrease over control (2.480 mg/g), and at the end of storage there was an increase in reducing sugar i.e. 2.600 mg/g with 4.839% increase over control. In surface sterilised seeds, a very small reduction in reducing sugar was observed as it was found to be 2.453 mg/g with a 1.089% decrease over control.

1.870 mg/g non-reducing sugar was recorded at the start of storage and after two months storage it showed a sharp depletion in infested seeds ie 0.810 mg/g with 56.684% decrease over control. The decrease in non-reducting sugar in infected seeds was found till 9th month of storage with an 18.021% decrease over control. At the end of storage at the 10th and 11th months, it shows an increase in non-reducing sugar ie 2.024 mg/g and 2.426 mg/g with 8.235% and 29.733% increase over control. In surface sterilized seed, there was reduction up to 8th month and after it showed an increase in non-reducing sugar. At the end of storage non-reducing sugar occurred as 1.953 mg/g with 4.439% increase over control. (Table 10.4)

From the seed germination study it has been concluded that seed germination decreases with an increase in the storage period of seed samples. This could happen due to the association of storage seed mycoflora with stored seeds. This finding was confirmed by earlier reports (Miles, 1961; Bhajbhuje, 1989; Barve, 1995; Ellis *et al.*, 1996). The samples of wheat, barley, corn, sorghum and peas stored at moisture contents and temperature favourable to the growth of storage fungi, but kept free of fungi, retained the germinability of 95–100% for some months, whereas similar samples inoculated with storage fungi, were reduced to zero or near zero germinability (Christensen, 1973).

The nitrogen requirement for the growth of fungi comes from nitrates, ammonium and organic sources, especially amino-acids. Proteins are the combination products of enzymes and amino acids. Nitrogen sources are good for growth and reproduction also, specially nitrate nitrogen, which is generally good for sporulation in several fungi. (Bilgrami and Verma, 1978)

The oil content decreased with an increase in storage period. Lalithakumari *et al.* (1971) reported that *Aspergillus flavus* reduced oil content of groundnuts remarkably. Similar were the findings of Rai and Saxena (1980) who reported that *Aspergillus flavus* was more effective in reducing oil content of Indian mustard. Nagel and Semeniuk (1947) reported the reduction in oil content of corn, and also reported that the reduction may be due to lipolytic activity of

Table 10.4: Changes in Starch, Total, Reducing and Non-Reducing sugars in *B. juncea* cv. Pusabold Seeds due to Fungal Association during Storage

S No.	Period of Incubation (Month)	Per cent Starch	Percentage Change in Starch Over Control	Total Sugars (TS) (mg/g)	% Change in TS over Control	Reducing Sugars (RS) (mg/g) (mg/g)	% Change in RS over Control	Non-reducing Sugars (NRS) (mg/g)	% NRS over Control
	0								
1	a*	62.921	0.000	4.350	0.000	2.480	0.000	1.870	0.000
	b*	62.921	0.000	4.350	0.000	2.480	0.000	1.870	0.000
2	1	62.921	0.000	4.350	0.000	2.480	0.000	1.870	0.000
		61.339	–2.514	3.690	–15.172	2.460	–0.806	1.230	–34.225
3	2	62.864	–0.091	4.093	–5.908	2.446	–1.371	1.647	–11.925
		59.521	–5.404	3.250	–25.287	2.440	–1.613	0.810	–56.684
4	3	62.821	–0.159	3.686	–15.264	2.406	–2.984	1.280	–31.551
		56.044	–10.930	3.140	–27.816	2.040	–17.742	1.100	–41.176
5	4	62.753	–0.267	3.633	–16.483	2.363	–4.718	1.270	–32.086
		50.806	–19.254	3.073	–29.356	2.000	–19.355	1.073	–42.620
6	5	62.624	–0.472	3.640	–16.322	2.333	–5.927	1.307	–30.107
		44.576	–29.156	3.256	–25.149	1.763	–28.911	1.493	–20.160
7	6	62.512	–0.650	3.793	–12.805	2.306	–7.016	1.487	–20.481
		40.460	–35.697	3.333	–23.379	1.850	–25.403	1.483	–20.695
8	7	62.379	–0.861	3.900	–10.345	2.326	–6.210	1.574	–15.829
		36.675	–41.713	3.450	–20.690	2.056	–17.097	1.394	–25.455
9	8	62.284	–1.012	4.206	–3.310	2.373	–4.315	1.833	–1.979
	32.272	–48.710	3.760	–13.563	2.246	–9.435	1.514	–19.037	
10	9	62.109	–1.291	4.306	–1.011	2.413	–2.702	1.893	1.230
		30.795	–51.058	3.806	–12.506	2.273	–8.347	1.533	–18.021
11	10	61.821	–1.748	4.393	0.989	2.440	–1.613	1.953	4.439
		27.009	–57.075	4.540	4.368	2.516	1.452	2.024	8.235
12	11	61.633	–2.047	4.406	1.287	2.453	–1.089	1.953	4.439
		24.453	–61.137	5.026	15.540	2.600	4.839	2.426	29.733

a*-control seeds.b* -infested seeds.

seed-borne fungi. Prasad and Singh (1983) observed a decrease in oil content due to fungi at higher relative humidity. Utilisation and reduction of starch content by *Aspergillus flavus* was shown by Premlata and Sinha (1985) in wheat seed, while Prasad and Pathak (1987) studied the quantitative biochemical change (Protein, Starch and Carbohydrate) in cereal seeds under different storage condition. Singh *et al.* (1973) showed a reduction in starch content in sunflower seeds and infected groundnut seeds.

In the present study, reduction in total sugar content was observed in the initial phase of storage and an increase was observed at the end of the storage period. This may be due to its preferential utilisation by some of the fungal species infesting the seeds. Preferential utilisation of sugars by fungi has also been reported by Cochrane (1958), Bilgrami *et al.* (1979), Bilgrami and Verma (1978). Similar results were obtained in the case of reducing and non-reducing sugar.

REFERENCES

Agrawal, P.K., Dadlani, M. and Shrawa, I. (1992). Techniques in Seed Science and Technology.' Eds. Agrawal, P.K. and M. Dadlani South Asian Publisher, New Delhi. p. 207.

Barve, Y.Y. (1995). Studies on storage fungi associated with seeds of some leguminous crops. Ph.D. Thesis, Nagpur University Nagpur, India.

Bhajbhuje, M.N. (1989). Investigation on seed mycoflora associated with vegetable seeds from Vidharbh region. Ph.D. Nagpur University, Nagpur. India.

Bilgrami, K.S. and Verma, R.N. (1978). Physiology of fungi. Vikas Publishing House Pvt. Ltd. Delhi. 505 pp.

Bilgrami, K.S., Prasad, T. and Sinha, R.K. (1979). International Biosciences monograph-9. Changes in nutritional components of stored seeds due to fungal association. Today and Tomorrows Printers and Publishers, New Delhi. p. 82.

Cochrane, V.W. (1958). Physiology of fungi. *John Wiley and Sons. Inc., New York.*

Christensen, C.M. and Kaufmann, H.H. (1965). Deterioration of stored grains by fungi. *Annual Rev. Phytopath.* 3: 69–84. Agrawal, U. (1965). Seed disorders of *Sesamum indicum* L., *Linum usitatissimum* L. and *Eruca Sativa* L. Ph.D. Thesis, Agra University.

Christensen, C.M. (1973). Loss of viability in storage: Microflora. *Seed Sci. and Technol.,* 1 : 547–562.

Davys, M.N.G. and Pirie, N.W. (1969). International Biological Programme (IBP). *Biotech. Bioengr.* 11: 528.

Dubois, M., K.A. Gills, J.K. Hamilton, P.A. Robers and E. Smith (1956). 'Methods in Microbiology' (eds.). J.K. Norris and D.W. Ribbons. Acad. Press London, N.Y. pp. 272.

Ellis, R.H., Hong, T.D., Astley, D., Pinnegar, A.E. and Kraak, H.L. (1996). Survival of dry and ultra-dry seeds of carrot, groundnut, lettuce, oilseed rape and onion during five year hermetic storage at two temperatures. *Seed Sci. and Technol.,* 24: 347–358.

International Seed Testing Association, (1966). International rules for seed testing. *Int. Seed Test. Assn.* 31: 1–157.

International Seed Testing Association, (1988). International rules for seed testing. *Proc. Int. Seed Test. Asso.* 63: 1–102.

Johnston, A. and Booth, C. (Eds.) (1983). 'Plant Pathologists Pocket Books.' The Combrain New Ltd. Aberystwyth, Wales. P.

Lalithkumari, D., Govindaswamy, C.V. and Vidhyasekaran, P. (1971). Effect of seed-borne fungi on the physiochemical properties of groundnut oil. *Indian Phytopathology*. 24: 283–289.

Miles, S.R. (1961). Germination variation and tolerances. Proceedings of the Association of Official Seed Analysts, 51: 86–91.

Musket, A.E. (1948). Technique for the examination of seeds for the presence of seed-borne fungi. *Trans. Br. Mycol. Soc.*, 30:74–83.

Nagel, C.M. and Semeniuk, G. (1947). Some mold induced changes in shelled corn. *Plant Physiol*. 22: 20–23.

Premlata, Singh and Sinha, K.K. (1985). Changes in the seed contents of Arhar infected with *Aspergillus parasiticus*. *Indian Phytopath*. 38 : 560–566.

Prasad, T. and Singh, B.K. (1983). Effect of relative humidity on oil properties of fungal infected sunflower seed. *Boil. Bull. India*, 5: 85–88.

Prasad, T., and Pathak, S.S. (1987). Impact of various storage systems on biodeterioration of cereals. *Indian Phytopath*. 40(1): 39–46.

Rai, J.N. and Saxena, A. (1980). Effect of some seed-borne fungi on the physio-chemical properties of the oil of Indian mustard. *Indian J. agric. Sci.* 50 : 769–772.

Sadasivam, S. and Manikam, A. (1992). 'Biochemical Method for Agricultural Science.' Wiley Eastern Ltd. New Delhi p. 246.

Sahai, A. and Mehrotra, B.S. (1982). Mycoflora associated with the seed of forest trees and their effect on germination. *Proc. Ind. Nat. Sci. Acad. Part B*. 48(5): 706–713.

Singh, B.P., Shukla, B.N. and Sharma, Y.K. (1973). Changes in oil, protein and carbohydrate contents of diseased safflower (*Carthamus tinctorius* L.) seeds. *P.K.V. Res. J.* 2 (1): 72.

11

Evaluation of Some Essential Oils for Their Fungitoxicity Against *Penicillium expansum* (Link)

PRAMILA TRIPATHI, N.K. DUBEY AND PRAFULL KUMAR

Centre of Advanced Study in Botany, Banaras Hindu University, Varanasi -221005, India

SUMMARY

Essential oils of Cajuputi (*Melaleuca leucodendron*),Caraway (*Anethum graveolens*), Cedarwood (*Cedrus deodara*), Lemongrass (*Cymbopogon citratus*), Lavender (*Lavandula officinalis*), Neem seed (*Azadirachta indica*), Palmarosa (*Cymbopogon martini*), Patchouli (*Pogostemon cablin*), Sandalwood (*Santalum album*), Valereian (*Valeriana officinalis*) and Vetiver (*Vetiver zizanoides*) were screened against the test pathogen *Penicillium expansum*. The cedarwood oil, lemongrass oil and palmarosa oil were found to be active against the test pathogen. The minimum inhibitory concentration of these essential oils was found to be 5000 ppm, 2000 ppm and 500 ppm respectively. The nature of fungitoxicity of the oils was static. The oils were found to withstand a high inoculum density. The cedarwood oil, lemongrass oil and palamarosa oils inhibited the most common fruit rotting fungi viz. *Aspergillus niger*, *Botryodiplodia theobromae*, *Botrytis cinerea*, *Penicillium digitatum* and *Penicillium italicum* at their respective MIC, showing a large antifungal spectrum.

Keywords: Fungitoxicity, essential oils, *Penicillium expansum*, Antifungal activity

Penicillium expansum (Link) is one of the most common decay causing fungi found on stored apples all over the world (Prusky *et al.*, 1985). It causes infection even at 0°C and the decay proceeds slowly at cold storage temperature (Buchanan *et al.*, 1974). The fungus not only causes fruit decay, but also produces the carcinogenic mycotoxin patulin. This toxin may rise due to unacceptable levels in fruit destined for processing, and may also result in off flavours.

This wound pathogen can be effectively controlled by benzimidazole fungicides and the commonly used post-harvest antioxidant, diphenylamine (Rosenberg, 1991). However, benzimidazole resistant isolates of the pathogens have been found in many packing houses (Reimann and Deising 2000, Dianz *et al.*, 2002) and continued post-harvest use of diphenylamine is being questioned as it is currently under review for certification. Further, the use of synthetic chemicals to control post-harvest deterioration of fruits has been restricted-due to their carcinogenicity, teratogenicity, high and acute toxicity, long degradation periods, environmental pollution and their effect on food and other side effects on human beings (Link, 1991; Caffarelli *et al.*, 1999; Unnikrishnan and Nath, 2002). In addition, many of the synthetic fungicides have been reported as residues in treated fruit commodities (Parmar and Devkumar 1993; Fernandez *et al.*, 2001, Doghein *et al.*, 2002).

In recent years, the study and use of chemicals (flavour compounds, jasmonates, glucosinolates, plant extracts and essential oils) from natural sources, which are non-toxic and specific in their action, is gaining considerable attention for exploitation as alternative chemical control measures to prevent the fungal rotting of fruits and vegetables (Tripathi and Dubey 2004). The essential oils produced by different plant species are in many cases biologically active, have antimicrobial, allelopathic, antioxidant and bioregulatory properties (French, 1985; Elakovich, 1988; Deans *et al.*, 1990; Vaughn and Spencer,1991; Caccioni and Guizzardi,1994). *In vitro* antifungal activity of the essential oils from higher plants has been evaluated by some workers (Bishop and Reagon, 1988; Dixit *et al.*, 1995; Bishop and Thornton, 1997; Chu, *et al.*, 1999,2001;Tsao and Zhou, 2000 Tripathi *et al.*, 2004). Keeping these facts in consideration, it has been thought desirable to find out the potency of some essential oils for the protection of apple fruits from post-harvest fungal deterioration caused by *Penicillium expansum*.

MATERIALS AND METHODS

Screening of Essential Oils

Essential oils of Cajuputi (*Melaleuca leucodendron*),Caraway (*Anethum graveolens*), Cedarwood (*Cedrus deodara*), Lemongrass (*Cymbopogon citratus*), Lavender (*Laveandula officinalis*), Neem seed (*Azadirachta indica*), Palmarosa (*Cymbopogon martini*), Patchouli (*Pogostemon cablin*), Sandalwood (*Santalum album*), Valereian (*Valeriana officinalis*) and Vetiver (*Vetiver zizanoides*) were procured from Genuine Chemical Co., Bombay. The test pathogen *Penicillium expansun* was obtained from IARI New Delhi. Its pathogenicity on fresh apple fruits was tested.

The fungitoxic activity of the essential oils was tested by the poisoned food technique of Grover and Moore (1962) and Perrucci *et al.*, (1994). Potato Dextrose Agar (Hi Media, 39 gm of medium dissolved in 1000 ml of distilled water) was used throughout the investigation. The medium was autoclaved at 120 °C for 30 minutes. After cooling the medium up to 40 °C, ten milligram of streptomycin was added to it and mixed thoroughly to prevent bacterial contamination as suggested by Gupta and Banerjee (1970). Requisite amounts of the oils were dissolved separately in 0.5 ml of 0.01 per cent aqueous solution of Tween 80 in a pre-sterilized petri plate (7cm diameter). 9.5 ml of PDA medium was pipetted to each petri plate and mixed thoroughly so as to obtain the requisite concentration 5000 ppm. For control sets, requisite amount of sterilised water was added to the medium in place of the oils. Each set of experiment was done in triplicate and was repeated twice. Discs of test fungus *P. expansum* (5 mm diameter), cut with the help of a sterilised cork borer from the periphery of a seven day old culture, were inoculated aseptically to the centre of each petri plate of the treatment and control sets. The petri plates were incubated at 27±1 °C for six days in an incubation chamber. The diameter of the fungal colony of treatment and control sets were measured in a mutually perpendicular direction on the seventh day. The percentage inhibition of growth was calculated by the mean value of colony diameters by the following formula:

$$\text{Percentage of mycelial inhibition} = \frac{Dc - Dt}{Dc} \times 100$$

Where Dc = Mean colony diameter of control sets.
Dt = mean colony diameter of treatment sets.

The results are recorded in Table 11.1, where the oils are arranged alphabetically. The oils of cedarwood, lemongrass and palmarosa were selected for further study in the present investigation, as these oils completely inhibited the radial colony growth of the test fungus, showing 100 per cent activity. These oils have been in use of traditional medicine since long and their fungitoxic properties would be having non-mammalian toxic nature, if applied in the management of storage fungi of fruits.

Fungitoxic Properties of the Essential Oils of Cedarwood, Lemongrass and Palmarosa

Minimum Inhibitory Concentration (MIC)

To find out the Minimum Inhibitory Concentration (MIC) at which the oils showed absolute fungitoxicity (complete inhibition of growth of test fungus), experiments were carried out by the usual poisoned food technique. Different concentrations (viz.200, 500, 1000, 2000, 3000, 4000, 5000 and 6000ppm) of the oils were prepared by separately dissolving their requisite amounts in 0.5 ml of 0.01 per cent of aqueous solution of tween-80, and then mixing with 9.5 ml potato dextrose agar medium. The medium of control sets contained requisite amounts of sterilised water dissolved in 0.5 ml Tween-80, in place of the oils. As usual, the prepared plates were inoculated upside down aseptically with the assay disc of the test fungus to the centre of the petri plates of the treatment and control sets. The petri plates were incubated at 27 ± 1°C for six days in a BOD incubator, and results were recorded as shown in Table 11.2.

Nature of Toxicity

Nature of toxicity (fungistatic/fungicidal) of the oils against the test fungus was determined, following Garber and Houston (1959) and Thompson (1989).

Requisite amounts of the oils were dissolved separately in 0.5 ml of 0.01 per cent of aqueous solution of Tween-80, and then mixed with 9.5 ml potato dextrose agar medium to get a final concentration of 5000 ppm for cedarwood oil, 2000 ppm for lemongrass oil and 500 ppm for palmarosa oil. Sterilised water was used in the control sets, in place of the oils. The plates were inoculated with the pathogen and incubated for six days at 27 ± 1°C. On the seventh day, the incubated discs were taken out from the plates washed with sterilised water and reinoculated to plates containing fresh potato dextrose agar medium. The revival of the growth of the fungal discs was observed, and the per cent inhibition of growth of the test fungus was calculated on the seventh day with respect to the control sets. The results of nature of the fungitoxicity, of cedarwood, lemongrass and palmarosa oils are presented in Table 11.3.

Effect of Increased Inoculum Density on Toxicity of the Oils

On increased inoculum density the effect of the test fungus on fungitoxicity of the cedarwood, lemongrass and palmarosa oils was studied, following Misra (1975) and Moleyar and Pattisapu (1987).

Requisite quantities of the oils were dissolved in 0.5 ml of 0.01 per cent aqueous solution of Tween 80, and then mixed to 9.5 ml liquid potato dextrose medium to make the final concentration of 5000 ppm for cedarwood, 2000 ppm for lemongrass and 500 ppm for palmarosa oil. Seven sets for each oil were thus prepared, and inoculated separately by the assay discs (5 mm in diameter) of test fungus in geometric progression of twos, ie 1, 2, 4, 8, 16, 32 and 64. The approximate number of spores of each set were recorded with the help of the Haemocytometer, by introducing a similar number of fungal discs in distilled water. For the control set, equal quantity of sterilised water was dissolved in 0.5 ml.Tween 80 was mixed with the liquid potato dextrose medium. All the sets were incubated for six days at 27 ± 1°C and the observations were recorded on the seventh day. The results are presented in Table 11.4.

Fungitoxic Spectrum

The range of fungitoxicity of the oils was determined at their toxic and hypertoxic concentrations, ie at 5000 ppm and 6000 ppm for cedarwood, 2000 ppm and 3000 ppm for lemongrass and 500 and 600 ppm for palmarosa oils against available fruit rotting fungi viz *Aspergillus niger, Botryodiplodia theobromae, Botrytis cinerea, Penicillium digitatum* and *Penicillium italicum* by the usual poisoned food technique, using potato dextrose agar medium. The observations and results are presented in Table 11.5.

RESULTS

It is evident from Table 11.1 that out of eleven essential oils tested against *P. expansum*, the three oils cedarwood, lemongrass and palmarosa showed absolute toxicity against the test fungus at 5000 ppm. The essential oils, cajuputi, cedarwood, lemongrass, lavender, neem seed and palmarosa exhibited moderate fungitoxicity (less than 100 per cent but more than 50 percent). Poor fungitoxicity (less than 50 per cent) was observed by the caraway, patchouli, sandal wood, valerian and vetiver essential oils. Therefore, due to absolute fungitoxicity and a long record of traditional medicinal use the essential oils of cedarwood, lemongrass and palmarosa were selected for further fungitoxic studies in the present investigation.

It is evident from Table 11.2 that the essential oils of cedarwood and lemongrass showed complete inhibition of the growth of the test fungus at their lowest concentration of 5000 ppm and 2000 ppm respectively. The essential oil of palmarosa completely inhibited the growth of the test fungus at the lowest concentration of 500 ppm.Therefore the minimum inhibitory concentration of the essential oils viz, cedarwood, lemongrass and palmarosa were recorded to be 5000 ppm, 2000 ppm and 500 ppm respectively.

Table 11.1: Screening of Some Essential Oils for Their Fungitoxocity Against *P. expansum*

Essential Oils	Plant Species	Family	Percentage inhibition at 5000 ppm
Cajuputi	Melaleuca leucodendronL.	Myrtaceae	50.0
Caraway	Anethum graveolensL	Apiaceae	44.5
Cedarwood	Cedrus deodara (Roxb.) Loud.	Pinaceae	100
Lemongrass	Cymbopogon citrates (DC.) Stapf.	Poaceae	100
Lavender	Lavandula officinalis Chaix et kitt	Lamiaceae	60.0
Neem seed	Azadirachta indica A.Juss	Meliaceae	70.0
Palmarosa	Cymbopogon martini (Roxb.) Stapf.	Poaceae	100
Patchouli	Pogostemon cablin (Blanco) Benth.	Lamiaceae	30.0
Sandalwood	Santalum album L.	Santalaceae	40.0
Valerian	Valeriana officinalis L.	Valerianaceae	10.0
Vetiver	Vetiver zizanoides (L.) Nash	Poaceae	20.0

Table 11.2: Minimum Inhibitory Concentration of Essential Oils

Concentration (ppm)	Per cent Inhibition of Test Fungus		
	Cedarwood Oil	Lemon Grass Oil	Palmarosa Oil
200	0	0	50
500	0	0	100
1000	0	50	100
2000	50	100	100
3000	80	100	100
4000	90	100	100
5000	100	100	100
6000	100	100	100

It is apparent from Table 11.3 that in case of all these oils there was revival of growth of the inhibited fungal discs at their respective minimum inhibitory concentrations. Therefore, the oils were found to be altogether fungistatic in nature.

The results of Table 11.4 show that the oils inhibited the fungal growth of the treatment sets, containing even 64 discs of test fungus indicating the potency of the essential oils of cedarwood, lemongrass and palmarosa to withstand high inoculum density of the test fungus.

Table 11.3: Nature of Toxicity of Essential Oils

MIC (ppm)	Per cent Inhibition	Nature of Toxicity	
	Treated	Reinoculate	
Cedarwood (5000ppm)	100	50	Fungistatic
Lemon grass (2000ppm)	100	40	Fungistatic
Palmarosa (500ppm)	100	60	Fungistatic

It is evident from the Table 11.5 that essential oils of cedarwood, lemongrass and palmarosa remained active at their toxic (5000,2000 and 500ppm respectively) and hypertoxic (6000,3000 and 600ppm respectively) concentrations, against all the five fungi viz *Aspergillus niger, Botryodiplodia theobromae, Botrytis cinerea, Penicillium digitatum* and *Penicillium italicum*, indicating their broad fungitoxic spectrum against fruit rotting fungi.

DISCUSSION

In the present investigation, it has been thought desirable to find out the efficacy of essential oils viz Cajuputi (*Melaleuca leucodendron*), Caraway (*Anethum graveolens*), Cedarwood (*Cedrus deodara*), Lemongrass (*Cymbopogon citratus*),

Table 11.4: Effect of Increased Inoculum Density on Fungitoxicity of Oils

Number of fungal discs	Approximate number of spores	Growth of the test fungus					
		Cedarwood		Lemongrass		Palmarosa	
		Treatment	Control	Treatment	Control	Treatment	Control
1	2358 × 10^3	–	+	–	+	–	+
2	47174 × 10^3	–	+	–	+	–	+
4	94348 × 10^3	–	+	–	+	–	+
8	188696 × 10^3	–	+	–	+	–	+
16	377392 × 10^3	–	+	–	+	–	+
32	754784 × 10^3	–	+	–	+	–	+
64	1509568 × 10^3	–	+	–	+	–	+

– indicates no growth of test fungus.
+ indicates growth of test fungus.

Table 11.5: Fungitoxic Spectrum of Essential Oils of Cedarwood, Lemongrass and Palmarosa

Fungi Tested	Per cent Inhibition of Growth of Test Fungus					
	Cedarwood		Lemongrass		Palmarosa	
	5000ppm	6000ppm	2000pppm	3000ppm	500ppm	600ppm
Aspergillus niger (Van Teigh)	100	100	100	100	100	100
Botryodiplodia theobromae (Pat.)	100	100	100	100	100	100
Botrytis cinerea (Pers:FR)	100	100	100	100	100	100
Penicillium digitatum (Pers:Fr)	100	100	100	100	100	100
Penicillium italicum (Whemer)	100	100	100	100	100	100

Lavender (*Lavandula officinalis*), Neem seed (*Azadirachta indica*), Palmarosa (*Cymbopogon martini*), Patchouli (*Pogostemon cablin*), Sandalwood (*Santalum album*), Valereian (*Valeriana officinalis*) and Vetiver (*Vetiver zizanoides*) as antifungal agents by standardising, through *in vitro* fungitoxic properties, by taking *P. expansum* as test fungus.

Out of the eleven essential oils screened by poisoned food technique, caraway, pathchouli, sandalwood, valerian and vetiver essential oils showed poor toxicity (less than 50 percent). Moderate fungitoxicity (less than 100 per cent but more than 50 per cent) was exhibited by the essential oils cajuputi, lavender, neemseed. 100 per cent activity has been shown by the essential oils viz cedarwood, lemongrass and palmarosa. Essential oils of cedarwood, lemongrass and palmarosa were selected for further study as these plants have been in use in traditional medicine, since long and their fungitoxic products would be having non-mammalian toxicity if applied in the management of the storage of fruits.

However, a perusal of literature shows that the most of earlier workers have merely tested the antimicrobial activity of a product against a specific test fungus. The potency of a product as an ideal fungitoxicant may be known if it has strong efficacy and activity to withstand the fungitoxic potency during adverse conditions. As such, in the present study, different fungitoxic properties viz, Minimum Inhibitory Concentration (MIC), nature of fungitoxicity, effect of increased inoculum density and fungitoxic spectrum of the oils of cedarwood, lemongrass and palmarosa were determined. The minimum inhibitory concentration of a fungitoxicant should be determined in order to prescribe an appropriate dosage for its exploitation, as a higher dosage may cause side effects and involve wastage. In the present investigation, the MIC of the essential oils of cedarwood, lemongrass and palmarosa was found to be 5000 ppm, 2000 ppm and 500 ppm respectively. The determination of the nature of a fungitoxicant (fungistatic/fungicidal) is of great practical value as it helps in regulating the time intervals of the treatment. A fungitoxicant may inhibit the growth of a fungus temporarily or permanently. In the former, the fungitoxicant is referred to as fungistatic while in the latter, as fungicidal. A perusal of literature shows that some of the oils show fungicidal or fungistatic nature. However, in the present study, the essential oils of cedarwood, lemongrass and palmarosa were altogether fungistatic in nature against the test fungus.

At a particular concentration, the potency of a fungitoxicant depends on the inoculum density of the test fungus. There are reports on the metabolism of secondary metabolites by different enzymes, produced by fungi (Jeffrey *et al.*, 1980).The antifungal activity of citral and menthol was found to vary with the inoculum density (Molyar and Pattisapou, 1987). In the present investigation, even the high inoculum density of the test fungus was unable to metabolise the essential oils into non-fungitoxic state. The findings indicate the possibility of the exploitation of the essential oils as fungitoxicants for fresh fruit commodities, which are heavily infested with the test fungus.

The fungitoxic spectrum of a plant product should be worked out to know its range of fungitoxicity. A fungitoxicant may exhibit a broad fungitoxic spectrum, inhibiting many fungi, or may be effective against some specific ones only. The fungitoxicants with broad fungitoxic spectrum offer greater promise to control fungal infestation of stored food commodities. A study of literature shows that plant products may possess a broad or narrow fungitoxic spectrum. In the present investigation the essential oils of cedarwood, lemongrass and palmarosa showed a broad fungitoxic spectrum, inhibiting five rotting fungi *Aspergillus niger, Botryodiplodia theobromae, Botrytis cinerea, Penicillium digitatum* and *Penicillium italicum,* indicating the potency of these oils to be exploited for the management of the most common fungal rotting of perishables during storage and transit.

REFERENCES

Bishop, C.D. and Regan, J. (1998).Control of the storage pathogen *Botrytis cinerea* on Dutch white cabbage (*Brassica oleracea* var *capitata*) by the essential oil of *Melaleuca alternifolia.J.Essent.Oil.Res.,* **10**:57–60.

Bishop, C.D. and Thornton, I.B. (1997). Evaluation of the antifungal activity of the essential oils of *Monarda citriodora* var *citriodora* and *Melaleuca alternifolia* on the post harvest pathogens, *Journal of Essential oil research,* **9**(1): 77–82.

Buchanan, J.R., Sommer, N.E., Fortlage, R.J., Maxie, E.C., Mitchel, F.G. and Hsieh, D.P.H. (1974). In Post harvest pathology of perishables edt. M.K. Dasgupta and N.C. Mandal Oxford and IBH Publishing Co. Pvt. Ltd. 3.

Caccioni, D.R.L. and Guizzardi, M. (1994). Inhibition of germination and growth of fruit and vegetable post harvest pathogenic fungi by essential oil components. *J. Ess. Oil. Res.,* **6**: 173–179.

Caffarelli, V., Rapagnani, M.R., Letardi, A., Triolo, L., Lancia, B. and Santaroni, P. (1999) Pesticide residues in fruit and vegetable products and the carcinogenic risk for consumers. *Rivista di Scienza dell' Alimentazione.* **28**(1): 11–24.

Chu, C.L., Liu, W.T. and Zhou, T. (2001).Fumigation of sweet cherries with thymol and acetic acid to reduce post-harvest brown rot and blue mold rot. *Fruits,* **56**: 123–130.

Chu, C.L., Liu, W.T., Zhou, T. and Tsao, R. (1999). Control of post-harvest gray mold rot of modified atmosphere packaged sweet cherries by fumigation with thymol and acetic acid.*Can. J. Plant Sci,* **79**: 685–689.

Deans, S.G., Svoboda, M., Gundudidza M. and Brechany, E.Y. (1990). Essential oils profile of several temperate and tropical plant: their antimicrobial and antioxidant activities. *Acta Horticulture Delturae* (In Press).

Dixit, S.N., Chandra, H., Tiwari, R. and Dixit, V. (1995). Development of botanical fungicide against blue mould of mandarins. *J. Stored. Prod. Res.,* **31**:165–172.

Dianz, F., Santos, M., Blanco, R. and Tello, J.C.(2002). Fungicide resistance in *Botrytis cinerea* isolated from strawberry crops in Huelva (South Western spain). *Phytoparasitica,* **30**(5): 529–534.

Dogheim, S.M., El-Marsafy, A.M., Salama, E.Y., Gadalla, S.A. and Nabil, Y.M. (2002). Monitoring of pesticide residues in Egyptian fruits and vegetables during 1997.*Food Additives and Contaminants,* **19**(11): 10156–1027.

Elakovich, S.D. (1988).Terpenoids as models for new agrochemicals. In Biologically active natural products potential use in agriculture.(ed.H.G.Cutler). ACS Symposium Series, **380**:250–261.

Fernandez, M., Pico, Y and Manes, J. (2001). Pesticide residues in orange from Valencia (Spain).*Food Additives and Contaminats,* **18** (70): 615–624.

French, R.C.(1985).The bioregulatory action of flavour compounds on fungal spores and other propagules, *Ann.Rev.Phytopathol.,* **23**:173–199.

Garber, R.H. and Houston, B.R. (1959). An inhibitor of *Verticillium albo –atrum* in cotton seed.*Phytopathol.,* **49**: 449–450.

Grover, R.K. and Moore, J.D. (1962). Toximetric study of fungicides against brown rot organism. *Sclerotinia fructicola* and *S.laxa.Phytopathol.,* **52:** 876–880.

Gupta, S. and Banerjee, A.B. (1970).A rapid method of screening antifungal antibiotic producing plants. Indian *J. Exp. Biol.,* **8**:148–149.

Jeffrey, J. Anderson and Stanley, Dagley (1980).Catabolism of Aromatic Acids in *Tricosporon cutaneum. Journal of Bacteriology,* 534–543.

Lingk, W. (1991). Health risk evaluation of pesticide contaminations in drinking water.*Gesunde pfangen.,* **43**: 21–25.

Misra, S.B. (1975). Antifungal activity of leaves of some higher plants. Ph.D. Thesis, Gorakhpur University, Gorakhpur.

Moleyar, V. and Pattisapu, N. (1987). Detoxification of essential oil components (Citral and Menthol) by *Aspergillus niger* and *Rhizopus stolonifer.J. Sci. Food. Agri.*, **39**: 239–246.

Parmar, B.S. and Devkumar, C. (1993). Pesticides: Future Scenario.In Botanical and Biopesticides. West Vill Publishing House. New Delhi. pp 197–199.

Perrucci, S., Mancianti, F., Ciont, P.L., Flamini, G., Morelli, I and Macchioni, G., (1994). *In vitro* antifungal activity of essential oils against some isolates of *Microsporum canis* and *M.gypseum.Plant.Med.*, 60 (**20**):184–187.

Prusky, D., Bazak , M. and Bein Arie, R. (1985). Development, persistence, survival and strategies for control of thiobendazole resistant strains of *Penicillium expansum* on pome fruits. *Phytopathology*, **75**: 877–822.

Rosenberger, D.A. (1991). Post harvest diseases blue mould and gray mould. In A.L. Jones and H.S. Aldwinckle (eds). *Compendium of apple and pear diseases.* APS press, St. Paul, Minn.

Reimann, S. and Deising, H.B. (2000).Fungicides : Risk of resistance development and search for new targets. *Archives of Phytopathology and Plant protection,* **33**(4) 329–349.

Thomposon, D.P. (1989). Fungitoxic activity of essential oil components on food storage fungi. *Mycologia*, **81**:151–153.

Tripathi, P. and Dubey, N.K. (2004). Exploitation of natural products as an alternative strategy to control post-harvest fungal rotting of fruits and vegetzables. *Post harvest Biology and Technology*. **32**(3): 235–245.

Tripathi, P. Dubey, N.K., Benergi, R. and Chansouria, J.P.N. (2004). Evaluation of some essential oils as a botanical fungitoxicant in the management of post-harvest rotting of citrus fruits. *World Journal of Microbiology and Biotechnology*, **20**: 317–321.

Tsao, R. and Zhou, T. (2000). Antifungal activity of monoterpenoids against post-harvest pathogens *Botrytis cinerea* and *Monilinia fructicola.J.essential oil res.*, **12**: 113–121.

Unnikrishnan, V. and Nath, B.S. (2000).Hazardous chemicals in foods. *Indian Journal of Dairy and Biosciences*, **11**: 155–158.

Vaughn, S.F. and Spencer, G.F. (1991).Volatile monoterpenes inhibit potato tuber sprouting. *Potato Journal*, **68**: 821–831.

12

Diversity of Stored Grain Pest and Their Role in Spreading Mould Contamination

NEETA PANDE

Botany Department, M.B.P.G. College Haldwani-263139 (U.A) India

SUMMARY

Stored insect is the most destructive biological agent of stored grain. *Sitophilus oryzae* Linn, *Rhyzopertha dominica* Fab. *Tribolium castaneum* Herbs *Oryzaephilus surinamensis* Linn and *Trogoderma granarium* Evert are the most common pest under typical Indian condition. Apart form visible damage pest attack may also initiate a chain of biochemical changes inside the grain kernel affecting the nutritive value and germinative ability of grain. Besides causing considerable losses on their host, these insects act as a vector or carrier for several pathogenic fungi. Many species of these fungi are well known to produce various kind of mycotoxins (a secondary metabolite produce by fungi) and can cause severe health hazard to consumer.

Keywords: Mould diversity, Mould contamination, Stored grain pest

Since the dawn of human evolution, cereal grains have always constituted an important source of food, whether it is wheat or any other grain. India is predominantly an agricultural country with nearly three-fourth of the people dependent on agriculture or rural economy. There the foodgrain production in the country increased from 50.8 million tons in 1950–51 to 152.37 million tons in 1983–84, the growth of Indian agriculture still continues to be linked with the vagaries of nature. Nearly, seventy per cent of total production of foodgrains in India is retained at the farm level where the unscientific and faulty storage conditions enhance the chances of insect and fungal attack, and thereby mycotoxin production (Goyal, 1989).

The storage of grains in India is dogged by the same problems, which are encountered throughout the world. During storage, various factors viz, physical (temperature and moisture), biological (properties of the produce, microorganisms, insects, rodents and birds), chemical (breakdown of the produce and pesticides), engineering (structural and mechanical), and socio-economic problems like finances, farming method, storage and marketing methods etc, play a crucial role in the upkeep and storage life of agricultural produts (Agrawal, 1977). Of these, insect infestation in by far the most troublesome and it is certainly a very significant and often a spectacular case of spoilage and loss. According to an estimate of FAO (1947), at least 10% of the harvested food corps were lost by biological attacks while they were under storage. Cereal grains are highly attractive to insects. The stored grains that are neglected or in properly handled, soon become populated and befouled by insects and mould. They contaminate either the inside or the outside of the kernel with their body, cast skin, faecal pellets and webbings. Grains once contaminated by insects can never be adequately cleared for the flour making process (Hall, 1954). Insects feeding on or in grain reduce part of the grain kernel, leading to a loss of nutrients, and consequently, decreased palatability of the grain (McGaughey, 1974; Sinha and Mc Cain, 1963; Vankat Rao *et al.*, 1960). According to Khare (1972), the odour eminating from the excrement was musty in the case of *Rhizopertha dominica* infestation, and in the case of *Tribolium castaneum*, the excrement emitted an annoying odour.

INSECT PESTS OF UTTARANCHAL

A number of insects have been found associated with stored grains under typical Indian conditions. Cereal constitutes the most important food and feed source of Uttaranchal. The environmental and storage conditions of this region are very conducive for the growth of stored pests. However, the incidence and extent of damage are location specific due to varied altitude, climatic storage conditions. Wheat, rice and maize are the main crops of Uttaranchal (Table 12.1). Though the climatic conditions, ie, range temperature, relative humidity etc. are favourable for a year-round agriculture practice; at the same time these two factors make a good environment for mould and insects to develop. Insect and mould proliferation leads to the discolouration of grains, heating and mustiness, loss in weight, decrease in germinability, and production of mycotoxin thus decreasing the quality and quantity of the foodgrains.

Most of the stored grain pests pass through the egg, larval and pupal stages before they finally become adult. An adult female, after reaching sexual maturity, copulates with the male and starts laying eggs, which vary in shape and size. The number laid, varies from few to several hundred. Insect development take place between 17°C to 30°C and relative humidity between 70% to 80%. Under ideal conditions of development, usually one cycle from egg to adult stage requires 4–6 weeks.

Most people believe that insects are generated de-novo in the grain, but this is far from the truth. All insects lay tiny eggs, which easily escape detection by the naked eyes. The manner and the place of laying eggs, by various types of insects infesting food grains differ. Some lay eggs inside the grain mass and others indiscriminately.

Stored grain pests are classified as major or minor, depending upon the severity of the damage they inflict on a particular commodity. Usually two types of insects are found to be infesting stored food grains (1) Hard bodied beetles and (2) Soft bodied flying insects, popularly known as 'moths'. These insects spend their life cycle either inside or outside the grain mass. In general, pests which complete their life cycle inside the grain kernel are more destructive than those which complete their life cycle outside in the grain mass.

Table 12.1: Diversity of Insect Pests in Different Cereals of Uttaranchal

Paddy	Rice Weevil (*Sitophilus oryzae* Linnaeus) Grain moth (Sitotroga cerealella Olivier) Lesser grain borer (*Rhyzopertha dominica* Fab)
Rice	Maize weevil (*Sitophilus zeamais* Motschulsky) Rice weevil (*S. oryzae* Linn.) Red flour beetle (*Tribolium castaneum* Herbs) Rice moth (*Corcyra cephalonica* Stainton) Saw toothed grain beetle (*Oryzaephilus surinamensis* Linn.) Flat grain beetle (*Cryptolestes pusillus* Schonherr)
Wheat	Red flour beetle (*Tribolium castaneum* Herbs) Saw toothed grain beetle (*Oryzaephilus surinamensis* Linn.) Lesser grain borer (*Rhyzopertha dominica* Fab) Khapra beetle (*Trogopderma granarium* Evert) Rice moth (*Corcyra cephalonica* Stainton)
Maize	Maize weevil (*Sitophilus zeamais* Motschulsky) Red flour beetle (*Tribolium castaneum* Herbs) Rice moth (*Corcyra cephalonica* Stainton)

ROLE OF INSECTS AS VECTORS OF FUNGI

Insects that infest stored grains rarely occur in the absence of fungi. It has long been known that there is a positive correlation between insect infestation and storage fungi invasion. The metabolic by-products and heat released by the metabolic activity of insects promote the development of the latter. Moreover, an insect could spread the fungus infestation in a certain stock (Pande, 1987). According to Agrawal *et al.* (1957), stored grains infested by *Sitophilus granarius* were heavily invaded by *A. restrictus*. This suggested that the infestation of grains by this insect might promote invasion of the seed by *A. restrictus* or

Table 12.2: Frequency of Occurrence of Fungi Associated Externally (E) as well as Internally (I) with Insects Present in Different Grain Samples

Name of fungi in grain sample	T. castaneum		S. oryzae		R. dominica	
In Wheat						
Aspergillus awamori Nakazawa	–	–	E	I	–	–
A. flavus Link	E	I	E	I	E	I
A. flavus var. *columnaris* Raper and Fennell	E	–	E	I	–	–
A. kambarensis Sugiyama	E	I	E	I	–	–
A. niger Tieghem	–	–	E	–	E	I
A. oryzae (Ahlb) Cohn	–	–	E	–	–	–
A. parasiticus Speare	E	I	E	I	E	–
A. sojae Sugiyama and Yamada	–	–	E	I	E	I
A. subolivaceus Fennell andRaper	E	I	–	–	–	–
A. tamarii Kita	E	I	E	–	E	–
A. terreus Thom	E	–	–	–	–	–
A. versicolor (Vaill) Tiraboschi	E	I	E	I	–	–
Penicillium citrinum Thom	E	–	E	–	–	–
P. crustosum Thom	E	I	E	I	–	–
P. cyclopium Westling	–	–	E	–	E	I
P. granulatum Bain	E	I	E	–	E	I
P. purpurogenum Stoll	–	–	E	–	–	–
Alternaria alternata (fr.) Keissler	–	–	–	–	E	–
A. tenuissima (Kunze ex Pers) Witts	E	–	–	–	–	–
Cladosporium cladosporioides (fres.) de Varies	E	I	E	I	E	–
Fusarium moniliforme Sheld	–	–	E	I	–	–
Humicola grisea Triaaen	E	–	–	–	–	–
In Rice						
Mucor racemosus Fres	–	–	E	–	–	–
Rhizopus arrhizus Fisch	–	–	E	–	E	–
Aspergillus awamori Nakazawa	–	–	E	–	E	–
A. candidus Link	–	–	E	I	–	–
A. flavus Link	E	I	E	I	E	I
A. foetidus (Naka) Thom and Raper	E	–	E	–	–	–
A. parasiticus Speare	E	I	E	I	–	–
A. sulphureus (Fres.) Thom and Church	E	–	–	–	E	–
A. terreus Thom	E	I	E	I	–	–
Penicillium citrinum Thom	E	–	E	–	–	–
P. chrysogenum Thom	–	–	E	I	E	I
P. cyclopium Westling	E	–	–	–	–	–
P. frequentans Westling	–	–	E	–	E	I
P. nigricans (Bain) Thom	E	–	–	–	–	–
Alternaria alternata (fr.) Keissler	E	–	E	–	–	–
A. tenuissima (Kunze ex Pers) Witts	E	I	E	I	–	–
Cladosporium cladosporioides (fres.) de Varies	E	I	E	I	E	–
C. lunata (Wakker) Boedijn	E	–	E	I	E	–

Fusarium lateritium Nees	–	–	E	–	E	–
Phorna eupyrena Sacc.	–	–	E	–	–	–
In Maize						
Mucor racemosus Fres	E	–	E	–	–	–
A. candidus Link	–	–	E	I	E	–
A. flavus Link	E	I	E	I	E	I
A. fumigatus Fres.	E	–	E	–	–	–
A. niger Tieghem	E	I	E	–	–	–
A. parasiticus Speare	E	I	E	I	–	–
A. tamarii Kita	–	–	E	–	E	–
A. terreus Thom	–	–	E	–	E	–
A. versicolor (Vaill) Tiraboschi	E	I	–	–	–	–
P. chrysogenum Thom	E	–	E	–	–	–
P. nigricans (Bain) Thom	–	–	E	I	E	–
P. simplicissimum (Oudem) Thom	E	–	E	I	E	–
A. alternata (fr.) Keissler	E	–	–	–	E	I
C. cladosporioides (fres.) de Varies	–	–	E	–	–	–
C. lunata (Wakker) Boedijn	E	–	–	–	–	–
Epicoccum purpurogenum Ehrend ex Schlecht	–	–	E	–	–	–
F. moniliforme Sheld	E	–	E	–	E	–
Verticillium albo–atrum Reinke and Berthold	E	–	E	I	E	–

other storage fungi. Studies conducted by Srinath *et al.* (1973) on the pupal micro flora of *Sitophilus oryzae* , *T. castaneum, T. granarium, C. chinensis, Corcyra cephalonica* St. and *Ephestia cautella* found that none of the pupae were microbiologically sterile and they acted as carriers of several toxigenic and saprophytic food spoilage microbes from their larval stage to the freshly emerging adults. Assessment of insect infestation in stored maize and their relationship to *Aspergillus flavus* contamination showed that there is a strong correlation between *A. flavus* and high densities of weevils, live or dead (Wright 1989). It is very alarming to note that there is a ubiquitous distribution of *A. flavus* (a potential carcinogen) outside and inside the insect body, and this fact reveals an even greater danger in the transmission of such fungi to non-infested healthy grain, specially since certain insects do transmit spores of mycotoxin producing fungi, such as *A. flavus, A. ochraceus, Penicillium citrinum* etc. (Srinath *et al.*, 1971 and 1973). In recent years, the interrelationship between insects and certain microorganisms found in stored grains is being investigated. All the post-embryonic stages of stored product insects, except eggs are known to carry (both externally and internally) several species of fungi (Table 12.2).

The activity of insects is interlinked with the activity of other biotic and abiotic factors. The physical conditions of the substrate also influence the growth and multiplication of insects. During the study it was found that *T. castaneum* flourished best in wheat flour, while *Soryzae* and *R. dominica* flourished best in whole wheat. Effect of fungal infestation of stored wheat on survival and multiplication of storage insects was also observed and it was found that there are many mould species like *A. tamarii*, *A. oryzae* and *Cladosporium cladosporioides* which favour insect survival and multiplication.

Insects are more critical in their response to high temperature. Most insect-pests are killed by temperatures above 40°C whereas most grain insects are able to withstand temperatures below freezing point for several days. However, when exposed to 65°C for a few minutes, all insects are killed. Thus, it is very clear that besides climate, the important parameters in the development and reproduction of storage pests are the kind and the condition of produce, its moisture content and relative humidity.

CONCLUSION

It may be concluded from the above review that stored insects are the most destructive biological agents of stored products. Apart from visible damage, pest attacks may also initiate a chain of biochemical changes inside the grain kernel, affecting the nutritive value and quality in general, a fact that is not quite realised by the common man. Food grains, which have been attacked, by insects, rodents, fungi and bacteria also become unfit for seed purpose because of the loss of germinative value. Besides causing considerable losses on their own storage insects act as a vector or carrier for several fungi viz *Aspergillus*, *Penicillium* sp. and *Fusarium* sp. Many species of these genera are well known to produce different kinds of mycotoxin. Thus the proper storage of food grains is a must so as to avoid mould and insect infestation. When the grains have no storage protection the insect population builds up rapidly thereby leading to mould infestation. Therefore, the loss and damage by insect pests is related to the storage duration and conditions.

ACKNOWLEDGEMENT

The author wishes to thank the Head, Department of Botany Kumaun University, Nainital for providing lab facility.

REFERENCES

Agrawal, N.S. 1977. Storage problems, losses, steps taken by the deptt. of food. p. 1–7. In Introduction to Action Prog. to Reduce farm storage losses in India. (ed. Drs N.S. Agrawal and G.K. Girish.) Deptt. of food, Ministry of Agri and Irrig. Kirishi Bhawan New Delhi.

Agrawal, N.S., Christensen, C.M. and Hodson, A.C.1957. Grain storage fungi associated with granary weevil. Jour. Ecom. Ent. 50(5), 659–663.

F.A.O. 1947. Agriculture Studies 2: 15–34.

Goyal, R.K. 1989. Prevention and control of mycotoxin in foodgrain in India. p. 108. In Mycotoxin prevention and control foodgrain ed. By R.L. Semple, A.S. Frio P.A. Hicks and J.V. Lozare.

Hall, D.W. 1954.Colonial Plant. Animal Proc. 4; 227.

Khare, B.P. 1972. Insect Pest of stored grain and their control in U.P.G.B. Pant Univ. of Agric. and Tech. Pantnagar Nainital. Bull. Grain Technol. P. 161.

Mc Gaughey, W.H. 1974. Insect development in milled rice, effect of variety, degree of milling, parboling and broken kernel. J. Stored Prood. Res. 10: 81–86.

Pande, Neeta. 1987. Studies on the relationship of some storage insects and storage fungi of grains. Ph.D. Thesis Kumaun University, Nainital. India.

Singh, D.N. and Mc Cain, F.S. 1963 .The relationship of some nutritional properties of the corn kernel to rice weevil infestation. Crop. Sci. 3: 259–610.

Srinath, D., Rajhunathan, A.N. and Majumdar, S.K. 1971. Stored Product insects are carrier of toxigenic fungi. Second International Symposium on Plant Pathology Ind. Phytopathol Soc. New Delhi, p. 32.

Srinath, D., Rajhunathan, A.N. and Majumdar, S.K. 1973. Mycoflora in the pupae of stored product insects. Bull.Grain Technol 12 (1): 32–35.

Vankal Rao, S., Jagrag, P., Bhagwan, R.K., Sankaran, A.N., Swaminathan , M. and Sriramchari, S. 1960. Ann Biochem. Exp. Med. 20: 135.

Wright, V.F. 1989. Assessment of Insect Infestation in stored maize and their relationship to *Aspergillus flavus* contamination p. 125. In mycotoxin prevention and control in food grain (eds R.L. Semple, A.S. Frio, A.P.A. Hicks and J.V. Lozare).

13

Disease Development and Disease Intensity of *Puccinia nepalensis* Barclay and Dietal at Nainital, Kumaun Himalaya

ANWESHA SAH AND UMA TIWARI PALNI

Department of Botany, Kumaun University, Nainital -263 002, India

SUMMARY

Puccinia nepalensis causes severe infection on *Rumex nepalensis*. Necrotic areas of the infected leaves showed 'green islands'. Basal leaves of the host plant showed severe infection, but the leaves higher up from the ground showed little or no infection. Histopathological studies showed uredospores and teleutospores. The per cent disease infection varied in different sites and localities.

Keywords: *Rumex nepalensis, Puccinia nepalensis,* disease development and intensity

The Uredinales or rust fungi of the class Basidiomycetes comprise a large group of obligate parasites containing 150 genera and over 7000 species; of these Pucciniaceae is the largest and the most important family of the order Uredinales, and the genus *Puccinia* is reported to have nearly 4000 species, causing a great variety of plant diseases in wild and several cultivated plants. It has a direct impact on their growth and development, leading in losses to crop yields. A large number of rust fungi are recorded from India, including the Himalayan region, but the reports are only in the form of enumeration or listing of rust fungi. The rust fungi of adjoining central Nepal and Pakistan have also been enumerated in a similar fashion (Ahmad 1956a,b 1969; Barclay 1890, 1891; Bisht *et al.*, 1994; Mitter and Tandon 1932, 1937; Okane *et al.*, 1992; Ono and Kakishima 1992; Ono and Okane 1995; Ono *et al.*, 1988, 1990, 1995; Palni and Pangtey 2000; Zhuang 1986). However, a detailed disease developmental study of many species of rust fungi causing diseases in the wild plants has not been investigated so far.

Rumex nepalensis Spreng is a common and economically important perennial herb that occurs throughout the Kumaun Himalaya, along roadsides, waste places, crop fields, field borders and camping sites. The plants of this species were severely infected by *Puccinia nepalensis* (Fig. 13.1). Therefore, the present study was undertaken to especially investigate the disease development and disease intensity on the host *R. nepalensis,* at Nainital, a disease hitherto least studied so far.

MATERIALS AND METHOD

Disease Development

Detailed field surveys were made in and around Nainital throughout the year, and on the basis of disease intensity, three different localities were selected *viz* Krishnapur (1,775m), Langham (2,000m) and Sleepy Hollow (2,100 m) to study the development and intensity of the disease.

Fig. 13.1: A leaf of *Rumex nepalensis* infected by *Puccinia nepalensis.*

Observations on disease development were recorded throughout the year, ie January to December in 2003 in three different selected sites around Nainital. The disease symptoms appeared in the form of small white spots full of mycelia, in the beginning of May, at all the sites, and they remained as such till mid-June on both the leaf surfaces (upper or adaxial side and lower or abaxial side). On the arrival of the monsoons (ie third week of June), the white mycelial spots gradually turned into small brownish patches (pustules) on both the surfaces of the leaves, and appeared widely scattered throughout the leaf surface. Later, these pustules started coalescing together forming larger pustules (patches) from the base to the apex or from the apex to the base or from the periphery to the centre. Finally the entire leaves were infected severely, leading to the yellowing of the leaves, with distinct brown spots, and gradually withered eventually leading to the death of the infected leaves.

The leaves showed a gradual change in the disease symptoms till January. The leaves close to the ground became heavily infected and senescent, while the ones away from the ground showed little or no symptoms.

Disease Indexing

For studying disease intensity, the scales used were according to the method of Elliott and Jenkins (1946) Ahmed (1976) and Pangtey (1979). The diseased leaves were classified into six categories on the basis of total area covered by the spots, namely (i) 0% (healthy), (ii) 10% (minimum infection), (iii) 25% (moderate infection), (iv) 50% (average infection), (v) 75% (heavy infection) and (vi) 100% (very heavy infection). Five sites from each locality were examined during 2003. Random samples were taken from each site and twenty leaves were taken even each site. Thus, 100 leaves were picked up from each locality. All these leaves were examined and classified according to their grade of infection. This was done by comparing with the diagrammatic scale which had been prepared and illustrated (Fig. 13.2).

The formula adopted during disease appraisal was based on those suggested by Naumov (1924), Ahmed (1976) and Pangtey (1979).

Degree of disease intensity per infected leaf $= d$

No. of infected leaflets studied per field $= x$

Average disease intensity per infected leaf, $F = \dfrac{d_1 + d_2 + d_3 \ldots d_x}{x}$

Total no. of leaflets studied per field $P = \dfrac{F.x}{m}$

Average disease intensity per field $I = \dfrac{P_1 + P_2 + P_3 \ldots P_x}{y}$

No. of fields examined in the locality $= y$

Average disease intensity in the locality $I = \frac{P_1 + P_2 + P_3 ... P_x}{y}$

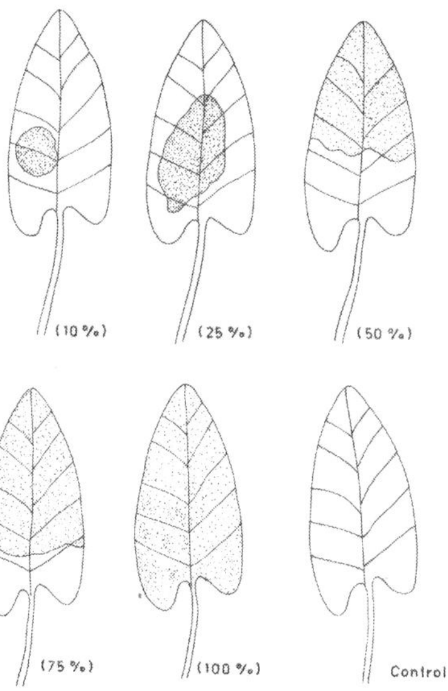

Fig. 13.2: Diagrammatic representation of disease intensity on the leaves of *Rumex nepalensis* infected.

Besides, histopathological studies were carried out by hand and microtome sectioning. For hand sectioning, the infected portions of the leaves were cut through with a razor and stained in cotton blue. Then slides were prepared for the study of mycelium and spores. The infected leaves were collected in the early developmental stages of the diseases, symptoms till severe infection developed. For the preparation of permanent slides of diseased leaf tissue showing fungal infection, mechanical microtomy was employed (Prasad and Prasad 1975). For this, Rotary microtome was used. Infected parts of the leaves were dehydrated in alcohol series before preparing the wax blocks.

RESULTS AND DISCUSSION

Puccinia nepalensis showed variable degrees of infection on the host (*R. nepalensis*) in three different selected localities, with three different climatic conditions, which supported the growth and development of the rust fungus. The symptoms of the disease changed as the infection became severe. The heavily infected leaves became yellow, senescent and finally died. The basal leaves, which are usually large, showed heavy infection. As the leaves were close to the ground, the severe infection could be due to sufficient moisture and shade which provided a congenial and favourable environment for disease development.

It was interesting to observe the necrotic areas of the infected leaves which showed the 'green islands', even when the rest of the leaf became yellow and senescent (Fig. 13.3). This phenomenon has been observed by Greene (1980), Thimann (1980) and Palni and Nandi (1990) in some fungi that cause rust diseases. They reported that these 'green islands' were rich in the plant growth substance, Cytokinin, synthesised by the rust fungus as food resource for the further progress of the disease symptoms of the disease.

The degree of infection recorded at the three selected localities has been given in Table 13.1, which shows that the maximum infection was recorded (22.02%) in Krishnapur locality followed by Sleepy Hollow (14.50%), and Langham (12.09%). Krishnapur locality showed the highest percentage infection due to a warmer climate, sufficient moisture and shade. Langham locality exhibited the minimum percent infection due to it being more open, sunny and without forest cover, whereas Sleepy Hollow showed medium infection as it had more moisture and sufficient shade due to the presence of evergreen oak forests. However, while the degree of infection varied considerably among sites as well as localities, the infection was found to be very severe during rainy season in all the sites and localities.

Histopathological studies revealed that during the early stages of disease development, the mycelium of the fungus spreads and invades the leaf tissues. As the disease progresses, uredospores develop abundantly. These spores appeared in the month of July when the moisture and humidity were suitable for their development and these spores were present on both sides (adaxial

Fig. 13.3: Green Island effect on the leaves of *Rumex nepalensis*.

Table 13.1: Per cent Disease Infection of *Puccinia nepalensis* at Three Different Localities (Krishnapur, Langham and Sleepy Hollow) during 2003.

Localities	Sites					Mean
	1	2	3	4	5	
Krishnapur	15.70	17.70	23.70	26.50	23.50	22.02
Langham	15.70	09.75	11.25	11.75	12.00	12.09
Sleepy Hollow	11.25	12.25	17.75	14.75	16.50	14.50

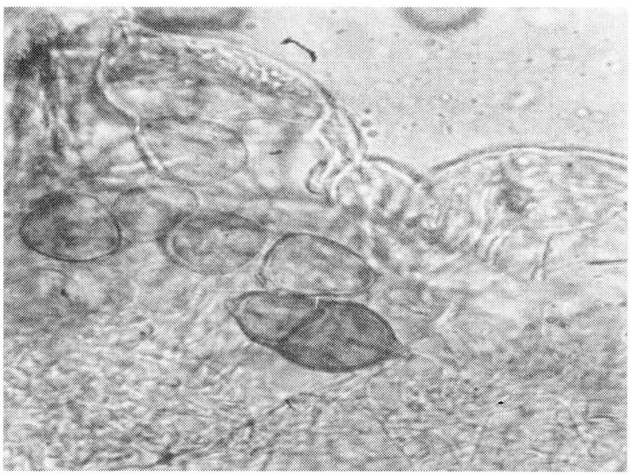

Fig. 13.4: Uredospores and teleutospores of *Puccinia nepalensis.*

and abaxial) of the leaves. However, the teleutospores were less common, as compared to the uredospores (Fig. 13.4).

ACKNOWLEDGEMENTS

The authors are thankful to Prof. S.P. Singh, Head, Department of Botany, D.S.B. Campus, Kumaun University, Nainital for encouragement and laboratory facilities. Thanks are due to Prof. Y.P.S. Pangtey for his help during the course of the present study.

REFERENCES

Ahmad, S. (1956). Uredinales of West Pakistan. *Biologia.* **2:** 26–101.

Ahmad, S. (1956). Fungi of West Pakistan. Monograph No. 1. *Biol. Soc. Pakistan.* Lahore.

Ahmad, S. (1969). Fungi of West Pakistan, Supplement 1. Monograph No. 5. *Biol. Soc. Pakistan,* Lahore.

Ahmed, I. (1976). Studies on the seed borne mycoflora and *Synchytrium* disease of til (*Sesamum indicum* L.). Ph.D. Thesis, Agra University, Agra.

Barclay, A. (1890). Descriptive list of Uredinales occurring in the neighbourhood of Simla (Western Himalaya). *J. Asiat. Soc. Bengal.* **9:** 75–112.

Barclay, A. (1891). Additional Uredinae from the neighbourhood of Simla. *J. Asiat. Soc. Bengal.* **60**: 211–230.

Bisht, G.S., Srivastava, S.L. and Singh, H. (1994). Survey of fungal plant diseases in a part of high altitude areas of Central Himalayas. In: *High Altitudes of the Himalaya (Biogeography, Ecology & Conservation)*, eds. Pangtey, Y.P.S. and Rawal, R.S., Gyanodaya Prakashan, Nainital. 281–297.

Elliott, C. and Jenkins, M.T. (1946). *Helminthosporium turcicum* leaf blight of corn. *Phytopath.* **36**: 660–666.

Greene, E.M. (1980). Cytokinin production by microorganisms. *Bot. Rev.*, **46**: 25–47.

Mitter, J.H. and Tandon, R.N. (1932). Fungi of Nainital Part-I, *J. Indian Bot. Soc.* **11**: 178–180.

Mitter, J.H. and Tandon, R.N. (1937). Fungi of Mussoorie. *Proc. Nat. Acad. Sci. India* **7B**: 175–180.

Naumov, N.A. (1924). On the question of the possibilities determining the degree of plant infection by fungus parasites Trudy IV. *Entom. Phytopath. Congr.*, Moscow, **1922**: 217–228.

Okane, I., Kakishima, M. and Ono, Y. (1992). Uredinales collected in the Murree hills, Pakistan. **In**: *Cryptogamic Flora of Pakistan, eds.*, Nakaike, T. and Malik, S. National Science Museum, Tokyo. **1**: 185–196.

Ono, Y., Adhikari, M.K. and Rajbhandari, K.H. (1988). Rust fungi of Kathmandu valley and adjacent areas, Nepal. In: *Cryptogams of the Himalayas*. eds. Nakaike, T. and Malik, S. National Science Museum, Tokyo, pp 197–216.

Ono, Y. Adhikari, M.K. and Rajbhandari, K.R. (1990). Uredinales of Nepal. *Rep. Tottori Mycol. Inst.* **28**: 57–75.

Ono, Y., Adhikari, M.K. and Kaneko, R. (1995). An annotated list of the rust fungi (Uredinales) of Nepal. In: *Cryptogams of the Himalaya (Nepal and Pakistan)*. eds., Watanabe, M. and Hagiwara, H. National Science Museum, Tsukuba, Japan. **3**: 69–125.

Ono, Y. and Kakishima, M. (1992). Uredinales collected in Swat valley, Pakistan. In: *Cryptogamic Flora of Pakistan*, eds., Nakaike, T. and Malik, S. National Science Museum, Tokyo. **1**:197–216.

Ono, Y. and Okane, J. (1995). Uredinales collected in Mt. Nanga Parbat area-Pakistan. In: *Cryptogamic of the Himalayas (Nepal & Pakistan)*. eds., Watanabe, M. and Haegiwara, H. National Science Museum, Tsukube, Japan. **2**: 61–68.

Palni, L.M.S. and Nandi, S.K. (1990). The biochemistry and molecular biology of cytokinin production and metabolism. **In**: Sinha, S.K., Sane, P.V., Bhargava, S.C. and Agarwal, P.K. eds., *Proc. Int. Congr. Plant Physiology*. Vol. II, New Delhi. pp. 750–759.

Palni, Tiwari Uma and Pangtey, Y.P.S. (2000). Rust fungi (Uredinales) of Indian Central Himalaya: An annotated list. In: *High Altitudes of the Himalaya-II (Biodiversity, Ecology & Conservation) ed.*, Pangtey, Y.P.S., Gyanodaya Prakashan, Nainital. pp. 223–246.

Pangtey, Y.P.S. (1979). Investigations on seed-borne fungi of *Dolichos biflorus* L. and *Glycine hispida* Maxim. two common legumes of Kumaun. Ph. D. Thesis, Kumaun University, Nainital.

Prasad, M.K. and Prasad, M. Krishna (1975). *Outlines of Microtechnique*. Emkay Publication. Delhi.

Thimann, K.V. (1980). *Senescence in Plants*. C.R.C.Press Inc.,Boca Raton, Florida.

Zhuang, J.Y. (1986). Uredinales from East Himalaya. *Acta Mycol. Sinica.* **5**:138–155.

Section C

Fungal Physiology/Ecology/Pollution

14

Colonisation and Successional Pattern of Aquatic Hyphomycetes in Freshwater Streams of Kumaun Himalaya

S.C. SATI AND N. (TIWARI) PANT

Department of Botany, Kumaun University, Nainital-263 002, India

SUMMARY

Colonisation and successional pattern of aquatic Hyphomycetes on submerged leaf litter of two different forest trees viz., oak (*Quercus leucotrichophora* A. camus) and pine (*Pinus roxburghii* Sarg.) of Nainital, Kumaun Himalaya were studied by using the nylon-mesh bag technique. Old leaves of oak and pine leaf litter were oven dried and placed in three sets—oak, pine and a mixture of these and submerged into a running freshwater stream for 15 months. Altogether, thirty species of aquatic hyphomycetes were colonised on the submerged leaf litter, the maximum being in oak (twenty-six species) followed by 18 and 22 species in pine and the mixture, respectively. The oak leaf litter was preferred to the pine leaf litter by the aquatic Hyphomycetes. The successional pattern varied from month to month and substrate to substrate. Colonisation was correlated with temperature and pH of the stream water.

Keywords: Aquatic hyphomycetes, Microbial colonisation, Decomposition, Succession

Ecological succession may be defined as an orderly sequence of different communities over a period of time in a particular area. The area which undergoes a major change in the physical or biological environment, there will be a sequence of communities, first a pioneer stage, then gradual changes and ultimately a relatively stable phase, the climax. Odum (1971) suggested three parameters to elaborate the sequence of ecological succession—(i) it is the orderly process of community changes which are directional and predictable, (ii) it results from the modification of the physical environment and population structure by the community, and (iii) it culminates in the establishment of a stable ecosystem, which is biologically possible, on the site.

The change in the species' composition is a fast process. Succession in aquatic fungi is derived from the alteration of a directional manner that becomes progressively more suitable for the development of different organisms and communities (Park, 1974a).

The succession of microorganisms like fungi, bacteria, actinomycetes etc., that occur within a microhabitat, has been called microsuccession. It is heterotropic in nature, begining on a substratum rich in organic matter. The substratum undergoes a process of decomposition and mineralisation and is gradually depleted in terms of organic content. Such a succession of microorganisms play a major role in the ecosystem, where the microbes are an important biotic component (Peterson and Cummins, 1974; Suberkropp et al., 1976).

Newton's (1971) study on the succession of aquatic fungi on natural leaf litter showed that the freshly fallen leaf material into the river quickly attracts a number of fungal colonists. Other and different fungi can also arrive and dominate with increasing degradation of substrate and biological changes. Willoughby and Archer (1973) found that the capacity of lotic freshwater environment to provide a microbial inoculum to initiate decomposition of allochthonous plant materials is always high. Of the species arriving on the bare litter substrate, only certain species, which are early successional species, establish themselves first. As the substrate is modified by the early colonisers, it becomes more suitable for the recruitment of 'late successional species'. Fungi dominating on decaying litter, from the beginning till the end, are known as 'regular' or resident fungi.

In the present investigation succession and colonisation pattern of aquatic Hyphomycetes on submerged leaf litter of two dominant forest trees of Nainital, Kumaun Himalaya has been studied by using a nylon-mesh bag technique.

MATERIALS AND METHODS

Old leaves of oak (*Quercus leucotrichophora* A. Camus) and pine (*Pinus roxburghii* Sarg.) were collected from trees before leaf fall and oven dried at 50°C for 3 days. 5 g. of dried leaves of each tree species-oak, pine and a mixture of these (oak+pine) were placed separately in 15 × 15 cm size nylon-mesh bags (1 mm^2 pore size) and sealed with a nylon thread. Three sets each consisting of 45 such bags (15 bags each of oak, pine and mixed litter) were kept in separate iron net boxes. One set each of these nylon mesh bags was placed into running freshwater streams at different sites.

Nine litter bags, three each of oak, pine and mixed (oak+pine) litter containing bags from the three sites were collected periodically at monthly intervals. These samples were kept in sterile polythene bags and brought to the laboratory. Samples of the litter bags were washed with running tap water to remove the microinvertebrates and sediments and then processed as described earlier in the taxonomic part of this study. Average temperature and pH were also recorded for each month. The incubated leaves were then screened directly under the microscope for the identification of aquatic Hyphomycetous conidia, and the presence of fungal species was recorded separately for each type of substrate, at different stages of submergence.

Organic carbon and nitrogen were recorded simultaneously from the leaf litter samples collected for the decomposition study, at different stages of submergence C/N ratio was also determined to observe its relation with the species number.

RESULTS AND DISCUSSION

Altogether thirty species of aquatic Hyphomycetes were recorded as colonisers of the submerged oak, pine and mixture (oak+pine) leaf litters at different stages of submersion during the present study. Oak litter substrate, having 36.46 initial C/N ratio had twenty-six species of aquatic Hyphomycetes, while pine (C/N ratio = 57.71) and mixed litter (C/N = 41.68) substrates had eighteen and twenty-two species, respectively (Table 14.1).

Colonisation of Oak Litter

The oak litter substrate was colonised by maximum number of aquatic Hyphomycetes at different stages of decomposition after submersion into the stream (Table 14.1). Eleven species were observed within 30 days of submersion and could be regarded as early colonisers of oak litter substrate. In the subsequent months of submersion, the species' composition slightly altered with an increase in species' number upto 4 months. Thereafter, the species' number gradually declined, with the exception of two middle months ie in the 7th –8th months (Fig. 14.2). There was a single species in the 9 month old submerged litter while in the following months, 6–7 species were observed with fewer conidia (Table 14.1). In the last collection (13 month), a little mass of the substrate remained in the bags and no aquatic Hyphomycetes were observed.

On oak leaf litter, *Clavariopsis aquatica, Alatospora acuminata* and *Anguillospora longissima* were found with maximum frequency of occurrence (> 80%). *Tricladium chaetocladium, Tetrachaetum elegans, Triscelophorus acuminatus, Heliscus lugdunensis, Dimorphospora foliicola, Lemonniera cornuta* and *Pleurophragmium sonum* accounted for 30–80% of occurrence (Table 14.1). The remaining species had the least frequency of occurrence (< 30%). It was interesting to note that the species which were found with high frequency of occurrence, had conidia formation early during 2–8 months of submergence.

The succession and colonisation pattern of different aquatic Hyphomycetes was not consistent on oak leaf litter. *Heliscus lugdunensis*, which was recorded in the first four months, disappeared from 5–10 months of submergence. Later, it was recorded with a few conidia (Table 14.1). *Tetrachaetum elegans* was also found as a frequently occurring species along with primary colonisers, but abundant conidia were observed only during the middle months. *Lunulospora cymbiformis, L. curvula, Lemonniera pseudofloscula, L. terrestris, Pestalotiopsis submersus, Flagellospora penicillioides, Trtracladium marchalianum, Camposporium pellucidum, Heliscella stellatacula, Tricladiopsis flagelliformis, Campylospora chaetocladia, Sytosynnema isthmospermum, Jaculispora submersa* and *Helicomyces roseus* were recorded only 1–3 times. Latter, six species were observed only once (in the 7 month old submerged oak litter). *Heliscella stellatacula*, however, occurred twice, but only in the 10th and 12th months, and could be regarded as true late colonist.

Colonisation of Pine Needle Litter

A total of 18 species of aquatic Hyphomycetes were observed on pine litter at different stages of submergence (Table 14.2). Of these, fourteen species were recorded within 30 days of submersion. In addition to *Lunulospora curvula, L. cymbiformis* and *Tetracladium marchalianum* the other species were similar to the early colonisers of oak litter. The species composition varied slightly in the following months and a gradual decline in species number was observed. However, three new colonisers viz, *Campylospora chaetocladia, Tricladium indicum* and *Heliscella stellatacula* were also recorded in 6th and 10th months of submergence. None of the species was observed in the 12th month but in the last collection (13th months old), when a small residue of litter remained, a few conidia representing three species were recorded (Table 14.2).

Lunulospora curvula and *Tetracladium marchalianum* were found in the first 4–5 months of submergence only. *Triscelophorus acuminatus* was less abundant in the initial stages, but it became prominent in the middle stages of the decomposing needles. *Flagellospora penicillioides* was absent in the first collection, but recorded in the 2nd to 6th months (old submerged pine needle litter) (Table 14.2).

Alatospora acuminata was found with the highest frequency of occurrence (92.31%) on pine needle litter. *Lemonniera pseudofloscula, Lunulospora cymbiformis, L. terrestris, Campylospora chaetocladia, Tricladium indicum* and *Heliscella stellatacula* had less than 30% frequency of occurrence as these were observed 1–3 times only. Amongst these fungi, the two species of *Lemonniera* and *Lunulospora*

Table 14.1: Occurrence of Aquatic Hyphomycetes on Oak Leaf Litter at Different Stages of Submersion

S.No.	Species	Collection Months 1	2	3	4	5	6	7	8	9	10	11	12	13	Total Occurrence	Frequency of Occurrence (%)
	C:N	21.32	26.68	21.09	16.55	18.59	21.89	19.15	19.33	20.53	19.74	19.12	19.25	19.14		
1	*Clavariopsis aquatica*	+	++	+++	+++	+	++	++	+		+	+	+		11	84.62
2	*Tricladium chaetocladium*	+	++	++	+++		+	++	+		+	+	+		10	76.92
3	*Tetrachaetum elegans*	+	+		++		++	+	++		+	+	+		9	69.23
4	*Triscelophorus acuminatus*	+	++	+	++	++	++	+++	++						8	61.54
5	*Heliscus lugdunensis*	+	++	+	+							+			5	38.46
6	*Alatospora acuminata*	+	+	+	+	+	+	+	++		+	+	+		11	84.62
7	*Anguillospora longissima*	+	++	+	++	+	++	+++	++	+	+		+		11	84.62
8	*Lemonniera cornuta*	+	+		+				+++						4	30.70
9	*L. pseudofloscula*	+													1	7.69
10	*L. terrestris*	+	+						+						3	23.02
11	*Dimorphospora foliicola*	+	+++	+++	+++	+	+++	++				+	+		9	69.23
12	*Lunulospora curvula*		+	+	+++										3	23.02
13	*L. cymbiformis*		+		+										2	15.32
14	*Pestalotiopsis submersus*		+												1	7.69
15	*Flagellospora penicillioides*			+	+										4	30.70
16	*Tetracladium marchalianum*			+											4	7.69
17	*Pleurophragmium sonam*				+		+	+	+						1	7.69
18	*Camposporium pellucidum*				+										1	7.69
19	*Unidentified* sp. 1				+										1	7.69
20	*Unidentified* sp. 2					+									1	7.69
21	*Tricladiopsis flagelliformis*							+							1	7.69
22	*Campylospora chaetocladia*							+							1	7.69
23	*Sytosynnema isthmosporum*							+							1	7.69
24	*Jaculispora submerse*							+							1	7.69
25	*Helicomyces roseus*							+							1	7.69
26	*Heliscella stellatacula*										+		+		2	15.32
	Total	11	13	10	15	6	7	13	10	1	6	6	7			

+ = Less conidial production++ = Moderate conidial production +++ = Abundant conidial production

Table 14.2: Occurrence of Aquatic Hyphomycetes on Pine Needle Litter at Different Stages of Submersion

S.No.	Species	Collection Months 1	2	3	4	5	6	7	8	9	10	11	12	13	Total Occurrence	Frequency of Occurrence (%)
	C:N	21.95	24.49	20.29	22.59	18.86	23.54	22.16	23.21	19.75	20.84	19.09	17.14	24.09		
1	Clavariopsis aquatica	++	++	+++	+++	+	++	+	+	+				+	9	69.23
2	*Tricladium chaetocladium*	++	++	++	++		++	+	+						7	53.85
3	*Lunulospora curvula*	+	++	+	++	+									5	38.46
4	*L. cymbiformis*	+	+		+										3	23.08
5	*Alatospora acuminata*	+	+	+	+	+	+	+++	+	+	+	+		+	12	92.31
6	*Anguillospora longissima*	+	++	++	++	+	+	+++	+						8	61.54
7	*Tetrachaetum elegans*	+	++	+	++		+++	++	+	+	+	+			10	76.92
8	*Heliscus lugdunensis*	++	++	++	+	+	+		+		+	+			9	69.23
9	*Triscelophorus acuminatus*	+	+		+		++	+++	++					+	7	53.85
10	*Lemonniera cornuta*	++	+		+				+					+	4	30.77
11	*L. pseudofloscula*	+													1	7.69
12	*L. terrestris*	+	+												2	15.38
13	*Tetracladium marchalianum*	+	+	+	+										4	30.77
14	*Dimorphospora foliicola*	++	+++	++	+++	+++	+++	+++	+++			+			9	69.23
15	*Flagellospora penicillioides*		+	+	++	+	+								5	38.46
16	*Campylospora chaetocladia*					+									1	7.69
17	*Heliscella stellatacula*										+				1	7.69
18	*Tricladium indicum*						+								1	7.69
	Total	14	14	10	13	8	10	7	9	2	4	4		3		

+ = Less conidial production ++ = Moderate conidial production +++ = Abundant conidial production *cymbiformis* were confirmed in the early stages of submergence, whereas the last three species were colonised in the later stages (Table 14.2).

As is evident from Table 14.2 pine needles were colonised by a sufficient number of aquatic hyphomycetes, with greater abundance of conidia, during the early 8 months, however, the species number as well as conidial production was low in the following months of submergence. This indicates that the early stages of decomposing needles were preferred more to the decomposed pine litter by these fungi.

Colonisation of Mixed Litter

Mixed litter ie mixture of 50% oak leaves (Angiosperm) and 50% pine needles (Gymnosperm) was taken to observe the combined effect on colonisation and the succession pattern of aquatic Hyphomycetes. Out of twenty-two species recorded on mixed litter substrate, a maximum of seventeen species were observed in the1st collection (after one month). In the 2nd month the number increased with the inclusion of two more species viz *Helicomyces roseus* and *Diplocladiella scalarioides* (Table 14.3). The species number gradually declined in the later stages of litter decomposition. However, three new species viz *Anguillospora longissima, Campylospora chaetocladia* and *Heliscella stellatacula* were also recorded. After 11 months very small amount of the litter was left in the bags, with no fungal species except for a few conidia of *Tetrachaetum elegans* (Table 14.3).

Table 14.3 shows that *Tetracladium marchalianum, Lunulospora curvula, Lemonniera terrestris, Camposporium pellucidum, C. antenatum, Tricladiopsis flagelliformis, Tricladium terrestre* and *Diplocladiella scalarioides* were restricted to early 24 months old submerged litter only and could be regarded as early colonists. Of these, the last three species were observed as new colonisers, which were absent in oak and pine. In mixed litter, *Heliscella stellatacula* was again observed in the later stages of submergence, thus confirming it as a late colonist.

On this substrate, *Alatospora acuminata* accounted for the maximum frequency of occurrence (84.6), while *Heliscus lugdunensis, Clavariopsis aquatica, Tetrachaetum elegans, Flagellospora penicillioides, Tricladium chaetocladium, Dimorphospora foliicola, Anguillospora longissima, Pleurophragmium sonam* and *Triscelophorus acuminatus* ranked second in frequency of occurrence (less than 30%). The remaining species had the least frequency of occurrence (Table 14.3).

The species number and conidia production were more prevalent in the early stages of decomposition than in the later stages. Perhaps, this might be due to the availability of suitable substrates in the early stages, in comparison to subsequent stages of decomposition.

COMPARISON AND DISCUSSION OF COLONISATION PATTERN

Park (1974) suggested that succession of fungal community in decomposable material may depend either on the substrate, interrelationship between

Table 14.3: Occurrence of Aquatic Hyphomycetes on Mixed (Oak+Pine) Litter at Different Stages of Submersion

S.No.	Species	Collection Months													Total Occurrence	Frequency of Occurrence (%)
		1	2	3	4	5	6	7	8	9	10	11	12	13		
	C:N	21.32	26.68	21.09	16.55	18.59	21.89	19.15	19.33	20.53	19.74	19.12	19.25	19.14		
1	*Heliscus lugdunensis*	++	++	+	+				+	+	+				7	53.85
2	*Triscelophorus acuminatus*	++	++	+	+	+	++	+	++						8	61.54
3	*Alatospora acuminata*	+	+	++	++	+	+	+	+++	+	+	+			11	84.62
4	*Clavariopsis aquatica*	+	++	++	+		++		+		+				7	53.85
5	*Tetracladium marchalianum*	+	++	+											3	23.08
6	*Tetrachaetum elegans*	+	+	++	+		++	+	++	+	+		+		10	76.92
7	*Flagellospora penicillioides*	+	+	+	+				+						5	38.46
8	*Lunulospora curvula*	+	+	+											3	23.08
9	*Tricladium chaetocladium*	+	+	++			++		++		+				6	46.15
10	*Lemonniera cornuta*	++	+						+						3	23.08
11	*L. terrestris*	+	+												2	15.38
12	*Dimorphospora foliicola*	++	+++	++	+		+	+	+++		+	+			9	69.23
13	*Camposporium pellucidum*	+	++												2	15.38
14	*C. antennatum*	+	++		+										3	23.08
15	*Tricladiopsis flagelliformis*	+	++												2	15.38
16	*Tricladium terrestre*	+	+												2	15.38
17	*Pleurophragmium sonam*	+	+	+	+	+									5	38.46
18	*Anguillospora longissima*			++	+	+	+	+	+	+	+	+			9	69.23
19	*Helicomyces roseus*		+		+		+								3	23.08
20	*Diplocladiella scalaroides*		+												1	7.69
21	*Campylospora chaetocladia*							+	+						2	15.38
22	*Heliscella stellatacula*									++	+	+			3	23.08
	Total	17	19	12	11	4	8	6	11	5	8	4	1			

+ = Less conidial production ++ = Moderate conidial production +++ = Abundant conidial production organisms, high inoculum availability and rapid growth or energy content of the substrate. When the course of succession follows, the early colonisers are replaced in the later stages by other colonisers, which are better adapted on the substrate. During the early stages of leaf litter processing in streams, aquatic Hyphomycetes are more dominant members of the microflora than bacteria (Kaushik and Hynes, 1968, 1971; Triska, 1970; Willoughby, 1974).

In the present investigation, colonisation and succession of aquatic Hyphomycetes on oak, pine and in mixed litter were studied. The oak litter substrate was colonised by the maximum number of aquatic Hyphomycetes at different stages of submersion, followed by mixed litter and needle litter (26, 22 and 18 species respectively).

This suggests that oak was preferred by the aquatic Hyphomycetes to mixed and pine litter, having a C/N ratio of 36.46, 41.68 and 57.71, respectively (Fig. 14.1).

In the first sample, which was recovered after 30 days of submergence, mixed litter had seventeen species while pine and oak were colonised by fourteen and eleven species, respectively. The C/N ratio of each of the substrates decreased at this stage to 20.0, 21.95 and 21.32, respectively. In the second month, when the C/N ratio slightly increased in each substrate (ie oak = 26.68; pine = 24.49; mix = 20.53), the species number also increased but gradually declined in subsequent collections, except in the case of oak (Table 14.4 and Fig. 14.1). A positive correlation was found with C/N ratio and pine litter species ($r = 0.576$, $P < .05$; $Y = 1.19–12.4x$), and oak litter species ($r = 0.139$ ns; $y = 0.26 –18.\ 06x$), while a negative correlation was observed with mixed litter species ($r = –0.355$; ns; $y = –1.6–34.15x$).

Oak and pine litter were found equally good substrate, for the colonisation of aquatic Hyphomycetes in the early stages of submersion. To compare the flora of different substrates in different stages of submergence, similarity indices (Sorensen's index) were calculated. This indicates that there was less than 90% similarity in the species of all three substrates used, in the 3rd month of submersion. But it declined in different substrates after 9 months of submersion. There were only 12 occasions when the similarity reached 90% in different substrates. Higher instances of species similarity (75–90%) were observed in between oak and pine substrates, followed by pine-mix and oak-mix substrates.

Most of the earlier workers (Ingold 1966, 1976; Manoharachary, 1986; Nilsson 1964; Ranzoni, 1953; Subramanian, 1983) assumed that these aquatic Hyphomycetes do not prefer conifer needles and remain absent from this substrate, but in the present study a large number of aquatic Hyphomycetes were collected on pine needles, in natural as well as in experimental conditions (Table 14.2 and 14.4). Such a high number of these fungi involved in the early colonisation of conifer needles confirm the findings of Barlocher and Oertli (1978), Barlocher (1981) and Barlocher and Rosett, (1981).

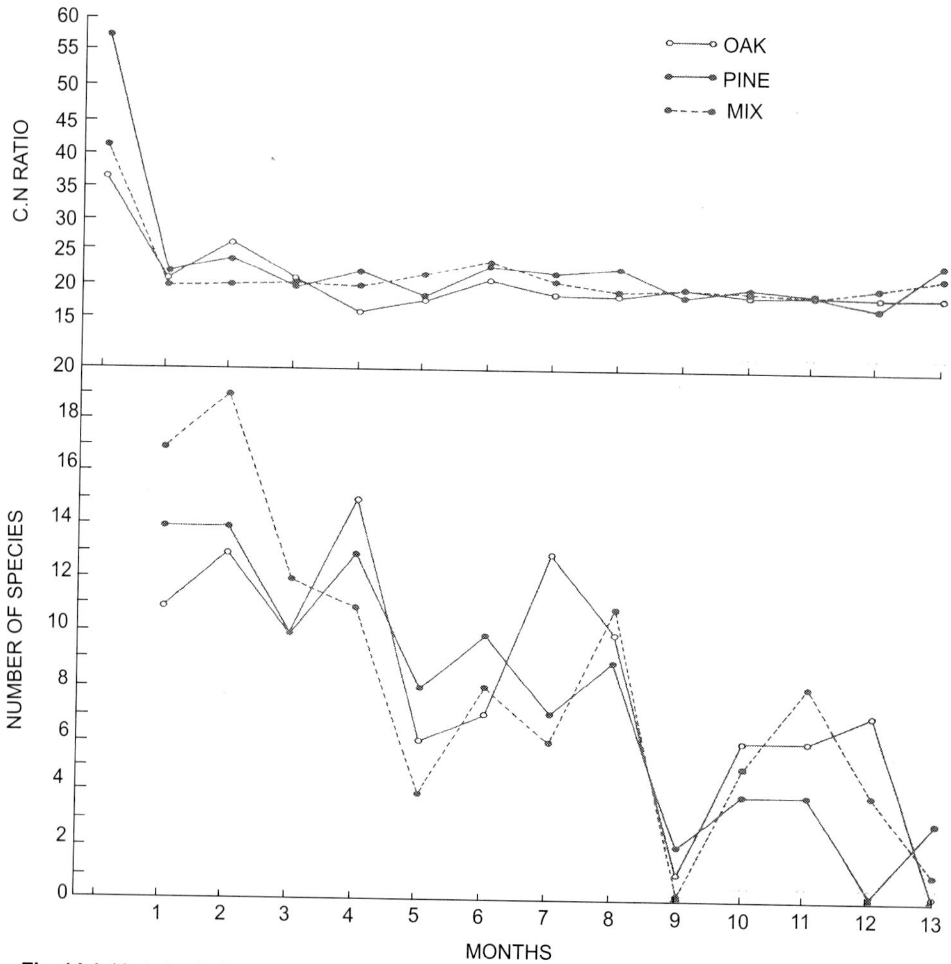

Fig. 14.1: Variation in Species Number and C/N ration in Oak, Pine and Mixed Litter at Different Stages of Submergence

Lunulospora curvula, L. cymbiformis, Lemonniera cornuta, L. terrestris, Tetracladium marchalianum, Tricladium terrestre, Tricladiopsis flagelliformis, Camposporium pellucidum and *C. antennatum* were restricted in the early months of submergence and regarded as early colonisers. Suberkropp and Klug (1976) reported *Flagellospora curvula, Lemonniera aquatica* and *Alatospora acuminata* as the early colonisers of submerged oak and hickory leaves, while *Anguillospora species* and *Clavariopsis aquatica* were recorded in the 4th week of submergence. *Tetracladium marchalianum* was absent in the early four weeks and also after 28 weeks of submergence. These observations also agree with Willoughby and Archer (1973) to some extent. *Clavariopsis aquatica* was present on the one month old submerged oak, pine and mixed litter, while *T. marchalianum* was present on pine and mixed litter only. Species of *Flagellospora, Lemonniera* and *Alatospora* were also found as early colonisers in the present investigation.

Table 14.4: Comparison in Species Composition of Oak, Pine and Mixed Litter Substrate (Q, P, M) in Different Stages of Submersion

		Collection Months												
		1	2	3	4	5	6	7	8	9	10	11	12	13
	Temp. (°C)	9.81	11	15.30	17.50	19.81	20.31	19.0	16.75	17.94	14.30	10.67	7.17	10.50
S. No.	Species pH	7.09	7.30	7.50	7.90	7.75	7.55	6.50	6.95	6.0	6.39	6.77	6.70	6.57
1.	*Alatospora acuminata*	QPM	QPM	QPM	QPM	QPM	QPM	QPM	QPM	P	QPM	QPM	QM	P
2.	*Anguillospora longissima*	QP	QP	QPM	QPM	QPM	QPM	QPM	QPM	Q	QM	M	QM	
3.	*Camposporium antennatum*	M	M		M									
4.	*Camposporium pellucidum*	M	M		Q									
5.	*Campylospora chaetocladia*					P		QM	M					
6.	*Clavariopsis aquatica*	QPM	QPM	QPM	QPM	QP	QPM	QP	QPM		Q	QM	Q	P
7.	*Dimorphospora foliicola*	QPM	QPM	QPM	QPM	QP	PM	QPM	QPM			QPM	QM	
8.	*Diplocladiella scalaroides*		M											
9.	*Flagellospora penicillioides*	M	PM	QPM	QPM	P	P		M					
10.	*Helicomyces roseus*		M		M		M	Q			PM	QPM		
11.	*Heliscella stellatacula*										QPM	M	QM	
12.	*Heliscus lugdunensis*	QPM	QPM	QPM	QPM	P	P		PM					
13.	*Jaculispora submerse*							Q						
14.	*Lemonniera cornuta*	QPM	QPM		QP				QPM					
15.	*L. pseudofloscula*	QP							Q					
16.	*L. terrestris*	QPM	QPM											
17.	*Lunulospora curvula*	PM	QPM	QPM	QP	P								
18.	*L. cymbiformis*	P	QP		QP									
19.	*Pestalotiopsis submerses*		Q											
20.	*Pleurophragmium sonam*	M	M	M	QM	M	Q	Q	Q					
21.	*Sytosynnema isthmosporum*							Q						
22.	*Tetrachaetum elegans*	QPM	QPM	PM	QPM		QPM	QPM	QPM	P	QPM	QPM	Q	M
23.	*Tetracladium marchalianum*	PM	PM	QPM	P									
24.	*Tricladiopsis flagelliformis*	M	M					Q						
25.	*Tricladium chaetocladium*	QPM	QPM	QPM	QP		QPM	QP	QPM		Q	QM	Q	
26.	*T. indicum*						P							
27.	*T. terrestre*	M	M											
28.	*Triscelophorus acuminatus*	QPM	QPM	QM	QPM	QM	QPM	QPM	QPM					P
29.	Unidentified sp. 1				Q									
30.	Unidentified sp. 2			Q										
	Total	Q=11 P=14 M=17	Q=13 P=14 M=19	Q=10 P=10 M=12	Q=15 P=13 M=11	Q=6 P=8 M=4	Q=7 P=10 M=8	Q=13 P=7 M=6	Q=10 P=9 M=11	Q=1 P=2 M=0	Q=6 P=4 M=5	Q=6 P=4 M=8	Q=7 P=0 M=4	Q=0 P=3 M=1

The least number of species were observed in 9-month-old decomposing substrates. In the 13th collection, a little substrate remained in all the bags and no species was found in oak, while pine and mixed litter had three and one species, respectively (Fig. 14.1). The conidial number was also least in those collections. *Alatospora acuminata, Anguillospora longissima, Clavariopsis aquatica, Dimorphospora foliicola, Tetrachaetum elegans, Tricladium chaetocladium* and *Triscelophorus acuminatus* were also among the early colonisers, but remained for a longer period on these substrates. Therefore, these could be categorised as regular colonisers of oak pine and mixed litter (Table 14.4).

The early colonisation of *Heliscus lugdunensis* on all the three substrates used in this study supports Willoughby and Archer (1973) who also found it as an early colonising species. However, Suberkropp and Klug (1976) reported this species as an occasionally sporulating species. *Dimorphospora foliicola,* which was observed as a late colonist (being present 5th month onwards) by Willoughby and Archer (1973), was an early colonizer of all the substrates used in the present study and agree with Newton (1971).

Many species occurred occasionally in different months of submergence. *Heliscella stellatacula* was always found to colonise in later stages of submergence in all the substrates and was regarded as a colonist (Table 14.4). The behaviours of colonisers as early or late has been well speculated by Park (1974b). The early colonists are replaced by late colonists, not because of changing nutritional conditions, but changes in the substratum, perhaps, due to autotoxin production or alteration of physical conditions by the early colonisers. In this study, no significant relation was found between species number and temperature, while pH showed a positive correlation. Linear regression equations were developed as given below:

Oak $-y = 0.069 - 7.59x$; $r = 0.542$, $p < 0.05$; Pine $-y = 0.089 - 6.91x$; $r = 0.723$, $p < .01$;
Mix $-y = -0.408$ ns; Y = species number; X = pH

It is generally assumed that microbial colonisation is greatly influenced by the temperature and pH of the microhabitat. In running water systems because of day to day fluctuations in the temperature and pH of stream water, colonisation cannot be based only on these factors (Barlocher and Rosset, 1981). However, temperature, availability of nutrient, pH and other chemical parameters of stream water all together may also play an effective role in the establishment of particular species and spore production (Barlocher and Rosset, 1981).

The differences found in successional pattern might be due to inherent differences in the mycoflora of the streams, method of placing the leaves in the stream or the methods used to determine the dominant species present. The preferential occurrence of aquatic Hyphomycetes on submerged leaves in freshwater streams has been recorded by Ingold (1975), Nilsson (1964), Ranzoni (1953) and Sridhar and Kaveriappa (1989). Therefore, other reasons for disparity in the findings of some earlier workers may be due to different substrate preferences by different investigators for the colonisation study.

REFERENCES

Barlocher, F. 1981. Fungi on the food and in the faeces of *Gamarus pulex*. Trans. Brit. Mycol. Soc., 76, 14–19.

Barlocher, F and Oertli, J.J. 1978. Inhibitors of aquatic hyphomycetes in dead conifer needles. Mycologia, 70: 964–974.

Barlocher, F. and Rosset. J. 1981. Aquatic Hyphomycetes Spora of two Black Forest and two Swiss Jura streams. Trans. Brit. Mycol. Soc., 76: 479–483.

Ingold, C.T. 1966. The tetraradiate aquatic fungal spore. Mycologia, 58, 43–56.

Ingold,C.T.1975. *An illustrated guide to aquatic and waterborne Hyphomycetes (Fungi imperfecti) with notes on their biology*. Freshwater Biol. Assoc. Scient. Publ. No. 30 England 96 pp.

Ingold, C.T. 1976. The morphology and biology of freshwater fungi excluding Phycomycetes. In 'Recent Advances in Aquatic Mycology'. (Ed. E.B. Gareth Jones) Publ. Paul Elek . Ltd. 335–357.

Kaushik, N.K. and Hynes, H.B.N. 1968. Experimental study on the role of autumn shed leaves in aquatic environment . J. Ecol., 56: 229–243.

Manoharacharya, C. 1986. Aspects and prospects of waterborne conidial fungi from India. Advances in Biological Research, 4, 160–183.

Newton, J.A. A mycological study of decay in the leaves of deciduous trees on the bed of a river.

Nilsson, S. 1964. Freshwater Hyphomycetes. Taxonomy, Morphology and Ecology. Symb. Bot. Upsal., 18:1–30.

Odum, E.P. 1971. 'Fundamentals of Ecology' W.B. Saunders Co., Philadelphia, 574

Park, D. 1974a. *Tricladium terrestre* sp. nov. Trans. Brit. Mycol . Soc., 63: 179–183.

Park, D. 1974b. Aquatic Hyphomycetes on non-aquatic habitats. Trans. Brit. Mycol. Soc., 63: 183–187.

Peterson, R.C. and Cummins, K. W. 1974. Leaf processing in a woodland stream. Freshwater Biol. , 4, 343–368.

Ranzoni, F.V. 1953. The aquatic Hyphomycetes of California. Farlowia ,4:353–398.

Sridhar, K.R. and Kaveriappa, K.M. 1989b. Colonisation of leaves by waterborne hyphomycetes in a tropical stream. Mycol.Res., 92: 392–396.

Suberkropp, K. and Klug, M.J. 1976. Fungi and bacteria associated with leaves during processing in a woodland stream. Ecology, 57: 707–719.

Suberkropp, K. and Godshalk, G.L. Klug, M.J. 1976. Changes in the chemical composition of leaves during processing in a woodland stream. Ecology, 57: 720–727

Subramanian, C.V., 1983. Hyphomycetes, Taxonomy and Biology. Acad. Press, London, 502 pp.

Willoughby, L.G. 1974. Decomposition of Litter in freshwater 659–681 in '*Recent Advances in Aquatic Mycology*' (Ed. E.B. Gareth-Jones) Elek Science London.

Willoughby, L.G. and Archer, J.F. 1973. The fungal spora of a freshwater stream and its colonisation pattern on wood. Fresh Wat. Biol., 3: 219–239.

Triska, F.J. 1970. Seasonal distribution of aquatic Hyphomycetes in relation to the disappearance of leaf litter from a woodland stream. Ph. D. thesis, University of Pittsburgh. Pittsburgh. PA.

15

Diversity, Biogeography and Ecological Functions of Filamentous Fungi in Mangrove Ecosystem

K.R. SRIDHAR AND K. ANANDA

Department of Biosciences, Mangalore University, Mangalagangotri, Mangalore 574 199, Karnataka, India

SUMMARY

The huge organic detritus generated by mangrove vegetation (110–2370 g/m^2) supports a wide variety of lives including diverse fungal communities. Fungi being the decomposers contribute significantly to the nutrient cycling and energy flow in mangrove ecosystem. To date, 55 mangrove tree species and their associates are known to support 191 filamentous mangrove fungi. The diversity of mangrove fungi dependent on the age of the mangrove, diversity of mangrove tree species, diversity of coastal trees and physicochemical features of the mangrove habitat. Mangrove fungi show several structural, functional and physiological adaptations, specificity to niche, substrate and zonation in mangrove habitats.

Species richness of mangrove fungi was consistently high on the detritus of *Rhizophora* spp. in several mangroves. The decomposition of leaves by fungi in mangrove habitats leads to a decline of organic carbon, increase in total nitrogen, organic matter, protein content and calorie of leaves. Up to 42% leachate of mangrove leaves gets incorporated into the food chain within 2–12 hr and hence, constitutes an important source of carbon, energy and nitrogen to the higher trophic levels. Fungal conditioning of mangrove litter results in the production of a large proportion of dissolved, rather than particulate organic matter. Mineralisation of lignocellulosic components of mangrove detritus is 10-fold lower than that of leachate fractions in the mangrove ecosystems. The enzyme equipment of the mangrove fungi further transforms such recalcitrant lignocellulosic fractions. In addition to typical marine fungi several geofungi and freshwater fungi have been recovered in mangrove habitats. A variety of modern techniques are available to assess the activities and functions of filamentous fungi in the mangrove ecosystem.

Keywords: Biodiversity, Biogeography, Colonisation, Decomposition, Mangroves, Mangrove fungi, Niche specificity, Nutrient cycling, Succession

'Marine Mycology' is one of the recent areas of research, which imparts knowledge on nutrient cycling, food web and energy flow in the marine ecosystem. 'Obligate marine fungi' are those, which grow and sporulate in marine or estuarine habitats, while 'facultative marine fungi', derived from freshwater or terrestrial substrates, can grow and possibly sporulate in the marine habitat (Kohlmeyer 1974). Although Desmazieres described the first facultative marine fungus (*Phaeosphaeria typhorum*) in 1849 and Durieu and Montagne discovered the first obligate marine fungus (*Halotthia posidoniae*) in 1869, for almost a century interest of mycologists on marine fungi was sporadic (Kohlmeyer and Kohlmeyer, 1979). The development of marine mycology was highly influenced by contributions by Barghoorn and Linder (1944). Subsequently (1951-1962), literature on marine fungi was generated from Australia, France, Germany, Great Britain and United States (Kohlmeyer and Kohlmeyer, 1979). In 1961, Johnson and Sparrow published the first monograph on marine mycology 'Fungi in Oceans and Estuaries'. A review published by G.C. Hughes (1975) has summarised the contributions on marine mycology since 1961.

Marine fungi are important in marine habitats as decomposers of dead organic materials and act as intermediaries of nutrient flow from organic matter to the higher trophic levels (Hyde and Lee, 1995). Marine fungi are well adapted to the saline environment and most of them require seawater for the successful completion of their life-cycle. Ascomycetes, in particular, show adaptation to the marine ecosystem in the production of appendages. These appendages help in the buoyancy of spores and entrapment and adherence to the surfaces. The habitats of fungi in marine and marine influenced ecosystem include: deep sea, estuaries, mangroves, sandy beaches, algal beds, marine lichens, tidal salt marshes and tropical shoreline trees (Kohlmeyer and Kohlmeyer, 1979). The monographs written by Kohlmeyer and Kohlmeyer (1979), Kohlmeyer (1984), Kohlmeyer and Volkmann-Kohlmeyer (1991a) and Hyde and Sarma (2000) are a major source for identifying and understanding the richness species and diversity of higher marine fungi. Recently, Alias (1996) listed 339 fungi known from mangroves, with 191 higher marine fungi (151 ascomycetes, 3 basidiomycetes and 37 deuteromycetes). The present review emphasises mainly on the biogeography, diversity and ecological functions of filamentous fungi in the mangroves of the Indian Ocean, Atlantic Ocean, Pacific Ocean, South-East Asia, Hong Kong and Japan. Niche specificity and litter decomposition by mangrove fungi, occurrence of geofungi and freshwater fungi in the mangrove ecosystem are also discussed.

THE MANGROVES

The term 'Mangroves' designates tidal wetland ecosystems formed by the association of plants and animals in intertidal low-lying areas of tropical and subtropical habitats. 'Mangroves' also designate tidal forests comprising trees, shrubs, palms, epiphytes, grasses and ground ferns (Chapman, 1977). About one-fourth of the world's tropical coastline is dominated by mangroves. According to Spalding *et al.* (1997), the world's total area of mangroves is about 181,000 km^2, out of which 41.5%, 26.5% and 22.6% mangroves are found in South-East Asia, America (including Caribbean) and Africa respectively. Mangroves have been classified into three broad categories (Tomlinson, 1986; Field, 1995): (i) *True mangroves,* which are restricted to intertidal areas between high water levels of neap of spring tides. (ii) *Minor species of mangroves,* which are inconspicuous elements of vegetation and rarely form pure communities. (iii) *Mangle associates,* which are salt tolerant plant species not found exclusively in the proximity of mangroves and occurring only in transitional vegetation, landwards and seawards. Chapman (1976) listed more than 100 mangrove tree species. About 80 species of true mangrove trees and shrubs have been recognised, of which 50-60 species make significant contributions to the structure of mangrove forests (Tomlinson, 1986; Field, 1995). High mangrove plant species diversity is seen in South-East Asian region (two-thirds of all species are found), while 15 and 10 species are known from the African and American coasts, respectively (Field, 1995). *Rhizophora* spp. are common in many coasts and islands. Mangrove trees have specific adaptations to their marshy habitat (e.g. prop roots, pneumatophores and viviparous seedlings) (Chapman, 1976). The organic detritus of mangrove vegetation (fallen leaves, twigs, bark, roots, seedlings and other dead parts) exposed to diurnal tidal activities, supports large animal and fungal communities.

MANGROVE FUNGI

Fungi and fungi-like organisms include: Oomycetidae and Thraustochytridae in the Chromista and the Ascomycota, Basidiomycota and the Chytridiomycetes in the Fungi. Cribb and Cribb (1955, 1956) in Australia reported the first account of mangrove fungi derived from *Avicennia marina*. In 1979 Kohlmeyer and

Kohlmeyer listed 42 species of higher mangrove fungi, worldwide (23 ascomycetes, 2 basidiomycetes and 17 mitosporic fungi). Subsequently, Hyde and Jones (1988) reported 89 species (62 ascomycetes, 2 basidiomycetes and 25 mitosporic fungi) from the Atlantic, Indian and Pacific Oceans. Although Kohlmeyer and Kohlmeyer (1979) considered mangrove fungi as the fourth largest group of marine fungi, recent studies have revealed that these fungi are the second largest group. Twenty-eight mangrove tree species consists of 120 species of higher fungi—87 ascomycetes, 2 basidiomycetes and 31 mitosporic fungi (Hyde, 1990b). *Rhizophora apiculata* harboured the maximum number of mangrove fungi—63 species (Sarma and Vittal, 2000). One-third of the species is known from the mangoves of South-East Asia. To date, fifty-five mangrove tree species and their associates are known to support 191 higher marine fungi (Alias, 1996).

Fungi being habitat-specific, a complete inventory can never be achieved especially due to habitat destruction by human interference, and other such activities which in turn lead to the extinction of fungi. Loss of mangrove fungi should not be overlooked due to their significant contribution to the functioning of marine and marine-influenced ecosystems. Considering the importance of mangroves in the food web, fisheries, land protection, recreation and tourism, little interest has been shown in the study of mangrove fungi.

BIOGEOGRAPHY AND DIVERSITY

Mangrove fungi have been incorporated in biogeographical maps published by Hughes (1974), Kohlmeyer (1983) and Jones (1993). Kohlmeyer (1984), Kohlmeyer and Volkmann-Kohlmeyer (1987a), Volkmann-Kohlmeyer and Kohlmeyer (1993) discussed the biogeography of specific fungi. Based on the studies conducted in South-East Asia, Indian, Pacific and Atlantic Oceans, Hyde and Lee (1995) revised the distribution maps of important mangrove fungal representatives: *Halosarpheia fibrosa, H. marina, Lignincola laevis* and *Lulworthia grandispora*. Drastic difference in mangrove fungal communities has been noticed between Oahu (Hawaii) and Belize (Caribbean) on the substrates of *Rhizophora mangle*. Many species were never found in Oahu due to the recent introduction of mangrove plant species. Further decline in mangrove fungal population was seen in Moorea Islands due to young mangrove stand. Jones (1993) has represented the geographical distribution of *Halorosellinia oceanica,* which is very common on mangrove wood and restricted to the tropics and subtropics. The distribution of this fungus mainly depends on temperature for growth and sporulation. The major factors which govern the distribution of mangrove fungi include the availability of substrates or hosts, water temperature and transport of fungi to the landmass across the seawater. The diversity of mangrove fungi is dependent on the age of the mangrove, diversity of mangrove tree species, the diversity of terrestrial tree flora and the physicochemical features of the mangrove habitat, (e.g. temperature, salinity and tidal range) (Hyde and Jones, 1988).

Indian Ocean

From the Indian Ocean, several mangrove locations have been investigated for the occurrence of mangrove fungi. Some important locations include: Indian Peninsula, Seychelles, Andaman Nicobar Islands, Atolls of Maldives, Islands of Lakshadweep, Mauritius and Aldabra. Hyde and Jones (1988) compiled the occurrence of seventy-five species of mangrove fungi, consisting of 57 ascomycetes, two basidiomycetes and 16 deuteromycetes, from the Indian Ocean. The total area of the Indian coast covered by the mangroves is estimated to be about 7% of the world mangroves (Untawale, 1987). Studies on mangrove fungi were initiated in Indian Peninsula during the 1980s. While Floristic studies have been conducted from the mangroves of Gujarat, Maharashtra, Goa, Tamil Nadu and Andhra Pradesh (see Table 1 and 2 for references), the mangroves of the rest of the coasts (Karnataka, Kerala, Orissa and West Bengal) are less explored or virtually unexplored for mangrove fungi. Similarly, studies on the mangroves of the islands adjacent to the Indian coast are also meagre. Thirteen new species, including three new genera have been described from the mangroves of the Indian Peninsula (Table 15.1). Most of these new fungi were described from the mangroves of Maharashtra coast. Although many floristic studies have been conducted on the mangroves of the Indian Peninsula, only a few quantitative studies are available (Borse, 1988; Ravikumar and Vittal, 1996, Sarma and Vittal, 2000; Borse *et al.*, 2000; Patil and Borse, 2001). In the mangroves of Maharashtra, 36 species were recovered, the dominant species being *Aigialus grandis, Massarina valatospora* and *Verruculina enalia* (Borse, 1988). Ravikumar and Vittal (1996) from the Pichavaram mangroves, Tamil Nadu recovered 48 species on the wood, prop roots and seedlings of *Rhizophora apiculata* and *R. mucronata*, with the maximum species (44) on prop roots. *Aigialus grandis, Asterosphaeriella mangrovis, Cirrenalia pygmea, Halocyphina villosa* and *Verruculina enalia* were the most dominant. From the deltas of Krishna and Godavari (Andhra Pradesh), Sarma and Vittal (2000) examined decaying wood, prop roots and seedlings of *Rhizophora apiculata* and recorded 63 species with maximum species on prop roots (61), then on wood (24) and seedlings (21). Sarma and Vittal (2000) also studied the occurrence of mangrove fungi on the wood, roots and pneumatophores of *Avicennia officinalis* and *A. marina* of Krishna and Godavari delta and recorded 65 species with a maximum of 61 species on wood, followed by 17 species on roots and 14 species on pneumatophores. On both *Rhizophora* and *Avicennia* spp., *Verruculina enalia* was the most frequent fungus. Detritus of nine host plant species of the mangroves of Godavari and Krishna also revealed the highest mangroves fungi in *R. apiculata* (Sarma and Vittal, 2001). A seasonal study of mangrove fungi was carried out at Godavari and Krishna delta (Sarma and Vittal, 1998–1999).

Though most fungi tend to occur during the monsoon season, no definite seasonal trend was evident. About 98 species of mangrove fungi (78 ascomyctes, 1 basidiomycete and 19 mitosporic fungi) have been reported from the mangroves of the Indian Coast (see Table 15.2 for checklist), with most of them confined to the mangroves of Gujarat, Maharashtra, Tamil Nadu and Andhra Pradesh. Several mangrove locations of the Indian coast are yet to be explored for mangrove fungi.

Table 15.1: New Genera and New Species of Mangrove Fungi Described from the Indian Peninsula

Fungus	Mangrove	Reference*
Ascomycetes		
**Acrocardiopsis patilii* Borse *et* Hyde	Malvan (Maharashtra)	Borse and Hyde, 1989
Aigialus mangrovei Borse	Dabhol (Maharashtra)	Borse, 1987e
Aigialus rhizophorae Borse	Dabhol (Maharashtra)	Borse, 1987e
Aniptodera indica Ananda *et* Sridhar	Udyavara (Karnataka)	Ananda and Sridhar, 2001
Asterosphaeriella mangrovis (Kohlm. *et* Vittal) Aptroot *et* Hyde	Bombay (Maharashtra)	Kohlmeyer and Vittal,1986
Bathyascus mangrovei Ravikumar *et* Vittal	Pichavaram (Tamil Nadu)	Ravikumar and Vittal, 1991
**Biatriospora marina* Hyde *et* Borse	Revadanda (Maharashtra)	Hyde and Borse, 1986a
Didymella avicenniae Patil *et* Borse	Revas (Maharashtra)	Patil and Borse, 1985b
**Julella avicenniae* (Borse) Hyde	Revas (Maharashtra)	Borse, 1987b (Hyde, 1992a)
	Pichavaram (Tamil Nadu)	Kohlmeyer and Vittal, 1986
Massarina armatispora Hyde, Vrijmoed, Chinnaraj *et* Jones	Kundapura (Karnataka)	Hyde, Vrijmoed, Chinnaraj and Jones, 1992
	Chorao Island (Goa)	Hyde, Vrijmoed, Chinnaraj and Jones, 1992
Massarina valatospora Hyde *et* Borse	Maharashtra	Hyde and Borse, 1986b
Passeriniella mangrovei Maria *et* Sridhar	Udyavara (Karnataka)	Maria and Sridhar, 2002
Anamorphs		
Cirrenalia basiminuta Raghukumar *et* Zainal	Orda (Goa)	Raghukumar, Zainal and Jones, 1988

*New genera

The Atolls of Maldives comprise about 1190 islands, which are known to harbour about 13 mangrove plant species, out of which *Avicennia marina*, *Bruguiera cylindrica*, *Ceriops tagal*, *Lumnitzera racemosa* and *Rhizophora mucronata* are dominant. Submerged and partially submerged dead wood from the trunks, branches, roots, prop roots and pneumatophores of mangrove plant species

Table 15.2: Check-list of Mangrove Fungi Reported from the Indian Peninsula

Fungus	Location*
Ascomycetes	
Acrocordiopsis patilii Borse *et* Hyde	2
Aigialus grandis Kohlm. *et* Schatz	1,2,5
Aigialus mangrovei Borse	2, 5
Aigialus parvus Schatz *et* Kohlm.	1,2,5
Aigialus rhizophorae Borse	2
Aniptodera chesapeakensis Shearer *et* Miller	6
Aniptodera haispora Vrijmoed, Hyde *et* Jones	6
Aniptodera indica Ananda *et* Sridhar	4
Aniptodera mangrovei Hyde	6
Antennospora quadricornuta (Cribb *et* Cribb) Johnson	1,2,5
Antennospora salina (Meyers) Yusoff, Jones *et* Moss	1
Ascochyta salicorniae Magnus	2
Ascocratera manglicola Kohlm.	2,5,6
Asterosphaeriella mangrovis (Kohlm. *et* Vittal) Aptroot *et* Hyde	1,2,5,6
Bathyascus avicenniae Kohlm.	5,6
Bathyascus mangrovei Ravikumar *et* Vittal	5
Belizeana tuberculata Kohlm.	5
Biatriospora marina Hyde *et* Borse	2,5
Ceriosporopsis cambrensis Wilson	2
Ceriosporopsis halima Linder	2
Chaetomastia typhicola (Karsten) Barr	6
Corollospora intermedia Schmidt	5
Corollospora maritima Werdermann	1
Corollospora pulchella Kohlm., Schmidt *et* Nair	2,6
Cryptosphaeria mangrovei Hyde	6
Dactylospora heliotrepha (Kohlm. *et* Kohlm.) Hafellner	1,2,5,6
Dictyosporium pelagicum (Linder) Hughes	2
Didymella avicenniae Patil *et* Borse	2,5
Eutypa bathurstensis Hyde *et* Rappaz	1,6
Halorosellinia oceanica Whalley, Jones, Hyde *et* Laessøe	1,3,4,6
Littispora abonnis (Kohlm.) Campb. Anderson *et* Shearer	2,6
Halosarpheia fibrosa Kohlm. *et* Kohlm.	2
Halosarpheia hamata (Höhnk) Kohlm.	5
Halosarpheia marina (Cribb *et* Cribb) Kohlm.	2,5,6
Halosarpheia minuta Leong	6
Littispora ratnagiriensis Patil *et* Borse	2,6
Natantispora retorquens (Shearer *et* Crane) Campb., Anderson *et* Shearer	2
Panorbis viscosus (Schmidt) Campb., Anderson *et* Shearer	6
Halosphaeria cucullata (Kohlm.) Kohlm.	5

Heleococcum japonense Tubaki	5,6
Helicascus kanaloanus Kohlm.	2
Julella avicenniae (Borse) Hyde	1,2,6
Kallichroma tethys (Kohlm. *et* Kohlm.) Kohlm. *et* Volkm.-Kohlm.	1,5,6
Lautospora gigantea Hyde *et* Jones	6
Lecanidion atratum (Hedw. *ex* Fr.) Endl.	6
Leptosphaeria australiensis (Cribb *et* Cribb) Hughes	1,2,5,6
Leptosphaeria marina Ellis *et* Everhart	2
Leptosphaeria oraemaris Linder	5
Leptosphaeria obiones (Crouan *et* Crouan) Saccardo	2
Leptosphaeria peruviana Speg.	5
Lignincola laevis Höhnk	2,6
Lignincola tropica Kohlm.	6
Lineolata rhizophorae (Kohlm. *et* Kohlm.) Kohlm. et Volkm.-Kohlm.	1,5,6
Lulworthia grandispora Meyers	1,2,5,6
Lulworthia medusa (Ellis *et* Everh.) Cribb *et* Cribb	2
Marinosphaera mangrovei Hyde	6
Massarina armatispora Hyde, Vrijmoed, Chinnaraj *et* Jones	3,4
Massarina thalassiae Kohlm. *et* Volkm.- Kohlm.	2,5,6
Massarina velatospora Hyde *et* Borse	1,2,5,6
Mycosphaerella pneumatophorae Kohlm.	5,6
Nais glitra Crane *et* Shearer	5,6
Neptunella longirostris (Cribb *et* Cribb) Pang *et* Jones	2,5,6
Ophiodeira monosemeia Kohlm. *et* Volkm.-Kohlm.	6
Passeriniellamangrovei Maria *et* Sridhar	4
Passeriniella obiones (Crouan *et* Crouan) Hyde et Mouzouras	6
Pedumispora rhizophorae Hyde *et* Jones	6
Quintaria lignatilis (Kohlm.) Kohlm. *et* Volkm.-Kohlm.	1,5,6
Rhizophila marina Hyde *et* Jones	4,6
Saccardoella marinospora Hyde	6
Saccardoella rhizophorae Hyde	6
Savoryella lignicola Jones *et* Eaton	2,6
Swampomyces armeniacus Kohlm. *et* Volkmn.-Kohlm.	5
Torpedospora radiata Meyers	1,2
Trematosphaeria lignatilis Kohlm.	2,5
Tubeufia setosa Sivanesan *et* Hsieh	6
Verruculina enalia (Kohlm.) Kohlm. *et* Volkm .-Kohlm.	1,2,5,6
Zopfiella latipes (Lundqvist) Malloch *et* Cain	1,2,6
Zopfiella marina Furuya *et* Udagowa	6
Basidiomycete	
Halocyphina villosa Kohlm. *et* Kohlm.	1,2,5,6
Anamorphs	
Ascochyta salicorniae Magnus	1

Bactrodesmium linderi (Crane *et* Shearer) Palm. *et* Stewart	2,6
Camarosporium palliatum Kohlm. *et* Kohlm.	2
Camarosporium roumeguerii Sacc.	1,2,6
Cirrenalia basiminuta Raghukumar et Zainal	3,5,6
Cirrenalia macrocephala (Kohlm.) Meyers *et* Moore	2,5,6
Cirrenalia pygmea Kohlm.	1,2,5,6
Cirrenalia tropicalis Kohlm.	5,6
Clavatospora bulbosa (Anast.) Nakagiri *et* Tubaki	2,4
Cytospora rhizophorae Kohlm. *et* Kohlm.	1,6
Ellisembia vagum (Nees *et* Nees) Subram.	6
Epicoccum purpurascens Ehrenb. *et* Schlecht.	6
Monodictys pelagica (Johnson) Jones	5
Periconia prolifica Anastasiou	1,2,5,6
Phomopsis mangrovei Hyde	1,6
Trichocladium achrasporum (Meyers et Moore) Dixon	1,2,5,6
Trichocladium alopallonellum (Meyers *et* Moore) Kohlm. *et* Kolkm.- Kohlm.	1,2,5,6
Zalerion maritimum (Linder) Anastasiou	1,2
Zalerion varium Anastasiou	1,2,5,6

*1. Gujarat: Borse *et al*, 2000; Patil and Borse, 2001
2. Maharashtra: Borse, 1984, 1987a-e, 1988; Borse and Hyde, 1989; Borse and Shrivastava,1988, 1994; Borse *et al*, 1988; Hyde and Borse, 1986a, b; Kohlmeyer and Vittal, 1986;Patil and Borse, 1983, 1985a, b; Shrivastava, 1989, 1994.
3. Goa: Hyde *et al*, 1992; Chinnaraj and Untawale, 1992; Raghukumar *et al*, 1988.
4. Karnataka: Ananda and Sridhar, 2001; Chinnaraj and Untawale, 1992; Hyde *et al*,1992; Maria and Sridhar, 2002; Sridhar and Kaveriappa, 1988.
5. Tamil Nadu: Raghukumar, 1973; Ravikumar and Purushothaman, 1988a, b; Ravikumar and Vittal, 1987, 1991, 1996.
6. Andhra Pradesh: Sarma and Vittal, 2000.

yielded 39 species of mangrove fungi (Chinnaraj, 1993a). The maximum number of fungal species was confined to *Rhizophora mucronata* (24 species). The dominant fungal species were *Asterosphaeriella mangrovis* and *Verruculina enalia*. Andaman and Nicobar consist of about 350 islands. Here the dead and decayed parts of six mangrove plant species yielded 49 ascomycetes, one basidiomycete and 13 mitosporic fungi (Chinnaraj, 1993b). Again, the highest number of species was recorded on *Rhizophora mucronata* (53). The most dominant fungi was *Verruculina enalia* and *Halocyphina villosa*. On the dead mangroves of Minicoy Island (Lakshadweep), 15 mangrove fungi comprising 11 ascomycetes, one basidiomycete and three mitosporic fungi were recorded (Chinnaraj, 1992). Hyde and Jones (1988) studied the mangrove litter accumulated on the floor of mangrove swamps of two islands at Seychelles. In Brillant Island 38 species were encountered; surprisingly the basidiomycete *Halocyphina villosa* was most dominant (27.3%), followed by two ascomycetes *Lulworthia grandispora* (22.7%) and *Antennospora quadricornuta* (12.4%). In Anse Boileau Island, 32 species were found, out of which the common species were *Halocyphina villosa* (13.3%) and

Halosarpheia marina (19.3%). Nine species were recorded from the mangroves of Aldabra (Kohlmeyer and Volkmann-Kohlmeyer, 1987a). Sixty-seven species (53 ascomycetes, 13 mitosporic fungi and 1 basidiomycete) of mangrove fungi were recovered from five mangrove stands of Mauritius (Poonyth *et al.*, 1999). Split wood blocks of *Bruguiera gymnorrhiza* and *Rhizophora mucronata* on submersion up to 78 weeks yielded in five mangroves of Mauritius 61 marine fungi, (Ponyth *et al.*, 2001). From the Indian Ocean, 34 mangrove fungi were recorded as more frequent (less than 10%) (Table 15.3).

Atlantic Ocean

The mangrove locations studied for fungi in the Atlantic coasts include: Sierra Leone, Bahamas, Belize, Florida, Sao Paulo, St. John, St. Thomas, Trinidad, Tobago and Virgin Islands. Newell (1976) recorded three frequent (less than 10%) mangrove fungi on *Rhizophora mangle* in Florida: *Etheirophora blepharospora, Lulworthia grandispora* and *Zalerion varium* (Table 15.3). Kohlmeyer (1980) compiled the occurrence of 54 species of mangrove fungi from the subtropical locations of the western Atlantic Ocean, the mangrove-bordered coasts of West Atlantic islands and the main land countries of North and South America (34 ascomycetes, 2 basidiomycetes and 18 mitosporic fungi). Aleem (1980) has studied the distribution and seasonal fluctuations of higher marine fungi in mangrove habitats of the coasts of Sierra Leone. Pneumatophores, prop roots, stems and twigs of *Avicennia africana, Rhizophora mangle* and *R. recemosa* were screened and 24 species of mangrove fungi (14 ascomycetes and 10 mitosporic fungi) were recovered. *Haligena viscidula* and *Torpedospora radiata* were frequent at the end of the rainy season (September-October), particularly on the roots of *Rhizophora recemosa.* These fungi were seen in association with the wood-boring organisms (Teredinidae). Mangrove wood and leaves of 10 locations of Sao Paulo (Brazil) yielded 32 species of margrove fungi (Booth, 1983). Kohlmeyer and Volkmann-Kohlmeyer (1987b) listed 46 fungi (37 ascomycetes, 2 basidiomycetes and 7 mitosporic fungi) from a few mangrove covered small islands of Belize. In Man-of-War Cay, Belize, the mangrove detritus was covered by lots of bird excreta (guano). Interestingly, at these enriched habitats, two basidiomycetes, *Halocyphina villosa* and *Nia vibrissa* and an ascomycete *Kallichroma tethys* were dominant. Sixty-five specics of mangrove fungi, comprising 44 ascomycetes, two basidiomycetes and 19 mitosporic fungi have been reported from the Atlantic Ocean by Hyde and Jones (1988). Twenty-two mangrove fungi have been listed from the Caribbean *Rhizophora mangle* (Volkmann-Kohlmeyer and Kohlmeyer, 1993). Although several mangrove locations of Atlantic Ocean have been screened for mangrove fungi, quantitative studies are scarce.

Table 15.3: Most Frequent Fungi (>10%) in Mangroves of Different Parts of the World*

Fungus	Indian Ocean	Pacific Ocean	Atlantic Ocean	South-East Asia	Hong Kong and Japan
Ascomycetes					
Acrocordiopsis patilii Borse *et* Hyde	5	–	–	–	–
Aigialus grandis Kohlm. *et* Schatz	4	–	–	1	–
Aigialus parvus Schatz *et* Kohlm.	4	–	–	1	–
Aniptodera chesapeakensis Shearer *et* Miller	–	1	–	3,4	–
Aniptodera mangrovei Hyde	–	–	–	2	–
Antennospora quadricornuta (Cribb *et* Cribb) Johnson	1,5	2	–	2	–
Antennospora salina (Meyers) Yusoff, Jones *et* Moss	–	2	–	–	–
Ascocretera manglicola Kohlm.	–	–	–	1	–
Asterosphaeriella mangrovis (Kohlm. *et* Vittal) Aptroot *et* Hyde	3,4	–	–	–	–
Astrosphaeriella nypae Hyde	–	–	–	3	–
Caryosporella rhizophorae Kohlm.	–	–	–	1,3	2
Cucullosporella mangrovei (Hyde *et* Jones) Hyde *et* Jones	–	–	–	2	–
Dactylospora haliotrepha (Kohlm. *et* Kohlm.) Hafellner	4,5	3	–	1,2,5	2
Dactylospora mangrovei Jones, Alias, Abdul-Wahab *et* Hsieh	–	–	–	2	–
Dictyosporium pelagicum (Linder) Hughes	–	–	–	2	–
Didymella avicenniae Patil *et* Borse	–	–	–	3	–
Etheirophora blepharospora (Kohlm. *et* Kohlm.) Kohlm. *et* Volkm.–Kohlm.	–	–	1	–	–
Eutypa bathurstensis Hyde *et* Rappaz	4	–	–	–	–
Halorosellinia oceanica (Schatz) Whalley, Jones, Hyde *et* Laessøe	–	–	–	2,3,5	–
Littisporaa abonnis (Kohlm.) Campb., Anderson *et* Sherrer	–	–	–	1	–
Halosarpheia fibrosa Kohlm. *et* Kohlm.	–	3	–	–	–
Halosarpheia marina (Cribb *et* Cribb) Kohlm.	1	1,3	–	2–5	–
Halosarpheia phragmiticola Poon *et Hyde*	–	–	–	–	1
Littispora ratnagiriensis (Patil *et* Borse) Campb., Anderson *et* Shearer	–	–	–	2,3	–
Natantispora retorquens (Shearer *et* Crane) Campb., Anderson *et* Shearer	–	–	–	2,4	–
Halosarpheia viscidula (Kohlm. *et* Kohlm.) Shearer *et* Crane	5	–	–	–	–
Helicascus kanaloanus Kohlm.	–	–	–	1	–
Hypoxylon hypomiltum Montagne	–	–	–	1	–
Julella avicenniae (Borse) Hyde	4	–	–	–	–
Kallichroma tethys (Kohlm. *et* Kohlm.) Kohlm. *et* Volkm.-Kohlm.	–	–	–	1–3	–
Leptosphaeria australiensis (Cribb *et* Cribb) Hughes	4	–	–	1–3	–
Lignincola laevis Höhnk	5	1,2	–	2–5	1
Lineolata rhizophorae (Kohlm. *et* Kohlm.) Kohlm. *et* Volkm.-Kohlm.	5	–	–	–	–
Lignincola tropica Kohlm.	–	–	–	2	–
Linocarpon appendiculatum Hyde	–	–	–	3	–
Linocarpon nypae (Henn.) Hyde	–	–	–	3	–

Lulworthia grandispora Meyers	1	1,3	1	1–3	–
Marinosphaera mangrovei Hyde	5	–	–	1–3	–
Massarina acrostichi Hyde	–	–	–	3	–
Massarina ramunculicola Hyde	–	–	–	1,2	–
Massarina velatospora Kohlm. etVolkm.–Kohlm.	5	–	–	1,3,6	–
Nais glitra Crane *et* Shearer	–	3	–	2	–
Neptunella longirostris (Cribb *et* Cribb) Pand *et* Jones	–	3	–	2	–
Oxydothis nypae Hyde *et* Nakagiri	–	–	–	3	–
Passeriniella savoryellopsis Hyde *et* Mouzouras	–	–	–	3	–
Quintoria lignatilis (Kohlm.) Kohlm. *et* Volkm.-Kohlm.	–	–	–	1	–
Rhizophila marina Hyde *et* Jones	1,4	–	–	5	–
Saccardoella rhizophorae Hyde	4	–	–	–	–
Savoryella lignicola Jones *et* Eaton	–	–	–	1–3	–
Savoryella longispora Jones *et* Hyde	–	–	–	1,2	–
Savoryella paucispora (Cribb *et* Cribb) Koch	–	–	–	2	–
Sphaeruline oraemaris Linder	–	–	–	3	–
Swampomyces triseptatus Hyde *et* Nakagiri	5	–	–	–	2
Thalassogena sphaerica Kohlm. *et* Kolkm.-Kohlm.	5	–	–	3	–
Tirispora unicaudata Jones *et* Vrijmoed	–	–	–	2	–
Torpedospora radiata Meyers	–	2	–	–	–
Trematosphaeria lineolatispora Hyde	–	1	–	–	–
Trematosphaeria mangrovei Kohlm.	5	–	–	–	–
Tunicatispora australiensis Hyde	–	3	–	–	–
Verruculina enalia (Kohlm.) Kohlm.	2–5	1	–	1,2,4	–
Basidiomycetes					
Calathella mangrovei Jones *et* Agerer	–	–	–	2	–
Halocyphina villosa Kohlm. *et* Kohlm.	1–5	–	–	1–4	–
Anamorphs					
Bactrodesmium linderi Crane *et* Shearer Palm *et* Stewart	–	–	–	2	1,2
Camarosporium roumeguerii Sacc.	4	–	–	–	–
Cirrenalia pseudomacrocephala Kohlm.	5	–	–	3	–
Cirrenalia pygmea Kohlm.	4	–	–	1–5	–
Cirrenalia tropicalis Kohlm.	–	–	–	2	–
Clavatospora bulbosa (Anast.) Nakagiri *et* Tubaki	4	–	–	–	–
Cumulospora marina Schmidt	5	–	–	–	–
Cytospora rhizophorae Kohlm. *et* Kohlm.	5	–	–	–	–
Dictyosporium pelagicum (Linder) Hughes	–	1	–	–	–
Monodictys pelagica (Johnson) Jones	–	1	–	–	–
Periconia abyssa Kohlm.	5	–	–	–	–
Periconia prolifica Anastasiou	5	3	–	2	–
Phialophorophoma litoralis Linder	–	–	–	1	–
Phomopsis mangrovei Hyde	4	–	–	1	–
Trichocladium acharasporum (Meyers *et* Meyers) Dixon	5	–	–	2	–
Trichocladium alopallonellum (Meyers *et* Moore) Kohlm. *et* Volkm.-Kohlm.	5	–	–	2,3	–
Zalerion varium Anastasiou	–	–	1	–	–
Total fungi	34	17	3	55	6

Indian Ocean: **1**, Seychelles (Hyde, 1986; Hyde and Jones, 1988); **2**, Andaman and Nicobar Islands (Chinnaraj, 1993b); **3**, Maldives (Chinnaraj, 1993a); **4**, Indian Peninsula (Borse, 1988; Borse *et al*, 2000; Patil and Borse, 2001; Ravikumar and Vittal, 1996; Sarma and Vittal, 2000; Sarma *et al*, 2001; Sridhar and Kaveriappa, 1988); **5**, Mauritius (Poonyth *et al*, 1999, 2001)
Pacific Ocean: **1**, Mexico (Hyde, 1992b); **2**, Hawaii (Volkmann-Kohlmeyer and Kohlmeyer, 1993); **3**,Australia (Hyde, 1990c)
Atlantic Ocean: **1**, Florida (Newell, 1976)
***South-East Asia**: **1**, Thailand (Hyde, 1989d; Hyde* et al, *1990, 1993); **2**, Malaysia (Alias* et al, *1995; Alias and Jones, 2000; Jones and Agerer, 1992; Jones and Kuthubutheen, 1989; Jones and Tan, 1987; Tan and Leong, 1992); **3**, Brunei: Hyde, 1988, 1989a, c, 1990b, 1992c); **4**, Singapore (Tan* et al, *1989); **5**, Sumatra (Hyde, 1989b); **6**, Philippines (Jones* et al, *1988)*
Hong Kong and Japan: **1**, Hong Kong (Virjmoed *et al*, 1994; Poon and Hyde, 1998); **2**,Japan (Nakagiri, 1993)

Pacific Ocean

The mangrove locations screened for marine fungi in the Pacific Ocean are the Australian mangroves, Fiji, French Polynesia, Galapagos, Hawaiian Islands and Mexico. Up to 1988, forty-seven mangrove fungi were known from the Pacific Ocean; the number was lower than the Atlantic (65) and Indian Oceans (75) (Hyde and Jones, 1988). Temperate mangroves of Australia revealed the occurrence of 25 species (Hyde, 1990c). Sixteen species of mangrove fungi are known from Mexico (Hyde 1992b). Out of them eight species were dominant on mangroves (11-46%). *Rhizophora mangle* in Hawaii yielded 21 species (Volkmann-Kohlmeyer and Kohlmeyer, 1993). From Moorea Islands, Volkmann-Kohlmeyer and Kohlmeyer (1993) recorded seven more species. On the mangroves of North Queensland, Hyde (1996) recorded 37 fungi. Based on a few quantitative studies on mangrove fungi of Pacific Ocean (Mexico and Australia), 17 fungi are frequent (14 ascomycetes and 3 mitosporic fungi) (Table 15.3).

A decline in mangrove fungi was seen from west to east of the Pacific Ocean and it has been speculated that this is due to the high plant species diversity in the Western Pacific. Small size and lack of diversity of mangrove plant species in the Central Pacific support this speculation (Hyde, 1990c; Kohlmeyer and Volkmann-Kohlmeyer, 1991b; Volkmann-Kohlmeyer and Kohlmeyer, 1993). For instance, although 22 species of mangrove fungi were recovered on *Rhizophora mangle* in Belize, none of them were found on the same host or any other plant species in the Hawaiian Archipelago.

South-East Asia

In South-East Asia, mangroves of Brunei, Malaysia, Thailand, Singapore and Sumatra attracted the attention of mycologists. In Kampong Danau mangrove (Brunei), Hyde (1988) recorded 41 species, out of which a maximum of 26 species were recorded at the mid-tidal levels. From Sersa mangrove (Brunei), 42 species were found (Hyde, 1989a), the dominant species being *Cirrenalia pygmea, Halocyphina villosa, Halosarpheia marina* and *Leptosphaeria australiensis*. Sersa mangroves (Brunei) were polluted with oil and consisted of only 23 species of mangrove fungi (Hyde, 1989a), out of which *Cirrenalia pygmea* and *Lulworthia grandispora* were dominant. The five mangrove tree species of Kampong Kapok and Tungit Api Api mangroves (Brunei) yielded 76 species of fungi (Hyde, 1990b), out of which *Rhizophora apiculata* harboured the maximum number of fungi (45 species).

Mangrove wood collected from Sungei Geyland Patah (Malaysia) revealed 32 fungi (Jones and Tan, 1987). In Pontian Besar mangrove (Malaysia), dead wood from the mid-tidal level yielded 30 species (Tan and Leong, 1992), the dominant species being *Halosarpheia marina, Natantispora retorquens* and *Lignincola laevis*. Alias and Jones (1994) compiled 125 mangrove fungi from *Avicennia marina* and *Bruguiera parviflora*. Sixty-one fungi were recorded on the test blocks of *A. marina* and *B. parviflora* exposed at Kaula Selangor mangrove (Malaysia) stand for 90 weeks (Alias and Jones, 2000). The highest number of species (38) occurred on *A. marina* during 72-90 weeks (late stage) of exposure, while on *B. parviflora*, the highest species (42) were found during 26–54 weeks (intermediate stage) of exposure. *Lignincola laevis* was the most dominant fungus on both kinds of wood blocks. Detritus from four mangrove plant species of Phang Nga Bay (Thailand) yielded 38 fungi (Hyde, 1989d). Among them, *Caryosporella rhizophorae, Cirrenalia pygmea, Leptosphaeria australiensis* and *Lulworthia grandispora* were dominant. *Rhizophora mucronata* harboured the highest number of fungi (26). The fungal flora of this study are similar to those recorded from other locations of South-East Asia (Brunei, Malaysia, Philippines and Sumatra) and New World mangroves (Hyde, 1989d). A total of 76 species of mangrove fungi were recovered from the root, bark and wood materials of three mangrove plant species from Thailand (Hyde *et al.*, 1990). From Ranong mangrove (Thailand) root, bark and wood samples of *Rhizophora apiculata* revealed the occurrence of 61 mangrove fungi, with *Savoryella longispora* being the most frequent fungus (Hyde *et al.*, 1993). The pattern of vertical distribution, succession and preference of mangrove fungi revealed a drastic difference in the species composition on bark, decorticated wood, roots and branches.

A few studies are available on the fungi of mangroves of Singapore. Woody debris of mangroves of Mandai yielded 44 fungi (Tan *et al.*, 1989). Wood blocks *of Avicennia alba* and *A. lanata*, submerged for 60 weeks in Mandai mangrove, resulted in the occurrence of 21 species (Tan *et al.*, 1989). The species' composition on both kinds of wood block was similar. The highest number of species occurred during 18–32 weeks (intermediate stage) submersion. The diversity was low during the initial as well as late stage of submersion. Studies on mangrove fungi of Sumatra are also meagre. Hyde (1989b) studied the decayed mangrove wood of *Rhizophora* and *Avicennia* from Kampong Nelayan mangrove and recorded 39 fungi. *Cirrenalia pygmea, Dactylospora haliotrepha, Halorosellinia oceanica, Halosarpheia marina, Lignincola laevis* and *Rhizophora marina* were dominant. *Rhizophora* wood yielded more fungi than *Avicennia* wood (20 vs 7 species).

In addition to typical mangrove vegetation, the mangrove ground fern, *Acrostichium speciosum* and the mangrove palm, *Nypa fruticans* in South-East Asia yielded interesting fungi (Hyde, 1989c; Hyde and Sutton, 1992). Intermittently submerged rachids of *Acrostichium* in Brunei yielded 22 species. *Aniptodera chesapeakensis, Cirrenalia pseudomacrocephala* and *Massarina acrostichi* were dominant. The fronds of *Nypa* in Brunei and Thailand yielded 43 fungi (Hyde and Sutton, 1992). On comparison of studies at Atlantic, Pacific and Indian Oceans, quantitative studies on mangrove fungi were highest at the South–East Asian region. Fifty-six species (45 ascomycetes, 2 basidiomycetes and 9 mitosporic fungi) were frequent (Table 15.3).

Hong Kong and Japan

A survey of the mangroves in the Pearl River Estuary, along the South China Coast viz, Hong Kong, Macau and Shenzhen revealed a rich fungal diversity, similar to that in the tropics (Vrijmoed and Jones, 1993). Common species recorded include *Bactrodesmium linderi, Halorosellinia oceanica, Lignincola laevis* and *Marinospora mangrovei.* Vrijmoed *et al.* (1994) studied the standing dead mangrove wood of Three Fathoms Cove (Hong Kong) and Taipa mangrove (Macau), and found 45 mangrove fungi (35 ascomycetes, 2 basidiomycetes and 8 mitosporic fungi). Four species were known to be frequent (less than 10%) in Hong Kong: *Bactrodesmium linderi, Halosarpheia phragmiticola, Lignincola laevis* and *Savoryella paucispora* (Vrijmoed *et al.*, 1994; Poon and Hyde, 1998) (Table 15.3), while in Macau *Lignincola laevis* was frequently (>10%) recorded. Mangrove leaves of South-West Islands of Southern Japan mainly consist of geofungi. Along with them a few typical mangrove fungi were recorded: *Cirrenalia pseudomacrocephala, Leptosphaeria neomaritima, Lulworthia grandispora, Varicosporina ramulosa* and *Zalerion varium.* A mangrove fungus, *Aniptodera salsuginosa* has been described from the mangroves of Iriomote Island (Japan) (Nakagiri and Ito, 1994). From Iriomote Island, Nakagiri (1993), several mangrove fungi were recovered, out of which five were very frequent (more than 10%) (Table 15.3).

FUNGAL RICHNESS ON WOOD

Several ecological factors affect the richness and diversity of mangrove fungi in mangrove ecosystems. The physicochemical features of the habitat, kind of substrata, chemical composition of the wood, presence or absence of bark, period of incubation, inhibition competition and salinity are some of the important factors which influence the ecology of mangrove fungi (Nakagiri, 1993; Hyde and Lee, 1995; Prasannarai and Sridhar, 1997; Jones, 2000). The mean number of fungi per wood sample of many mangrove locations has been compiled in Table 15.4. Most observations are confined to the mangroves of South-East Asia. The highest number of fungi per sample was 3.4 at Beau Champ Mangrove, Mauritius (Poonyth *et al.*, 2001). The least was 0.92 per sample at Sersa Mangrove, Brunei; incidentally this location is polluted with oil (Hyde, 1989a). Regression analysis between total wood samples analysed and total isolates, revealed a significant dependence of the number of isolates on the number of samples ($r^2 = 0.8746$; $p<0.001$). Additional data from the Indian, Atlantic and Pacific Oceans are lacking for a fair comparision. Such quantitative assessment might reveal the status of mangroves (young, intermediate, mature or impoverished by human interference). Table 15.5 reveals the mean species richness per sample of nine mangrove woody materials. This data largely comes from the mangroves of South-East Asia. A highest of 3.27 species per sample was found on the decorticated wood blocks of *Avicennia marina,* while it was least (0.95 per sample) on the roots and trunks of *Bruguiera parviflora* with intact bark. Detailed methodology on fungal inventory is required to recognise species' richness and diversity of a specifc mangrove plant, species and its part in order to link the host and substrate preference of mangrove fungi.

NICHE SPECIFICITY

Many factors are known to affect the occurrence of mangrove fungi: oceanic region, nature of mangrove floor (mud or sand), position of intertidal region, nature of substrate (driftwood, root and branches), water salinity and pH (Hyde and Jones, 1988; Jones, 2000). Within mangrove ecosystems further ecological niches are apparent: mangrove floor (dead leaves, wood and seedlings), dead wood intact with plants (roots and branches) and bark (roots and branches). Intertidal parts of mangroves (roots, stems and branches) represent extreme microhabitats. Drying at low tides, with formation of salt crystals on the surface of bark and wood alternate with wetting by salt or rainwater, such prevailing conditions of mangroves influence the fungal activity and that in turn affects the nutrient cycling. Fungi show many adaptive features such as structural and physiological adaptation, occupation of specific zones of mangroves, specificity for host or substrate, and succession in litter decomposition.

Adaptation

The extreme conditions of mangroves might have resulted in the evolution of mangrove fungi, with specific structural and functional adaptations (Kohlmeyer and Volkmann-Kohlmeyer, 1987b). Some form black clypei (shield-like stroma overlying one or more perithecia) (e.g. genera: *Adonia, Aigialus, Massarina* and *Swampomyces*), superficial melanized ascomata (e.g. genera: *Belizeana, Hypoxylon* and *Trematosphaeria*), and some superficial fruit bodies: carbonaceous or thick walled ascomata (e.g. genera: *Ascocratera, Caryosporella* and *Lophiostoma*). *Kallichroma tethys* with soft fruit bodies adapted to mangrove roots has orange-coloured ascomata containing oil droplets with carotenoids which protect it from insulation (Kohlmeyer and Volkmann-Kohlmeyer, 1987b; Kohlmeyer *et al.*, 1995). Many mangrove-inhabiting fungi are bitunicate forms with large perithesia, ascospores without elaborate appendages like true marine fungi. They possess a gelatinous sheath, hamate polar appendages which unfurl into long threads, with some being without appendages (e.g. genera: *Massarina valatospora, Cucullospora mangrovei* and *Verruculina enalia* respectively). Nakagiri and Ito (1994) demonstrated that the appendage of *Aniptodera salsuginosa* adapted to mangrove brackish water ecosystem. *Aniptodera indica* also showed such adaptations (Ananda *et al.*, 2001). Unappandaged spores posses sticky cell walls, which aid in attachment (e.g. *Caryosporella rhizophorae* and *Verruculina enalia*). Fungal fructifications also seem to be adapted to the mangrove intertidal levels. The spore colour, wall texture of ascoma and the mechanisms of ascospore release are some evidences of this fact.

Actively spore-releasing fungi were located at the upper intertidal zone, while passively spore-releasing fungi were found throughout the tidal range (Hyde *et al.*, 1993). Biochemical and physiological adaptations of a typical mangrove fungus *Cirrenalia pygmea* have been studied (Ravishankar and Suryanarayanan, 1998; Ravishankar *et al.*, 1994, 1995, 1996). Impact of salinity on fatty acids, amino-acids, enzymes and melanin production of *Cirrenalia pygmea* showed adaptation to the mangrove ecosystem. *Halocyphina villosa* also showed morphological and physiological adaptations to mangrove habitats (Rohrmann and Molitoris, 1986).

Kohlmeyer (1984) opined that species richness and number of fruit bodies were highest in eutrophic habitats. At the mangroves of Man-of-War Cay the prop roots, pneumatophores and fallen branches were covered with bird excreta (Guano). Two basidiomycetes *Halocyphina villosa* and *Nia vibrissa* and an ascomycete *Kallichroma tethys* were in great numbers in this habitat. The nutrient enrichment due to Guano at the Man-of-War Cay resulted in higher number of mangrove fungi than in Twin Cays (Kohlmeyer *et al.*, 1995). *Swampomyces armeniacus* is an obligate mangrove fungus found on the roots of submerged mangroves along with *Limnoria* sp., shipworms, *Spirorbis* sp. and *Bostrichia* sp. (Kohlmeyer and Volkmann-Kohlmeyer, 1987b). *Caryosporella rhizophorae*, the obligate marine fungus always occurs in association with marine algae, cirripeds, fungi and molluscs (Kohlmeyer, 1985).

Vertical Zonation

A distinct vertical zonation of mangrove fungi have been noticed in Sierra Leone (Aleem, 1980), Belize (Kohlmeyer *et al.*, 1995), Seychelles (Hyde and Jones, 1988), Andaman Nicobar Islands (Chinnaraj, 1993b), Malaysia (Jones and Tan, 1987), Thailand (Hyde *et al.*, 1990, 1993), Brunei (Hyde, 1988) and Philippines (Jones *et al.*, 1988). Mangrove fungal zonation has been studied on some mangrove plant species: *Rhizophora apiculata* (Hyde, 1988; Hyde *et al.*,

Table 15.4: Species' Richness of Mangrove Fungi on Wood Samples Studied at Different Mangrove Locations (*Polluted with oil).

Mangrove location	Mean fungi/wood	Reference
Beau Champ, Mauritius	3.4 1.1–1.4	Poonyth *et al*, 2001 Poonyth *et al*, 1999
Beau Rivage, Mauritius	3.3 1.1–1.3	Poonyth *et al*, 2001 Poonyth *et al*, 1990
Kaula Selangor, Malaysia	3.27	Alias and Jones, 2000
Macondé, Mauritius	2.7 1–1.1	Poonyth *et al*, 2001 Poonyth *et al*, 1999
Mandai, Singapore	2.63	Tan *et al.*, 1989
Poudre d'Or, Mauritius	2.5 1–1.1	Poonyth *et al*, 2001 Poonyth *et al*, 1999
Mandai, Singapore	2.2	Tan *et al.*, 1989
Ile d'Ambre, Mauritius	2.1 1.2–1.4	Poonyth *et al*, 2001 Poonyth *et al*, 1999
Phang Nga Bay, Thailand	1.81	Hyde, 1989d
Taipa, Macau	1.7	Vrijmoed *et al*, 1994
Kampong Kapok and	1.63	Hyde, 1990b
Tungit Api Api, Brunei		
Pontian Besar, Malaysia	1.51	Tan and Leong, 1992
Kampong Nelayan, Sumatra	1.5	Hyde, 1989b
Brillant, Seychelles	1.46	Hyde and Jones, 1988
Port Dickson, Malaysia	1.46	Jones and Agerer, 1992
Sungei Geyland Patah, Malaysia	1.33	Jones and Tan, 1987
Seria, Brunei	1.32	Hyde, 1989a
Three Fathms Cove, Hong Kong	1.2	Vrijmoed *et al*, 1994
Anse Boileau, Seychelles	1.1	Hyde and Jones, 1998
Gujarat, India	1	Patil and Borse, 2001
Pirotan Island, Gujarat	1	Borse *et al.*, 2000
Maharashrra, India	1	Borse, 1988
*Sersa, Brunei	0.92	Hyde, 1989a

1990, 1993), *R. mucronata* (Hyde, 1988) and *Sonneratia griffithii* (Hyde et *al.*, 1990). Hyde (1988) has compared the most frequent fungi found at lower-, mid- and upper-littoral levels in mangrove trees of tropical and subtropical locations. Some fungi have a broad distribution pattern and are found throughout the tidal range (eg *Cirrenalia pygmea, Leptosphaeria australiensis* and *Lulworthia*

Table 15.5: Species' Richness of Mangrove Fungi on Wood Samples of Different Mangrove Tree Species

Plant species	Nature of wood	Mean fungi/wood	Reference
Aegiceras corniculatum	Roots and trunks	1.95	Hyde, 1989d
Avicennia alba	Decorticated wood	2.67	Tan et al, 1989
	Branches	2.22	Hyde, 1990b
Avicennia lanata	Decorticated wood	2.60	Tan *et al*, 1989
Avicennia marina	Decorticated wood	3.27	Alias and Jones, 2000
Bruguiera parviflora	Decorticated wood	2.57	Alias and Jones, 2000
	Roots and trunks	0.95	Hyde, 1989d
Rhizophora apiculata	Prop roots and branches	1.55	Hyde, 1990b
	Prop roots	1.46	Jones and Agerer, 1992
Rhizophora mucromata	Split wood	2.1–3.4	Poonyth *et al*, 2001
	Roots and trunks	2.18	Hyde, 1989d
	Prop roots and branches	1.50	Hyde, 1990b
	Prop roots	1.1–1.4	Poonyth *et al*, 1999
Sonneratia alba	Twigs and branches	1.83	Hyde, 1990b
Xylocarpus granatum	Roots and trunks	1.65	Hyde, 1989d
	Roots	1.07	Hyde. 1990b

grandispora), while others are specific to the upper-littoral (e.g. *Cytospora rhizophorae, Halorosellinia oceanica* and *Savoryella lignicola*), mid-littoral zones (e.g. *Caryosporella rhizophorae*) and lower-littoral zone (e.g. *Antennospora quadricornuta* and *Thalassogena sphaerica*) (Aleem, 1980; Hyde, 1988, 1989b; Hyde and Jones, 1988). The stakes of mangrove plants, artificially attached to prop roots showed that many fungal species sporulated below the water line (i.e., at 2–2.4 m depth from the surface or 0.6–1.2 m above the sediment) (Kohlmeyer *et al.*, 1995). Three fungi sporulated on the stacks close to the sediment (e.g. *Antennospora quadricornuta, Lulworthia* sp. and *Phoma* sp.). *Halocyphina villosa, Lulworthia* sp. and *Trichocladium alopallonellum* occurred above the high water line. *Nia vibrissa* preferred the submerged part of the stakes.

Based on the studies on mangrove fungal adaptations and pattern of zonation (Hyde and Lee, 1995), the following observations have been made: mangrove fungi show vertical zonation; greater diversity occurs at above mean-tide level; the loculoascomycetes, fungi with carbonaceous ascomata, show active spore release, possess coloured or ornamented ascospores with superficial ascomata confined to above mean-tide levels. Ascomycetes with passive spore dispersal, membranous walls and immersed ascomata occur throughout the tidal range. At the zone between high intertidal regions and terrestrial regions, many unusual fungi sporulate; they might have adapted for decaying the mangrove wood, which is not severely affected by salinity.

Table 15.6: Pattern of Dominanence of Mangrove Fungi on Mangrove Wood During Decomposition in Different Locations.

Plant species	Status of wood	Fungus* and location	Reference
Avicennia alba	Without bark (Singapore)	*Aigialus parvus* (L) *Dactylospora haliotrepha* (L) *Lignincola laevis* (E) *Verruculina enalia* (T)	Tan *et al*, 1989
Avicennia lanata	Without bark (Singapore)	*Aigialus parvus* (L) *Lignincola laevis* (E) *Verruculina enalia* (T)	Tan *et al*, 1989
Avicennia marina	Without bark (Malaysia)	*Antennospora quadricornuta* (L) *Dictyosporium pelagicum* (I, L) *Halocyphina villosa* (I, L) *Halosarpheia marina* (E, I, L) *Littispora ratnagiriensis* (E, I) *Natantispora retorquens* (E, I) *Lignincola laevis* (E, I, L) *Neptunella longirostris* (E, I, L) *Lulworthia grandispora* (E, I) *Perconia prolifica* (I) *Savoryella lignicola* (I, L) *Savoryella paucispora* (I) *Trichocladium achrasporum* (I, L) *Bactrodesmium linderi* (L) *Verruculina enalia* (L)	Alias and Jones, 2000
Avicennia sp.	Branches (Belize)	*Halosphaeria cuculata* (E) *Lulworthia grandispora* (L) *Ophiodera monosemeia* (L) *Periconia prolifica* (E)	Kohlmeyer *et al*, 1995
Bruguiera gymnorrhiza	Split wood (Mauritius)	*Cirrenalia pygmea*(E) *Cumulospora marina* (E) *Dactylospora haliotrepha* (L) *Lignincola laevis* (I) *Periconia prolifica* (T) *Trichocladium achrasporum* (I) *Verruculina enalia* (L)	Poonyth *et al*, 2001
Bruguiera parviflora	Without bark (Malaysia)	*Dictyosporium pelagicum* (L) *Halosarpheia marina* (E, I) *Littispora ratnagiriensis* (E) *Natantispora retorquens* (I, L) *Lignincola laevis* (E, I) *Neptunella longirostris* (I) *Lulworthia grandispora* (E) *Savoryella lignicola* (I, L) *Savoryella longispora* (L) *Savoryella paucispora* (I) *Trichocladium achrasporum* (I, L) *Verruculina enalia* (E, I, L)	Alias and Jones, 2000
Rhizophora apiculata	Without bark (Singapore)	*Dactylospora haliotrepha* (L) *Halorosellinia oceanica* (L) *Lignincola laevis* (E) *Verruculina enalia* (E)	Leong *et al*, 1991

	With bark (Brunei)	*Halosarpheia minuta* (E) *Hypophloeda rhizospora* (L) *Lulworthia grandispora* (L) *Manglicola guatemalensis* (L)	Hyde, 1991
Rhizophora mangle	Prop roots (Belize)	*Dactylospora haliotrepha* (L) *Etheirophora blepharospora* (E) *Littispora abonnis* (L) *Halosphaeria cucullata* (E) *Kallichroma tethys* (E) *Passeriniella savoryellopsis* (L) *Periconia prolifica* (E)	Kohlmeyer *et al*, 1995
Rhizophora mucronata	Split wood (Mauritius)	*Cirrenalia pygmea* (E) *Cumulospora marina* (E) *Dactylospora haliotrepha* (L) *Lignincola laevis* (I) *Periconia prolifica* (T)	Poonyth *et al*, 2001
Xylocarpus granatum	With bark (Brunei)	*Halosarpheia marina* (L) *Halorosellinia oceanica* (L) *Leptosphaeria australiensis* (L) *Periconia prolifica* (L)	Hyde, 1991

*Stage of colonisation: **E**, early; **I**, intermediate; **L**, late; **T**, throughout

Studies on vertical zonation have been largely confined to the locations of South-East Asia. It is necessary to understand the pattern of fungal zonation at narrow tidal ranges (e.g. Belize, 0.15 m; Florida, 0.8 m) and large tidal ranges (e.g. South China Sea, 2.6 m; Brunei, 3 m; northern territory of Australia, 7 m) (Hyde and Lee, 1995; Kohlmeyer *et al.*, 1995). Some more studies at mangroves of different parts of the world may reveal more information on the preference and pattern of fungal zonation.

Host and Substrate Specificity

A total of 120 mangrove fungi have been recorded on 28 mangrove plant species, out of which 37 species are confined to any one of the plant species, suggesting the possibility of host specificity among some mangrove fungi (Hyde *et al.*, 1990). Many studies have addressed the host specificity of mangrove fungi (e.g. Kohlmeyer and Kohlmeyer, 1979; Hyde, 1990a; Hyde and Jones, 1988; Hyde *et al.*, 1990). Hyde *et al.* (1990) found some mangrove fungi specific to the mangrove tree species (e.g. *Rhizophora*: *Rhizophila marina*); some were very common on a single tree species (e.g. *Sonneratia griffithii*: *Halorosellinia oceanica*); some were common to all tree species (e.g. *Dactylospora haliotrepha*). Greater preferences of mangrove fungi on *Rhizophora* spp. have been noticed than other plant species. To date, the *Rhizophora apiculata* is the only mangrove plant species known to harbour the maximum number of mangrove fungi (63 species) (Sarma and Vittal, 2000, 2001a). Many fungi were unique to *Rhizophora* as host (e.g. *Capillatispora corticola, Caryosporella rhizophorae, Hypophloeda rhizospora, Rhizophila marina* and *Etheirophora blepharospora* (Hyde *et al.*, 1993; Sarma and Vittal, 2000). The occurrence of mangrove fungi also depends on the kind of mangrove wood. The mangrove wood types differ from plant to plant. *Xylocarpus granatum* has hard wood, *Rhizophora* sp. has medium type and *Avicennia* and *Sonneratia* has softwood. Hence, the fungal flora differs on these niches (Hyde, 1990b). *Passeriniella savoryellopsis* and *Sphaerulina oraemaris* were common on hard wood. *Halorosellinia oceanica* and *Littispora ratnagiriensis* were common on softwood, while *Cirrenalia pygmea, Caryosporella rhizophorae* and *Leptosphaeria australiensis* were common on medium wood. Besides the specificity of fungi on mangrove plant species, they were also found to prefer bark, wood, twigs, pneumatophores and seedlings. Some fungi were most common on the pneumatophores of *Sonneratia griffithii* (e.g. *Aigialus grandis* and *Massarina valatospora*), while others are specific on twigs (e.g. *Saccardoella mangrovei* and *Savoryella lignicola*) (Hyde *et al.*, 1990). Some fungi were exclusively found on the bark of mangrove roots and branches (Kohlmeyer and Kohlmeyer, 1979). Some were confined to the bark of *Rhizophora apiculata* (e.g. *Hypophloeda rhizospora, Phomopsis mangrovei* and *Rhizophila marina*), while others were confined to woody tissues (e.g. *Caryosporella rhizophorae*) (Hyde *et al.*, 1993). *Eutypa bathurstensis* was confined to the wood of *Avicennia* spp. (Sarma and Vittal, 2000). Many studies support the above observations (Kohlmeyer and Kohlmeyer, 1979; Hyde and Jones, 1988; Hyde *et al.*, 1990; Sarma and Vittal, 2000).

Although host and substrate specificity is evident in many studies, the preference and frequency of occurrence of mangrove fungi is dependent on several factors: water temperature, water salinity, vertical zonation, pattern of succession and kind of wood (soft, medium and hard) (Hyde *et al.*, 1993). Besides these factors, the age of the wood (young or old part of the tree), presence or absence of bark, period of submersion and stage of colonisation greatly influence the type of fungi and their colonisation pattern (Tan *et al.*, 1989).

Pattern of Succession

Succession leads to replacement of species at each stage and the species modify the environment less suitable for themselves or more suitable for others (Hudson, 1962). A few studies discuss the succession of fungi on mangrove leaves, wood and seedlings. In *Rhizophora,* bark is an important factor (a niche) in determining mangrove fungi (Hyde *et al.*, 1993). In the first year, the bark preferring fungi developed without decay of wood beneath, but subsequent colonisers were more wood decomposers than bark (Hyde, 1991; Kohlmeyer *et al.*, 1995). The prop roots may be first colonised by bark inhabiting fungi (Hyde, 1991), after the bark decomposed or detached, the wood decaying fungi colonise (Leong *et al.*, 1991). *Caryosporella rhizophorae* grows only on hard woods, which are dry or submerged. The ascocarps require more time to develop to maturity and can not survive on bark or softwood. Marine woodborers facilitate the colonisation of wood decomposers (Kohlmeyer *et al.*, 1995). Interestingly, severely damaged roots were simultaneously colonised by bark as well as wood decomposing fungi (Hyde *et al.*, 1993). A distinct pattern of succession has been noticed on the colonisation of mangrove fungi on the decorticated wood blocks of *Avicennia alba* and *A. lanata* on submersion for 60 weeks at Mandai mangrove (Singapore) (Tan *et al.*, 1989). The early coloniser was *Lignincola laevis* (6 and 26–32 weeks); some fungi occurred throughout the submersion period without any peaks (eg *Halosarpheia marina*), while others were found throughout with a peak during the middle period, but some were intermediate to late colonisers. The highest number of species occurred within 18–32 weeks (intermediate period). The species diversity declined in both substrates during the initial and final periods of submersion. On immersion of wood blocks of *Avicennia marina* and *Bruguiera parviflora* at Kaula Selangor mangrove (Malaysia), no distinct pattern of fungal succession was seen (Alias and Jones, 2000). However, a clear pattern of colonisation was evident: early (6–18 weeks), intermediate (26–54 weeks) and late (60–96 weeks). The number of species per sample was highest during the intermediate stage of exposure (6–8) than early (1.8–4.2) and late (4–7) stages which corroborates with the earlier study at Singapore (Tan *et al.*, 1989). Fungal succession was also found on the seedlings of *Rhizophora mangle* (Newell, 1976). In Newell's (1976) study, the majority of the higher fungi were terrestrial. At the third stage of succession, the mangrove fungi *Lulworthia grandispora* and *Zalerion varium* were dominant. Based on similarity index, Alias and Jones (2000) compared fungal species diversity on the mangrove wood exposed at four mangrove locations, for different periods of immersion (early, intermediate and late). Fungal diversity was lower at all stages for Belize and Brunei (Hyde, 1991; Kohlmeyer *et al.*, 1995) than at Singapore and Malaysia (Tan *et al.*, 1989; Leong *et al.*, 1991; Alias and Jones, 2000). In most locations, species diversity increased with exposure period, except for Belize. Table 15.6 shows the pattern of dominance of mangrove fungi on mangrove wood in different mangrove locations. Most fungal succession studies have been conducted at the mangroves of South-East Asia. For a comparative analysis, studies from other locations are essential with specific woody substrata. Based on Table 15.6, *Lignincola laevis* was consistantly found to be an early to intermediate coloniser of mangrove woody materials in South-East Asian mangroves. This fungus might have a better competitive and resource capturing ability.

Interaction of three mangrove fungal species on mangrove wood was studied by Tan *et al.* (1995). The growth and ascocarp formation of *Aigialus parvus*, *Lignincola laevis* and *Verruculina enalia*, cultured individually and in combination on three kinds of wood blocks (*Avicennia alba* and *Bruguiera cylindrica* and *Rhizophora apiculata*) was studied. Based on the mycelial encroachment, *Verruculina enalia* was considered to be the most aggressive species. Mixed culture of fungi generally delayed the onset of sporulation and also affected the abundance of ascocarp formation. The presence of *Lignincola laevis* enhanced ascocarp formation by *Verruculina enalia* on all types of wood. It has been predicted that the metabolites secreted by *Lignincola laevis* have enhanced the sporulation of *Verruculina enalia*. This evidence supports the view of Hyde *et al.*, (1993): 'the speed of fungal growth and sporulation might determine the competition'. There is a likelihood that the first fungus, due to its activities, created ideal conditions for the development of a second fungus, although both were colonised simultaneously.

LITTER DECOMPOSITION

Annual litter production in mangroves ranges from 110-2370 g / m^2 (Bunt, 1995; Saenger and Snedaker, 1993). The diversity of mangrove plant species being higher in the Indo-West Pacific than in the New World, it affects the quality and diversity of both leafy and woody litter available for decomposition. Early studies on mangrove litter decomposition were confined to the mangroves of upper latitudes (Blum *et al.*, 1988; Fell and Master, 1980; Steinke and Ward, 1987; Steinke *et al.*, 1990). The rate of decomposition depends on the relative position of the litter along the flooding duration. Hence, the submerged leaves tend to decompose more rapidly than the leaves that exist on intertidal zones (Robertson *et al.*, 1992; Steinke and Ward, 1987). Newell (1996) is of the opinion that in salt marshes and mangroves, balance among the major groups of decomposers (eumycotes, oomycotes and bacteria) differs with regard to their utilisation of vascular plant litter. Since Oomycota are potentially important in mangrove leaf litter decomposition, ergosterol (largely restricted to Eumycotic fungi and lacking in Oomycota and Chitridiomycota) was extremely low in leaf litter (Newell and Fell, 1992). However, electron microscopic study revealed hyphal growth in a variety of leaf tissues. Besides, Oomycetes and Eumycotic fungi, the unicellular Heterotrophic Protists, Labyrinthulomycota (Labyrinthulales and Thraustochitridiales) were actively involved in mangrove litter decomposition.

Leaf Decomposition

The decomposition of leaves of *Avicennia marina* was studied at the mangroves at Bulwer Island, Queensland (Australia) at high and low tidal levels during summer and winter (Mackey and Smail, 1996). The rate of mass loss of leaf litter was significantly higher at low tide in summer than at high tide in winter. The duration required for 50% dry mass loss during summer was 59 and 44 days in high and low tide respectively, while during winter it was 98 and 78 days. The exponential relationship of leaf litter decay (t_{50}) and latitude indicated the importance of temperature (hence the season) in organic matter cycling in mangroves. Incubation of *Avicennia marina* leaves in estuarine water for three months in the laboratory by Chale (1993) resulted in rapid weight loss in the first 24 hrs (19%), thereafter at six weeks it was up to 30%. Nitrogen decreased initially and increased rapidly at six weeks, while phosphate and potassium decreased with time. Leaching and microbial activities were responsible for changes in nutrients.

Raghukumar *et al.* (1995) observed three distinct phases of leaf litter decomposition of *Rhizophora apiculata*: (i) rapid loss of detrital weight, including the reduction of proteins, carbohydrates, reducing sugar, phenolics and cellulose; (ii) attainment of peak fungal biomass and rapid decline of most of the organic constituents including C/N ratio almost to the lowest levels within 21 days; (iii) fungal and bacterial biomass decreased after 3 and 5 weeks, respectively with further decline being seen in cellulose and lignin. The decomposition of *Rhizophora mucronata* leaves in laboratory and field conditions showed a decline of organic carbon and ash, with an increase in total nitrogen, organic matter, protein and calories (Sumitra-Vijayaraghavan *et al.*, 1980). The C/N ratios decreased during decomposition of mangrove leaves (Robertson *et al.*, 1992; Steinke and Ward, 1987). At the advanced stages of decay, nitrogen exceeded the initial values, indicating immobilisation from external sources and it is strongly believed that fungi enrich the nitrogen in decomposing mangrove leaf litter (Benner *et al.*, 1990). Tannins and other inhibitory phenolic compounds are the major fractions of dissolved organic matter (18%) in *Rhizophora mangle* leaf leachate in Bahamas (Benner *et al.*, 1986). Except for high concentrations, they did not inhibit microbial uptake or microbial degradation of mangrove leaves. Up to 42% leachate of the mangrove leaves was incorporated into the food web within 2–12 hr and hence it constitutes an important source of carbon, energy and nitrogen to the higher trophic levels in mangroves. The bulk of the leachates of mangrove leaves was mineralised rapidly into microbial biomass (30%) (Benner and Hodson, 1985). In contrast, the mineralisation of lignocellulosic component of mangrove leaves and wood was 10-fold lower than the mineralisation of leachate fraction. Studies on the production of enzymes by mangrove fungi on mangrove litter are scarce. The cellulase of decomposing *Kandelia candel* leaves at Three Fathoms Cove (Hong Kong) attained a peak at six weeks decomposition (Hodgkiss and Leung, 1986). Recently, Pointing *et al.* (1998, 1999) studied the cellulolytic activities of seventeen mangrove fungi in relation to pH, temperature and salinity.

Wood Decomposition

The decomposition of twigs of *Avicennia marina* in the mangroves of Bulwer Island, Australia revealed a higher mass loss during summer than in winter (t_{50}: 383 and 179 days vs 1327 and 1207 days in high and low tide) (Mackey and Smail, 1996). In the Man-of-War Cay, Belize, high borer activity of shipworms (*Teredo bratschi*) on mangrove stakes was due to the abundance of nutrients (ammonia, nitrate and phosphate) from guano deposits (Kohlmeyer *et al.*, 1995). Such shipworm damage facilitates further decomposition of wood by mangrove fungi. Hyde *et al.* (1993) observed that the mangrove fungi occupy small zones on woody tissue (5-10 cm) and the change in the texture and colour of wood is dependent on the kind of fungal species. For instance, wood colonised by *Massarina thalassiae* is usually soft and bleached white, while by *Leptosphaeria australiensis* it is soft and gray. The surface of the wood is blackened if colonised by *Halorosellinia oceanica*.

Nutrient Cycling

The activities of sesarmid crabs, leaf quality and environmental conditions lead to the rapid disappearance of leaf litter at high intertidal zones (Robertson *et al.*, 1992; Steinke *et al.*, 1990). Feeding choice experiment (*Avicennia marina, Bruguiera exaristata, Ceriops tagal* and *Rhizophora stylosa*) to crabs *Sesarma messa* and *S. smithii* showed significant preference of decayed leaves over newly fallen leaves irrespective of the plant species. Though choice of leaves by both crabs was not influenced by leaf chemical properties; *S. smithii* chose *Rhizophora stylosa* over other plant litter. The crude lipid extract of *Rhizophora mucronata* leaves when incorporated at 0.5, 1 and 2% (w/w) with a basal diet (42% protein; 10% lipid, 4% carbohydrates and 2% vitamins and minerals) of penaeid prawn, *Penaeus indicus*, promoted assimilation efficiency, growth and biochemical constituents (Ramesh and Kathiresan, 1992). Senescent leaves of *Rhizophora apiculata* and *A. marina* allowed to decompose in the Pichavaram mangrove in the East coast of India during pre-monsoon, monsoon, post-monsoon and summer seasons (Rajendran and Kathiresan, 1999). During the pre-monsoon and post-monsoon seasons, prawns, crabs and fishes were attracted towards the decomposing leaves. The number of fauna attracted towards the decaying litter increased from the 10th day onwards and became maximum between 30 and 40 days of decomposition. High assimilation efficiency was found in shrimps *Metapenaeus manoceros* (86%) on the leaf litter of *Acanthus ilicifolius, Avicennia officinalis* and *Rhizophora mucronata* in combination with rice bran as diet (Ramadhas and Sumitra-Vijayaraghavan, 1979). It is presumed that fungal processing of litter in mangroves results in a large proportion of dissolved rather than Particulate Organic Matter (POM) (Hyde and Lee, 1995). But the quantity, quality and fate of Dissolved Organic Matter (DOM) in the trophic structure of mangrove ecosystem are not well understood.

GEOFUNGI AND FRESHWATER FUNGI

Terrestrial fungi have been neglected on the assumption that they are from terrestrial origin. However, terrestrial fungi have been isolated from marine habitats viz, estuaries and coastal waters, decomposing plants and animals, mangrove and deep-sea sediments. They have been isolated from water and mud samples of mangroves of Red Sea, Egypt (Khallil *et al.*, 1991) on baited wood in Pearl River Estuary, Hong Kong (Vrijmoed *et al.*, 1982a, 1982b, 1986), mangrove wood at Sierra Leone (Aleem, 1980), seedlings of *Rhizophora mangle* at Florida (Newell, 1976) and mangrove leaves at Chorao Island, Goa, India (Raghukumar *et al.*, 1995). Rhizosphere fungi of *Avicennia officinalis* and *Rhizophora mucronata*, Pichavaram mangrove, India, consists of several terrestrial fungi with the exception of *Cirrenalia pygmea* (Venkatesan and Natarajan, 1985; Salique *et al.*, 1985). Mangrove mud, rhizosphere, rhizoplane and non-rhizosphere zones of Sunderban mangrove swamps, India, yielded many terrestrial fungi (Garg, 1983; Chowdhary *et al.*, 1982). The rhizosphere and non-rhizosphere soils of *Avicennia officinalis*, Maharashtra, India, revealed the occurrence of terrestrial fungi (Nair *et al.*, 1991). Mangrove mud at Vellar Estuary, Pichavaram mangroves, Cochin backwaters and Andaman Nicobar Islands consists of several terrestrial fungi (Silique *et al.*, 1985; Prabhakaran and Gupta, 1990; Misra, 1986). In Cochin backwaters mangrove leaves, stem, roots and pneumatophores of *Avicennia* and *Acanthus* yielded terrestrial fungi along with mangrove fungi. *Cirrenalia pseudomacrocephala* and *Trichocladium alopallonellum* were found on the dead *Avicennia* stem during pre-monsoon season with high salinity (Prabhakaran *et al.*, 1987). Recently, terrestrial fungi were recovered from the deep sea environment of the Arabian Sea (Raghukumar and Raghukumar, 1998). Seawater, sea foam and beach soil of Arabian Gulf Coast and Saudi Arabia yielded terrestrial fungi along with marine fungi and freshwater hyphomycetes (Bokhary *et al.*, 1992) on direct examination and baiting. Similarly, seasonal sampling of the leaf litter from the Nethravathi estuary, India also revealed the occurrence of many freshwater Hyphomycetes and a typical mangrove fungus, *Clavatospora bulbosa* (Sridhar and Kaveriappa, 1988). It is likely that terrestrial and freshwater fungi are involved in plant detritus degradation in mangrove habitats under certain conditions and seasons. For instance, leaf, bark and wood decay in mangrove canopies or decomposition of litter on mangrove floor (pneumatophores, prop roots and twigs) during low saline or rainy season or air-water interphase by terrestrial fungi cannot be ruled out. Similarly, aquatic Hyphomycetes are active during low saline conditions of mangroves and backwaters (Sridhar and Kaveriappa, 1988).

Occurrence of different groups of fungi in mangroves depends on the selection of substrates and the techniques employed. Inhibition of spore germination of non-marine fungi by seawater (Kirk, 1980) and marine soil (Lockwood, 1977) was seen. The inhibitory effects of seawater are reduced or partly removed by autoclaving. Addition of glucose to seawater also helped to overcome the inhibitory effects (Kirk, 1980). Evidence is accumulating on the adaptation and activity of terrestrial fungi under marine/mangrove ecosystem as facultatives or indwellers or residents (Chandralata, 1999; Raghukumar and Raghukumar, 1998). Such controversies are also known with regard to actinomycetes in marine sediments (Goodfellow and Williams, 1983).

TECHNIQUES

A variety of techniques have been employed to study the activities of mangrove fungi. Detailed procedures and protocols have been outlined (Jones and Hyde, 1988; Sarma and Hyde, 2001). For inventing and monitoring fungi, examination of drift and intertidal substrates, baited materials, foam samples, direct plating of water and sediments are required. Based on literature, a wide variety of media have been listed by Hyde *et al.* (1987) for isolations and sporulation of marine or mangrove fungi. The isolation media consists of a variety of natural substrates (e.g. balsa wood bait and balsa broth) and nutrient agar media (e.g. cornmeal-seawater agar and yeast-peptone-glucose-seawater agar). The culturing techniques have to be employed by simulating the natural habitats (natural substrate, desiccation, agitation, temperature, pH and salinity). Methods to detect the substrate mass loss, changes in C/N ratio, estimation of fungal biomass, enzyme activities, radiolabelling techniques (^{14}C) have been briefly discussed by Jones and Hyde (1988). Recently, Fell and Newell (1998) outlined the biochemical and molecular methods to study marine mangrove fungi. The techniques include ergosterol, glucosamine and phospholipid-bound fatty acids, DNA fingerprinting through PCR and nucleotide sequence analysis. Besides ecological observations, biochemical and molecular techniques provide a clearer picture on the role and importance of filamentous fungi in the mangrove ecosystem.

CONCLUSIONS

'Mangrove mycology' is a recent field of research, opening up possibilities of studing the ecosystem function and exploiting fungal power. Initial studies on mangrove fungi dealt with a list of fungi on mangrove litter, indicating the presence or absence at specific geographical locations. Subsequently, quantitative analysis was employed to find out the frequency of occurrence and extent of colonisation on mangrove substrate. Later, for uniform assessment of mycoflora, baiting of specific mangrove leaves or woody litter was employed. However, biogeographic and diversity studies on mangrove fungi are inadequate. Relatively, mangroves of South-East Asia have been studied more than the tropical and subtropical parts. Due to lack of sufficient quantitative surveys, it is difficult to decide the dominant, frequent and rare species among mangrove fungi. Based on the data (see Table 15.3), *Dactylospora haliotrepha, Halosarpheia marina, Lulworthia grandispora* and *Verruculina enalia* are common and frequent (more than 10%) in the mangroves of Indian Ocean, Pacific Ocean and South-East Asia. Alias and Jones (2000) also recognised the above fungi (except for *Dactylospora haliotrepha*) as being highly dominant and playing an important role in mangrove wood decomposition in a Malaysian mangrove. These fungi were also dominant during wood decomposition in other mangroves of South-East Asia (see Table 15.6). Possibly they are the 'core group' which drive the process of litter decomposition in mangroves .

Mangroves being the most productive ecosystem, contributions of filamentous fungi gain utmost importance in the decomposition, biogeochemical cycles, nutrient transfer and energy flow in mangrove and marine food webs. Investigations on the geographical distribution and diversity of mangrove fungi are inadequate. In addition to quantitative studies on mangrove fungi, the status and physicochemical features of mangrove habitats need to be documented to link the fungal activities to the environmental factors. It is evident that marine fungi, geofungi and freshwater fungi overlap in the mangrove ecosystem. Except for occurrence, the role of geofungi and freshwater fungi in mangrove habitats is yet not clear. Similarly, studies on the occurrence and role of endophytic fungi in the mangrove ecosystem are inadequate. Possibilities of the production and role of phyto-hormones (e.g. *Phaeosphaeria* sp. synthesise gibberellins on decomposing smooth cardgrass in salt marshes, see Newell, 1993) by mangrove fungi in mangrove habitat is yet to be seen or experimented. According to Raghukumar *et al.* (1995) and Newell (1996) the fungal biomass actually greater than what is so far recognised in the mangrove ecosystem. Hence, precise techniques to estimate the total fungal biomass in mangroves help to understand their contribution. Studies on the supply of essential nutrients viz, fatty acids, sterols, vitamins and growth factors to the detritivores by the mangrove fungi is not clearly known. Significance of mangrove fungi in the dynamics of particulate organic matter and dissolved organic matter in mangrove habitat has to be addressed further. Fungi are essentially habitat specific. Although mangroves are the richest source of mycoflora, their destruction is occuring at an alarming rate (Ong, 1995). Human interference, such as encroachment of mangrove habitats and addition of pollutants, is resulting in the extinction of mycoflora the essential component of the food web. It is necessary to conserve the fungal genetic resources *in situ* and *ex situ* to understand the mangrove ecosystem function and for the future requirements of natural products and biotechnological inventions.

REFERENCES

Aleem, A.A. 1980. Distribution and ecology of marine fungi in Sierra Leone (Tropical West Africa). *Bot. Mar.* 23: 679-688.

Alias, S.A. 1996. *Ecological and Taxonomic Studies of Lignicolous Marine Fungi in Malaysian Mangroves*. Ph.D. Thesis, University of Portsmouth, UK.

Alias, S.A. and Jones, E.B.G. 1994. Biodiversity of microfungi in tropical mangroves. In: *International Mycological Congress 5*. Vancouver, Canada, pp. 3.

Alias, S.A. and Jones, E.B.G. 2000. Colonisation of mangrove wood by marine fungi at Kuala Selangor mangrove stand, Malaysia. *Fungal Diversity*, 5: 9–21.

Alias, S.A., Kuthubutheen, A.J. and Jones, E.B.G. 1995. Frequency of occurrence of fungi on wood in Malaysian mangroves. *Hydrobiologia* 295: 97–106.

Ananda, K. and Sridhar, K.R. 2001. *Aniptodera indica*, a new species of mangrove inhabiting ascomycete from the west coast of India. *J. Environ. Biol. 22*: 283–286.

Barghoorn, E.S. and Linder, D.H. 1944. Marine Fungi: Their Taxonomy and Biology. *Farlowia 1*: 395-467.

Benner, R. and Hodson, R.E. 1985. Microbal degradaton of the leachable and lignocellulosic components of leaves and wood from *Rhizophora mangle* in a tropical mangrove swamp. *Mar. Eco. Prog. Ser. 23*: 221–230.

Benner, R., Hatcher, P.G. an Hedges, J.I. 1990. Early diagnisis of mangrove leaves in a tropical estuary: bulk chemical characterisation using solid-state ^{13}C NMR and elemental analysis. *Geochim. Cosmochem. Acta 54*: 2003–2013.

Benner, R., Peele, E.R. and Hodson, R.E. 1986. Microbial utilisation of dissolved organic matter from leaves of the red mangrove, *Rhizophora mangle*, in the French Creek Estuary, Bahamas. *Estuarine, Coastal and Shelf Science 23*: 607–619.

Blum, L.K., Mills, A.L., Zieman, J.C. and Zieman, R.T. 1988. Abundance of bacteria and fungi in seagrass and mangrove detritus. *Mar. Ecol. Prog. Ser. 42*: 73–78.

Bokhary, H.A., Moslem, M.A. and Parvez, S. 1992. Marine fungi of the Arabian Gulf Coast and Saudi Arabia. *Microbiologica 15*: 281–290.

Booth, T. 1983. Lignicolous marine fungi from Sao Paulo, Brazil. *Can. J. Bot. 61*: 488–506.

Borse, B.D. 1984. Marine fungi from India I. *Indian Bot. Reptr. 3*: 156–157.

Borse, B.D. 1987a. Marine fungi from India IV. *Trans. Mycol. Soc. Japan 28*: 55–61.

Borse, B.D. 1987b. Marine fungi from India V. *Current Science 56*: 1109–1111.

Borse, B.D. 1987c. Marine fungi from India VI. *Indian Bot. Reptr. 6*: 108–110.

Borse, B.D. 1987d. Marine fungi from India VII. *Indian Bot. Reptr. 6*: 62–64.

Borse, B.D. 1987e. New species of *Aigialus* from India. *Trans. Br. Mycol. Soc. 88*: 424–426.

Borse, B.D. 1988. Frequency of occurrence of marine fungi from Maharashtra coast, India. *Indian J. Mar. Sci. 17*: 165-167.

Borse, B.D. and Hyde, K.D. 1989. Marine fungi from India III. *Acrocordiopsis patilii* gen. *et* sp. nov. from mangrove wood. *Mycotaxon 30*: 535–540.

Borse, B.D. and Shrivastava, A.D. 1988. Marine fungi from India VIII. *Indian Bot. Reptr. 7*: 26–30.

Borse, B.D. and Shrivastava, A.D. 1994. Marine fungi from India IX: Loculoascomycetes. *Indian Bot. Reptr. 7*: 18–25.

Borse, B.D., Kelkar, D.J. and Patil, A.C. 2000. Frequency of occurrence of marine fungi from Pirotan Island (Gujarat), India. *Geobios 27*: 145–148.

Borse, B.D., Ramesh, Ch. and Shrivastava, A.D. 1988. Marine Fungi from Maharashtra (India). *Indian Bot. Reptr. 7*: 18-25.

Bunt, J.S. 1995. Continental scale patterns in mangrove litter fall. *Hydrobiologia 259*: 135–140.

Chale, F.M.M. 1993. Degradation of mangrove leaf litter under aerobic conditions. *Hydrobiologia 257*: 177–183.

Chandralata, R. 1999. What are geofungi doing in the marine environment? *Asian Mycological Congress*. chennai, India, 12.

Chapman, V.J. 1976. *Mangrove Vegetation*. Cramer Vadyz, Liechtenstein, Germany.

Chapman, V.J. 1977. *Wet Coastal Ecosystem*. Elsevier Science, UK.

Chinnaraj, S. 1992. Higher marine fungi of Lakshadweep Islands and a note on *Quintaria lignatilis*. *Cryptogamie, Mycol. 13*: 312–319.

Chinnaraj, S. 1993a. Manglicolous fungi from the atolls of Maldives, Indian Ocean. *Indian J. Mar. Sci. 22*: 141–142.

Chinnaraj, S. 1993b. Higher marine fungi from the mangroves of Andaman and Nicobar Islands. *Sydowia 45*: 109–115.

Chinnaraj, S. and Untawale, A. G. 1992. Manglicolous fungi from India. *Mahasagar 25*: 25-29.

Chowdhary, H.J., Garg, K.L. and Jaitly, A.K. 1982. Occurrence of fungi in rhizosphere, rhizosplane and non–rhizosphere zones of some mangroves. *Indian J. Mar. Sci. 11*: 138–142.

Cribb, A.B. and Cribb, J.W. 1955. Marine fungi from Queensland-I. Univ. Queensl. Pap., *Dep. Bot. 3*: 77–81.

Cribb, A.B. and Cribb, J.W. 1956. Marine fungi from Queensland-II. Univ. Queensl. Pap., *Dep. Bot. 3*: 97–105.

Desmazieres, J.B.H.J. 1849. *Plantes Cryptogames de France*. 2nd Edition, No. 1778, Lille.

Durieu de Maisonneuve, C. and Montagne, J.F.C. 1869. Pyranomycetes Fr. pp. 443–608. In: Bory de Saint-Vincent, J. and Durieu de Maisenneuve, C. (Editors) *Exploration Scientifique de Algerie*. Botanique, Paris.

Fell, J.W. and Master, I.M. 1980. The association and potential role of fungi in mangrove detritus system. *Bot. Mar. 23*: 257–263.

Fell, J.W. and Newell, S.Y. 1998. Biochemical and molecular methods for the study of marine fungi. pp. 259–283 In: Cooksey, K.E. (Editor), *Molecular Aproaches to the Study of the Ocean*. Chapman and Hall, London.

Field, C. 1995. *Journey Amongst Mangroves*. International Society for Mangrove Ecosystems, Okinawa, Japan, South China Printing Co., Hong Kong.

Garg, K.L. 1983. Vertical distribution of fungi in Sunderban Mangrove mud. *Indian J. Mar. Sci. 12*: 48–51.

Goodfellow, M. and Williams, S.T. 1983. Ecology of actinomycetes. *Ann. Rev. Microbiol. 37*: 189–216.

Hodgkiss, I.J. and Leung, H.C. 1986. Cellulase associated with mangrove leaf decomposition. *Bot. Mar. 29*: 467–469.

Hudson, H.J. 1962. Succession of microfungi on aging leaves of *Sacharum officinarum*. *Trans. Br. Mycol. Soc. 45*: 395–423.

Hughes, G.C. 1974. Geographical distribution of the higher marine fungi. *Veroff. Inst. Meeresforsch. Bremenh. Suppl. 5*: 419–441.

Hughes, G.C. 1975. Studies of fungi in oceans and estuaries since 1961. I. Lignicolous, caulicolous and foliicolous species. *Oceanogr. Mar. Biol. Annu. Rev. 13*: 69–180.

Hyde, K.D. 1986. Frequency of occurrence of lignicolous marine fungi in the tropics. In: Moss, S.T. (Ed.) *The Biology of Mangrove Fungi*. Cambridge Univ. Press, UK, 311–322.

Hyde, K.E. 1988. Observations on the vetical distribution of marine fungi on *Rhizophora* spp., at Kampong Danau mangrove, Brunei. *Asian Mar. Biol. 5*: 77–81.

Hyde, K.D. 1989a. Ecology of tropical marine fungi. *Hydrobiologia 178*: 199-208.

Hyde, K.D. 1989b. Ecology of tropical marine fungi from North Sumatra. *Can. J. Bot. 67*: 3078–3082.

Hyde, K.D. 1989c. Intertidal fungi from mangrove fern, *Acrostichium speciosum*, including *Massarina acrostichi* sp. nov. *Mycol. Res. 93*: 435–438.

Hyde, K.D. 1989d. *Caryospora mangrovei* sp. nov and notes on marine fungi from Thailand. *Trans. Mycol. Soc. Japan 30*: 333–341.

Hyde, K.D. 1990a. A study of the vertical zonation of intertidal fungi on *Rhizophora apiculata* at Kampong Kapok mangrove, Brunei. *Aquat. Bot. 36*: 255–262.

Hyde, K.D. 1990b. A comparison of the intertidal mycota of five mangrove tree species. *Asian Marine Biology 7*: 93–107.

Hyde, K.D. 1990c. Intertidal fungi from the warm temperate mangroves of Australia, including *Tunicatispora australiensis* gen. *et* sp. nov. *Aust. Syst. Bot. 3*: 711–718.

Hyde, K.D. 1991. Fungal colonisation of *Rhizophora apiculata* and *Xylocarpus granatum* poles in Kampong Kapok mangrove, Brunei. *Sydowia 43*: 31–38.

Hyde, K.D. 1992a. *Julella aviceniae* (Borse) comb. (Thelenellaceae) from intertidal mangrove wood and miscellaneous fungi from the N.E. coast of Queensland. *Mycol. Res. 96*: 939–942.

Hyde, K.D. 1992b. Intertidal mangrove fungi from the west coast of Mexico, including one new genus and two new species. *Mycol. Res. 96*: 25–30.

Hyde, K.D. 1992c. Fungi from decaying intertidal fronds of *Nypa fruticans*, including three genera and four new species. *Bot. J. Linn. Soc. 110*: 95–110.

Hyde, K.D. 1996. Biodiversity of microfungi in North Queensland. *Australian Systematic Botany 9*: 261–271.

Hyde, K.D. and Borse, B.D. 1986a. Marine fungi from Seychelles V. *Biatriospora marina* gen *et* sp. nov. from mangrove wood. *Mycotaxon 26*: 263–270.

Hyde, K.D. and Borse, B.D. 1986b. Marine fungi from Seychelles VI. *Massarina velatospora*, a new marine Ascomycotina from mangrove wood. *Mycotaxon 27*: 161–167.

Hyde, K.D. and Jones, E.B.G. 1988. Marine mangrove fungi. *P.S.Z.N.I. Marine Ecology 9*: 15-33.

Hyde, K.D. and Lee, S.Y. 1995. Ecology of mangrove fungi and their role in the nutrient cycling: what gaps occur in our knowledge? *Hydrobiologia 295*: 107-118.

Hyde, K.D. and Sarma, V.V. 2000. Pictorial key to higher marine fungi. In: *Marine Mycology –A Practical Approach* (Fungal Diversity Research Series I). Fungal Diversity Press, Hong Kong, pp. 201–264.

Hyde, K.D. and Sutton, B.C. 1992. *Nypaela frondicola* gen. *et* sp. nov., *Plectophomella nypae* sp. nov. (Coelomycetes) from intertidal fronds of *Nypa fruticans*. *Mycol. Res. 96*: 210–214.

Hyde, K.D., Chalermongse, A. and Boonthavikoon, T. 1990. Ecology of intertidal fungi at Rangong mangrove, Thailand. *Trans. Mycol. Soc. Japan 31*: 17–27.

Hyde, K.D., Chalermongse, A. and Boonthavikoon, T. 1993. The distribution of intertidal fungi on *Rhizophora apiculata*. pp. 643–652 In: Morton, J.B. (Editor) *The Marine Biology of South China Sea*. Hong Kong Univrsity Press, Hong Kong.

Hyde, K.D., Farrant, C.A. and Jones, E.B.G. 1987. Isolation and culture of marine fungi. *Bot. Mar. 30*: 291–303.

Hyde, K.D., Vrijmoed, L.L.P, Chinnaraj, S. and Jones, E.B.G. 1992. *Massarina armatispora* sp. nov., a new intertidal Ascomycete from mangroves. *Bot. Mar. 35*: 325–328.

Johnson, T.W. and Sparrow, F.K. 1961. *Fungi in Oceans and Estuaries*. J. Cramer, Weinheim, Germany.

Jones, E. B. G. 1993. Tropical marine Fungi. pp. 73–89, In: Isaac, S., Frankland, J.C., Watling, R. and Whalley, J.S. (Editors) *Aspects of Tropical Mycology*. Cambridge University Press, Cambridge, UK.

Jones, E.B.G. 2000. Marine fungi: some factors influencing biodiversity. Fungal Diversity 4: 53–73.

Jones, E.B.G. and Agerer, 1992. *Calathella mangrovei* sp. nov. and observations on the mangrove fungus *Halocyphina villosa*. *Bot. Mar. 35*: 259-265.

Jones, E.B.G. and Hyde, K.D. 1988. Methods for the study of mangrove marine fungi. pp. 9–27. In: Agate, A.D., Subramanian, C.V. and Vennucci, M. (Editors) *Mangrove Microbiology: Role of Microorganisms in Nutrient Cycling of Mangrove Soils and Waters*. UNDP / UNESCO, New Delhi, India.

Jones, E.B.G. and Kuthubutheen, A.J. 1989. Malaysian mangrove fungi. *Sydowia 41*: 160–169.

Jones, E.B.G. and Tan, T.K. 1987. Observations on manglicolous fungi from Malaysia. *Trans. Br. Mycol. Soc. 89*: 390–392.

Jones, E.B.G., Uyenco, F.R. and Follosco, M.P. 1988. Fungi on driftwood collected in the intertidal zone from the Philippines. *Asian Mar. Biol. 5*: 103–106.

Khallil, A.-R.M.A., El-Hissy, F.T. and Bagy, M.M.K. 1991. Mycoflora of mangroves of Red Sea in Egypt. *Folia Microbiol.* *36*: 456–464.

Kirk, P.W. 1980. The mycofloristic effect of seawater on spores of terrestrial and marine higher fungi. *Bot. Mar.* *23*: 233–238.

Kohlmeyer, J. 1974. On the definition and taxonomy of higher marine fungi. *Veroff. Inst. Meeresforsch. Bremerh. Suppl.* *5*: 263-286.

Kohlmeyer, J. 1980. Tropical and subtropical filamentous fungi of the Western Atlantic Ocean. *Bot. Mar.* *23*: 529-544.

Kohlmeyer, J. 1983. Geography of marine fungi. *Aust. J. Bot. Suppl. Ser.* *10*: 67–76.

Kohlmeyer, J. 1984. Tropical marine fungi. P.S.Z.N.I.: *Marine Ecology* *5*: 329-378.

Kohlmeyer, J. 1985. *Caryosporella rhizophorae* gen. *et* sp. nov. (Massariaceae), a marine Ascomycete from *Rhizophora mangle. Proc. Indian Acad. Sci. (Plant Sci.)* *94*: 355–361.

Kohlmeyer, J. and Kohlmeyer, E. 1979. *Marine Mycology: The Higher Fungi.* Academic Press, New York, USA.

Kohlmeyer, J. and Vittal, B.P.R. 1986. *Lophiostoma mangrovei*, a new marine ascomycete from the tropics. *Mycologia* *78*: 485–489.

Kohlmeyer, J. and Volkmann-Kohlmeyer, B. 1987a. Marine fungi from Aldabra, Galapagos and other tropical islands. *Can. J. Bot.* *65*: 571–582.

Kohlmeyèr, J. and Volkmann-Kohlmeyer, B. 1987b. Marine fungi from Belize with a description of the two new genera of Ascomycetes. *Bot. Mar.* *30*: 195-204.

Kohlmeyer, J. and Volkmann-Kohlmeyer, B. 1991a. Illustrated Key to the filamentous higher marine fungi. *Bot. Mar.* *34*: 1-61.

Kohlmeyer, J. and Volkmann-Kohlmeyer, B. 1991b. Marine fungi of Queensland, Australia. *Aust. J. Mar. Freshwater Res.* *42*: 91-99.

Kohlmeyer, J. and Bebout, B. and Volkmann-Kohlmeyer, B. 1995. Decomposition of mangrove wood by marine fungi and Teredinids in Belize. *P.S.Z.N.I. Mar. Ecol.* *16*: 27–39.

Leong, W.F., Tan, T.K. and Jones, E.B.G. 1991. Fungal colonisation of *submerged Bruguiera cylindrica* and *Rhizophora apiculata* wood. *Bot. Mar.* *34*: 69–76. Lockwood, J.K. 1977. Fungistatisis in soils. *Biol. Rev.* *52*: 1–43.

Makey, A.P. and Smail, G. 1996. The decomposition of mangrove litter in a subtropical Mangrove forest. *Hydrobiologia* 332: 93–98.

Maria, G.L. and Sridhar, K.R. 2002. A new ascomycete *Passeriniella mangrovei* sp. nov. from the mangrove forest of India. *Indian Journal of Forestry* *25*: 319–322.

Misra, J.K. 1986. Fungi from mangrove muds of Andaman Nicobar Islands. *Indian J. Mar. Sci.* *15*: 185–186.

Nair, L.N., Rao, V.P. and Chaudhuri, S. 1991. Micorflora of *Avicennia officinalis* Linn. pp. 52–55 In: Agate, A.D., Bonde, S.D. and Kumaran, K.P.N. (Editors) *Proceedings of the Symposium on Significance of Mangroves.* Pune, India.

Nakagiri, A. 1993. Intertidal fungi from Iriomote Island. Institute of Fermentation of Osaka *Research Communcations (Osaka)* *16*: 24–62.

Nakagiri, A. and Ito, T. 1994. *Aniptodera salsuginosa*, a new mangrove-inhabiting ascomycete, with observations on the effect of salinity on ascospore appendage morphology. *Mycol. Res.* *98*: 931–936.

Newell, S.Y. 1976. Mangrove fungi: the succession in the mycoflora of red mangrove (*Rhizophora mangle*) seedlings. pp. 51–91 In: Jones, E.B.G. (Editor) *Recent Advances in Aquatic Mycology.* Eleck Science, London.

Newell, S.Y. 1996. Establishment and potential impacts of eukaryotic mycelial decomposers in marine/terrestrial ecotones. *J. Exp. Mar. Biol. Ecol.* *200*: 187–206.

Newell, S.Y. and Fell, J.W. 1992. Ergosterol content of living and submerged decaying leaves and twigs of red mangrove. *Can. J. Bot.* *38*: 979–982.

Ong, J.E. 1995. The ecology of mangrove conservation and management. *Hydrobiologia 295*: 343–351.

Patil, S.D. and Borse, B.D. 1983. Marine fungi from Maharashtra (India) III. Some fungi from mangroves. *Indian Bot. Reptr. 2:* 56–58.

Patil, S.D. and Borse, B.D. 1985a. Marine fungi from Indian mangroves. pp. 151–152. In: Bhosale, L.J. (Editor) *The Mangroves: Biology, Utilisation and Conservation of Mangroves*. Maharashtra, India.

Patil, S.D. and Borse, B.D. 1985b. Marine fungi from Maharashtra (India) IV: Some Loculoascomycetes. *Trans. Mycol. Soc. Japan 26*: 271–276.

Patil, K.B. and Borse, B.D. 2001. Studies on higher marine fungi from Gujarat coast (India). *Geobios 28*: 41–44.

Pointing, S.B., Vrijmoed, L.L.P. and Jones, E.B.G. 1998. A qualitative assessment of lignocellulose degrading enzyme activity in marine fungi. *Bot. Mar. 41*: 293–298.

Pointing, S.B., Buswell, J.A., Jones, E.B.G. and Vrijmoed, L.L.P. 1999. Extracellular cellulolytic enzyme profiles of lignocolous mangrove fungi. *Mycol. Res. 103*: 696–700.

Poon, M.O.K. and Hyde, K.D. 1998. Biodiversity of intertidal estuarine fungi on *Phragmites* at Mei Po Marshes, Hong Kong. *Bot. Mar. 41*: 141–155.

Poonyth, A.D., Hyde, K.D. and Peerally, A. 1999. Intertidal fungi in Mauritian mangroves. *Bot. Mar. 42*: 243–252.

Poonyth, A.D., Hyde, K.D. and Peerally, A. 2001. Colonisation of *Bruguiera gymnorrhiza* and *Rhizophora mucronata* wood by marine fungi. *Bot. Mar. 44*: 75–80.

Prabhakaran, N. and Gupta, R. 1990. Activity of soil fungi of Mangalvan, the mangrove ecosystem of Cochn backwater. *Fishery Technology 27*: 157–159.

Prabhakaran, N., Gupta, R. and Krishnankutty, M. 1987. Fungal activity in Mangalavan: An estuarine mangrove ecosystem. *Proc. Natn. Sem. Estuarine Management*. Trivandrum, India, 458–463.

Prasannarai, K. and Sridhar, K.R. 1997. Effect of incubation period of driftwood on the occurrence of marine fungi. *Indian J. Mar. Sci. 26*: 380–382.

Raghukumar, S. 1973. Marine lignicolous fungi from India. Kavaka 1: 73-85.

Raghukumar, C. and Raghukumar, S. 1998. Barotolerance of fungi isolated from deep-sea sediments of the Indian Ocean. *Aquatic Microbial Ecology 15*: 153–163.

Raghukumar, S., Zainal, A. and Jones, E.B.G. 1988. *Cirrnalia basiminuta*: a new lignicolous marine Deuteromycete from the tropics. *Mycotaxon 31*: 163–170.

Raghukumar, S., Sathe-Pathak, V., Sharma, S. and Raghukumar, C. 1995. Thraustochytrid and fungal component of marine detritus III. Field studies on the decomposition of leaves of the mangrove *Rhizophora apiculata*. *Aquat. Microb. Ecol. 9*: 117–125.

Rajendran, N. and Kathiresan, K. 1999. Do decomposing leaves of mangroves attract fishes? *Current Science 77*: 972–976.

Ramadas, V. and Vijayaraghavan, Sumitra-. 1979. Efficiency of energy utilisation in the shrimp, *Metapenaeus monoceros* fed mangrove leaves. *Indian J. Mar. Sci. 8*: 114–115.

Ramesh, M.X. and Kathiresan, K. 1992. Mangrove cholesterol in the diet of panaeid prawn *Penaeus indicus*. *Indian J. Mar. Sci. 21*: 164–166.

Ravikumar, D.R. and Purushothaman, A. 1988a. *Cirrenalia tropicalis*, an addition to Indian marine fungi. *Current Science 57*: 674.

Ravikumar, D.R. and Purushothaman, A. 1988b. *Corollospora intermedia*, a lignicolous marine fungus from India. *Current Science 57*: 898–899.

Ravikumar, D.R. and Vittal, B.P.R. 1987. Studies on mangrove fungi of India I. Note on some Ascomycetes. *Kavaka 15*: 99-103.

Ravikumar, D.R. and Vittal, B.P.R. 1996. Fungal diversity on decomposing biomass of mangrove plant *Rhizophora* in Pichavaram estuary, East Coast of India. *Indian J. Mar. Sci. 25*: 142-144.

Ravikumar, D.R. and Vittal, B.P.R. 1991. A new species of *Bathyascus* from Indian mangroves. *Mycol. Res. 95*: 370–371.

Ravikumar, D.R. and Vittal, B.P.R. 1996. Fungal diversity on decomposing biomass of mangrove plant *Rhizophora* in Pichavaram estuary, East Coast of India. *Indian J. Mar. Sci. 25*: 142–144.

Ravishankar, J.P. and Suryanarayanan, T.S. 1998. Influence of salinity on the activity of polyol metabolism enzymes and peroxidase in the marine fungus *Cirrenalia pygmea* (Hyphomycetes). *Indian J. Mar. Sci. 27*: 237–238.

Ravishankar, J.P., Muruganandam, V. and Suryanarayanan, T.S. 1994. Effect of salinity on fatty acid composition of *Cirrenalia pygmea*, an obligate marine fungi. *Bot. Mar. 37*: 479–481.

Ravishankar, J.P., Muruganandam, V. and Suryanarayanan, T.S. 1995. Isolation and characterisation of melanin from a marine fungus. *Bot. Mar.38*: 413–416.

Ravishankar, J.P., Muruganandam, V. and Suryanarayanan, T.S. 1996. Effect of salinity on amino acid composition of the marine fungus *Cirrenalia pygmea*. *Current Science 70*: 1086–1087.

Robertson, A.I., Alongi, D.M. and Boto, K.G. 1992. Food chains and carbon fluxes. pp. 293–325 In: Robertson, A.I., and Alongi, D.M. (Editors) *Tropical Mangrove Ecosystems: Coastal and Estuarine Studies*. Volume 41, American Geophysical Union, Washington DC, USA.

Rohrmann, S. and Molitoris, H.P. 1986. Morphological and physiological adaptions of the cyphellaceous fungus *Halocyphina villosa* (Aphyllophorales) to its marine habitat. *Bot. Mar. 29*: 239–247.

Saenger, P. and Snedaker, S.C. 1993. Pantropical trends in mangrove above-ground biomass and annual litter fall. Oecologia 96: 293–299.

Salique, S.M., Krishnamurthy, K. and Palaniappan, R. 1985. Ecology of fungal population in the Vellar Estuary and Pichavaram Mangroves, South India. pp. 153–162. In: Bhosale, L.J. (Editor) *The Mangroves: Biology, Utilisation and Conservation of Mangroves*. Maharashtra, India.

Sarma, V.V. and Vittal, B.P.R. 1998–1999. Studies on the ecology of manglicolous fungi from the East Coast of India, observations on the seasonal occurrence. *Kavaka* 26–27: 105–120.

Sarma, V.V. and Hyde, K.D. 2001. A review on frequently occurring fungi in mangroves. *Fungal Diversity 8*: 1–34.

Sarma, V.V. and Vittal, B.P.R. 2000. Biodiversity of mangrove fungi on different substrata of *Rhizophora apiculata* and *Avicennia* spp. from the Godavari and Krishna deltas, and East Coast of India. *Fungal Diversity 5*: 23–41.

Sarma, V.V. and Vittal, B.P.R. 2001. Biodiversity of manglicolous fungi on selected plants in the Godavari and Krishna deltas, east coast of India. *Fungal Diversity 6*: 115–130.

Sarma, V.V., Hyde, K.D. and Vittal, B.P.R. 2001. Frequency of occurrence of mangrove fungi from the East Coast of India. *Hydrobiologia 455*: 41–53.

Spalding, M., Blasco, F. and Field, C. 1997. World Mangrove Atlas. Cambridge Samara Pub. Co., Cambridge, UK.

Shrivastava, A.D. 1989. Marine fungi from Bombay III: Some loculoascomycetes. *Indian Bot. Reptr. 8*: 16–19.

Shrivastava, A.D. 1994. Marine fungi from Bombay IV. *Indian Bot. Reptr. 13*:92–93.

Sridhar, K.R. and Kaveriappa, K.M. 1988. Occurrence and survival of aquatic hyphomycetes in brackish and sea water. *Arch. Hydrobiol. 113*: 153–160.

Steinke, T.D. and Ward, C.J. 1987. Degradation of mangrove leaf litter in the St. Lucia Estuary, as influenced by season and exposure. *S. Afr. J. Bot. 53*: 323–328.

Steinke, T.D., Barnabas, A.D. and Somaru, R. 1990. Structural changes and associated microbial activity accompanying decomposition of mangrove leaves in Mgheni Estuary. *S. Afr. J.Bot. 56*: 39–48.

Vijayaraghavan, Sumitra-, Ramadas, V., Kumari, K.L. and Royan, J.P. 1980. Biochemical changes and energy content of the mangrove, *Rhizophora mucronata* leaves during decomposition. *Indian J. Mar. Sci. 9*: 120–123.

Tan T.K. and Leong, W.F. 1992. Lignicolous fungi of tropical mangrove wood. *Mycological Research* 96: 413–414.

Tan, T.K., Leong, W.F. and Jones, E.B.G. 1989. Succession of fungi on wood *Avicennia alba* and *A. lanata* in Singapore. *Can. J. Bot. 67*: 2686-2691.

Tan, T.K., Teng, C.L. and Jones, E.B.G. 1995. Substrate and microbial interactions as factors affecting ascocarp formation by mangrove fungi. *Hydrobiologia 295*: 127–134.

Tomlinson, P.B. 1986. The Botany of Mangroves. Cambridge University Press, Cambridge, UK.

Untawale, A.G. 1987. Country Reports: India. pp. 51–87. In: *Mangroves of Asia and the Pacific: Status and Management*. Technical Report of the UNDP / UNESCO Research and Training Pilot Programme on Mangrove Ecosystems.

Venkatesan, T. and Natarajan, R. 1985. The Mangroves. pp. 163–169 In: Bhosale, L.J. (Editor) *Biology, Utililzation and Conservation of Mangroves*. Maharashtra, India.

Volkmann-Kohlmeyer, B. and Kohlmeyer, J. 1993. Biogeographic observations on Pacific marine fungi. *Mycologia 85*: 337-346.

Vrijmoed, L.L.P. and Jones, E.B.G. 1993. Observations on subtropical mangrove fungi. pp. 310 In: Isaac, S., Frankland, J.C., Walting, R. and Whalley, A.J.S. (Editors) *Aspects of Tropical Mycology*. Cambridge University Press, Cambridge, UK.

Vrijmoed, L.L.P., Hodgkiss, I.J. and Thrower, L.B. 1982a. Factors affecting the distribution of lignicolous marine fungi in Hong Kong. *Hydrobiologia 87*: 143–160.

Vrijmoed, L.L.P., Hodgkiss, I.J. and Thrower, L.B. 1982b. Seasonal patterns of primary Colonisation by lignicolous marine fungi in Hong Kong. *Hydrobiologia 89*: 253-262.

Vrijmoed, L.L.P., Hodgkiss, I.J. and Thrower, L.B. 1986. Occurrence of fungi on submerged pine and teak blocks in Hong Kong. *Hydrobiologia 135*: 109–122.

Vrijmoed, L.L.P., Hyde, K.D. and Jones, E.B.G. 1994. Observations on mangrove fungi from Macau and Hong Kong, with descriptions of two new ascomycetes: *Diaporthe salsuginosa* and *Aniptodera haispora*. *Mycol. Res. 98*: 699–704.

16

Species Composition of Waterborne Conidial Fungi in Two Altitudinally Different Streams of Kumaun Himalaya

M. Belwal, N. Pargaien and S. Bisht

Department of Botany, Kumaun University, Nainital-263002, India

SUMMARY

Species composition of waterborne conidial fungi were studied in two altitudinally different streams viz Jageshwar (1775 m. asl) and Jeoli (1150 m. asl). Sampling of different submerged leaf litter and water foam was done over a period of two years at monthly intervals. The maximum number of species (fifty-four species) occurred in Jageshwar stream with only 35 species in the Jeoli stream during the study period. Difference in occurrence was attributed to altitudinal difference of these two streams. Similarity Index calculated for these two streams mycoflora was 74.15%. Two species, *Clavariopsis azlanii* and *Clavatospora tentacula* were found restricted to Jeoli.

Keywords: Stream mycoflora/Occurrence and distribution/Aquatic ecosystem

In the recent past, many species other than the zoosporic fungi, ie, conidial fungi have also been described from aquatic habitat (Sati and Tiwari, 1997). Waterborne conidial fungi occur throughout the year in any fast flowing stream, however their abundance relates to the availability of leaf litter in the stream (Webster and Descals 1981). In a temperate stream which passes through deciduous forests, the community of conidial fungi reaches its peak sometime in autumn. Low temperature and sufficient availability of substrates results in higher production of waterborne conidial fungi (Iqbal and Webster 1973). As fresh and senescent leaves fall into clean and well aerated water bodies, the leaf surface gets colonised by waterborne conidial fungi (Ingold, 1975; Manoharachary and Madhusudan Rao 1983). Barlocher and Kendrick (1974) and Kaushik and Hynes (1971) have reported that these fungi act as intermediaries of energy flow in the aquatic ecosystem. Madhusudan Rao and Manoharacharya (1982) stated that leaf material liberated into the aquatic environment quickly attracts a number of fungal species in a preferential manner.

Waterborne conidial fungi represent the major microbial element in the decay of leaves. Increasing attention has been given to the role of these fungi in the decaying of leaves (Barlocher and Kendrick, 1974, 1976, 1981; Gareth Jones, 1976; Suberkropp, 1984; Suberkropp and Klug, 1976 and Triska, 1970). Normal brown coloured leaves associated with aerobic decay and softened tissue, especially bear rich crops of water-borne conidial fungi. Barlocher and Kendrick (1974) and Kaushik and Hynes (1971) have reported that, these fungi act as intermediaries of energy flow in the aquatic ecosystem. Their occurrence and taxonomy have been studied by a number of workers all over the world (Ingold, 1975, 1976 and Webster *et al.*, 1976).

In India, though very little work has been done on aquatic conidial fungi (Subramanian and Bhatt, 1981) yet, the Himalayan region is not untouched. Aquatic Hyphomycetous flora of Kumaun Himalaya has been described by Sati and Tiwari (1997) while earlier a few species were also recognised by Mer and Khulbe, (1981). More than 50 species belonging to different genera of these fungi have been recorded from various habitats by Sati and his co-workers. (Sati *et al.*, 1989, 2002; Sati and Tiwari, 1990, 1992 a, b; Tiwari, 1992).

MATERIALS AND METHODS

In the present investigation, waterborne conidial fungi, occurring in running freshwater bodies of Nainital and Almora districts located in the outer part of central Himalaya ie, Kumaun Himalaya (29° 5′ N to 31° 25′ N latitude and 77° 48′ to 81° 0′longitude) were studied. Periodical samplings were done from two streams of selected sites, as mentioned earlier. Collection of materials viz, submerged decaying leaf litter and water foam accumulated at the barriers were done and processed within 12 hours of each collection.

The collected plant materials were washed thoroughly under running tap water for 3–4 hours to remove extraneous materials, including plankton and soil particles. The leaf litters were then cut into small pieces (4–5 cm) and placed in sterile petridishes containing sterile water, and kept for incubation at 20 ±2°C. After 3–4 days, the dishes containing incubated leaf litter were examined under the low power of a microscope for the presence of water-borne conidial fungi.

The collected foam samples were fixed in 5% FAA at the spot to avoid conidial germination and examined, placing drop by drop on a glass slide, directly under the low power of a microscope. Presence of conidia was used as an index for the species present. Data was analysed for each study site to determine species composition of waterborne conidial fungi, and similarity index was made by using following formulae (Mishra, 1968).

RESULTS

Species Composition in Jeoli Stream

At the Jeoli stream, thirty-five species of waterborne conidial fungi belonging to twenty-three genera viz. *Alatospora, Anguillospora, Articulospora, Campylospora, Camposporium, Clavariopsis, Clavatospora, Diplocladiella, Flabellospora, Flagellospora, Helicomyces, Helicosporium, Heliscus, Lemonniera, Lunulospora, Pestalotiopsis, Setosynnema, Speiropsis, Tetracladium, Tetraploa, Tripospermum* and *Triscelophorus* were recorded from different submerged decaying leaf litters and water foam samples (Table 16.1). Of these, 28 species viz., *Alatospora acuminata, A. flagellata, Anguillospora crassa, A. longissima, Campylospora chaetocladia, C. purvula, Camposporium pellucidum, Clavariopsis aquatica, C. azlanii, Clavatospora tentacula, Flabellospora acuminata, Flagellospora penicillioides, Helicomyces roseus, Heliscus lugdunensis, Lemonniera cornuta, L. pseudofloscula, Lunulospora curvula, L. cymbiformis, Pestalotiopsis submersus, Setosynnema isthmosporum, Speiropsis scopiformis Tetrachaetum elegans, Tetracladium apiense, T. marchalianum, T. setigerum, Tripospermum cameleopardus, Triscelophorus acuminatus* and *T. monosporus* were found common in their occurrence in both the habitats ie water foam samples and decaying submerged leaf litter samples. Three species *Diplocladiella longibrachiata, D. scalaroides* and *Tetraploa aristata* were exclusively recorded from foam samples and were absent in submerged leaf litter. Whereas four species *Articulospora tetracladia, Lemonniera terrestris, Helicosporium lumbricoides* and *Triscelophorus konajensis* were found rare in occurrence and restricted to decaying submerged leaf litter samples only. On the basis of similarity index calculated for leaf litter and foam occurring species, 90% similarity was found.

Species Composition in Jageshwar Stream

Species composition of waterborne conidial fungi in the higher altitudinal stream, Jageshwar is presented in Table 16.2. At this stream, a total of 54 species belonging to 33 genera of waterborne conidial fungi viz, *Acaulopage, Alatospora, Anguillospora, Articulospora, Camposporium, Campylospora, Clavariopsis, Dimorphospora, Diplocladiella, Dwyaangam, Flabellocladia, Flabellospora, Flagellospora, Helicosporium, Helicomyces, Heliscina, Heliscus, Isthmotricladia,*

Table 16.1: Species Composition of Waterborne Conidial Fungi in Freshwater Stream Jeoli.

S.No	Name of Fungi	Habitat			
		Spore Types	Foam	Leaf Litter	Substrate
1.	Alatospora acuminata *Ingold*	Staurosporae	+	+	Sc,Mi
2.	A. flagellata *(Gonczol) Marvanova*	Staurosporae	+	+	Pn,
3.	Anguillospora crassa *Ingold*	Sigmoid	+	+	Pn,
4.	A. longissima *(Sacc and Therry) Ingold*	Sigmoid	+	+	Sc, Mi
5.	*Articulospora tetracladia* Ingold	Staurosporae	–	+	Sc
6.	*Camposporium pellucidum* (Grove) Hughes	Sigmoid	+	+	Sc
7.	*Campylospora chaetocladia* Ranzoni	Staurosporae	+	+	N, Pn
8	*Campylospora purvula* Kuzuha	Staurosporae	+	+	Pn,B
9.	*Clavariopsis aquatica* de Wildeman	Staurosporae	+	+	B,Pn
10	*Clavariopsis azlanii* Nawawi	Staurosporae	+	+	Sc, Mi
11	*Clavatospora tentacula* (Umphlet) Nilsson	Staurosporae	+	+	Sc, Mi, B
12.	*D. longibrachiata* Nawawi and Kuthubutheen	Staurosporae	+	–	–
13.	*D. scalaroides* Arnaud	Staurosporae	+	–	–
14.	*Flabellospora acuminata* Descals and Webster	Staurosporae	+	+	B
15.	*Flagellospora penicillioides* Ingold	Sigmoid	+	+	Sc, Mi
16.	*Helicosporium lumbericoides* (Sac) Linder	Helicoid	–	+	Pn
17.	Helicomyces roseus *Link*	Helicoid	+	+	Pn
18.	*Heliscus lugdunensis* Ingold	Staurosporae	+	+	Sc, B
19.	*Lemonniera cornuta* Ranzoni	Staurosporae	+	+	Mi
20.	L. pseudofloscula *Dyko*	Staurosporae	+	+	B, Pn
21.	*L. terrestris* Tubaki	Staurosporae	–	+	Mi
22.	*Lunulospora curvula* Ingold	Sigmoid	+	+	Sc, Mi, B
23.	*L. cymbiformis Miura*	Sigmoid	+	+	Sc, Mi, B
24.	*Pestalotiopsis submersus* Sati and Tiwari	Coelomycetae	+	+	Sc
25.	*Setosynnema isthmosporum* Shaw and Sutton	Sigmoid	+	+	Pn
26.	*Speiropsis scopiformis* Kuthubutheen and Nawawi	Sigmoid	+	+	Mi, Sc
27.	*Tetrachaetum elegans* Ingold	Staurosporae	+	+	Sc
28.	*Tetracladium apiense* Sinclair and Eicker	Staurosporae	+	+	Mi
29.	*T. marchalianum* de Wildeman	Staurosporae	+	+	Mi, Sc, Pn
30.	*T. setigerum* (Grove) Ingold	Staurosporae	+	+	Sc, B, L
31.	*Tetraploa aristata* Barkily and Broome	Staurosporae	+	–	–
32.	*Tripospermum cameleopardus* Ingold, Dann and Mc. Dougall	Staurosporae	+	+	L
33.	*Triscelophorus acuminatus* Nawawi	Staurosporae	+	+	Sc, Mi
34.	*T. monosporus* Ingold	Staurosporae	+	+	Sc, Mi, B
35.	*T. konajensis* Sridhar	Staurosporae	–	+	B
	Total		31	32	

+ = Present, –= Absent
B= *Bauhinia* sp, Sc = *Syzigium cumuni*, Mi = *Mangifera indica*, Pn = *Polygonum nepalensis* , L = *Lantana* sp

Table 16.2: Species Composition of Waterborne Conidial Fungi in Freshwater Stream Jageshwar

S.No.	Name of Fungi	Spore Types	Habitat		Substrate
			Water Foam	Leaf Litter	
1	Acaulopage tetraceros *Drechsler*	Staurosporae	–	+	Ps.
2	*Alatospora acuminata* Ingold	Staurosporae	+	+	Ps, Pr, Ra, Cd, Ao, Ba, Lo
3	*A. flagellata* (Gonczol) Marvanova	Staurosporae	–	+	Ps Ra Ao,
4	*A. pulchella* Marvanova	Staurosporae	+	+	Ps, Pr, Ra, Ao, Ba, Lo
5	*Anguillospora crassa* Ingold	Sigmoid	+	+	Pn
6	*A. filliformis* Greathhead	Sigmoid	+	+	Ps, Ba
7	*A. furtiva* Descals	Sigmoid	+	+	Ql, B
8	*A. longissima* (Sacc and Therry) Ingold	Sigmoid	+	+	Ps, Ra, Cd, Ao, Lo,
9	*Articulospora tetracladia* Ingold	Staurosporae	–	+	Ps, Ra, Ao, Ba, Lo
10	*Camposporium pellucidum (Grove) Hughes*	Sigmoid	+	+	Ao
11	*Campylospora chaetocladia* Ranzoni	Staurosporae	+	+	Ra
12	*Clavariopsis aquatica* de Wildeman	Staurosporae	+	+	Ps, Pr, Ra, Cd, Ao, Ba,
13	*Dimorphospora foliicola* Tubaki	Amerosporae	–	+	Ra, Cd, Ao, Ba, Lo
14	*Diplocladiella appendiculate* Nawawi	Staurosporae	+	+	Pn
15	*D. longibrachiata* Nawawi and Kuthubutheen	Staurosporae	+	+	Pn
16	*D. scalaroides* Arnaud	Staurosporae	+	–	–
17	*Dwyaangam Cornuta* Descals	Staurosporae	+	–	–
18	*Flabellocladia gigantea* Nawawi	Staurosporae	+	–	–
19	*F. tetracladia* (Nawawi) Nawawi	Staurosporae	+	+	Ql, Lo
20	*Flabellospora acuminata* Descals and Webster	Staurosporae	+	+	Ql, Qf
21	*Flagellospora penicillioides* Ingold	Sigmoid	+	+	Ps, Pr, Ra, Cd, Ao, Ba, Lo
22	*Helicosporium lumbericoides* (Sac.) Linder	Helicoid	+	+	Pn
23	*Helicomyces roseus* Link	Helicoid	+	+	Ra
24	*Heliscella campanulata* Marvanova	Staurosporae	+	+	Ps, Lo
25	*Heliscus lugdunensis* Ingold	Staurosporae	+	+	Ps, Pr, Cd, Ao
26	*Isthmotricladia gombakensis* Nawawi	Staurosporae	+	+	Pn
27	*Lemonniera alabamensis* Sinclair and Jones	Staurosporae	–	+	Pn, N
28	*L. aquatica* de Wildeman	Staurosporae	+	+	N
29	*L. cornuta* Ranzoni	Staurosporae	–	+	Ps, Pr, Ra, Cd, Ao, Ba, Lo
30	*L. pseudofloscula* Dyko	Staurosporae	+	+	Ra, Ba, Lo

31	*L. terrestris* Tubaki	Staurosporae	–	+	Ps, Pr, Ra, Ao, Ba, Lo
32	*Lunulospora curvula* Ingold	Sigmoid	+	+	Ps, Pr, Ra, Cd, Ao, Ba, Lo
33	*L. cymbiformis* Miura	Sigmoid	+	+	Ps, Pr, Ra, Cd, Ao, Ba, Lo
34	*Lateriramulosa uni inflata* Matsushima	Staurosporae	+	–	–
35	*Magdalannaea monogramma* Arnaud	Staurosporae	+	–	–
36	*Naidella fluitans* Marvanova and Bandoni	Staurosporae	–	+	Pn, N
37	*Pestalotiopsis submersus* Sati and Tiwari	Coelomycetae	+	+	Pr, Ra
38	*Setosynnema isthmosporum Saw and Sutton*	Sigmoid	+	+	Pr
39	*Speiropsis pedatospora* Tubaki	Staurosporae	+	+	Pn
40	*Speiropsis scopiformis* Kuthubutheen and Nawawi	Sigmoid	+	+	Pr, Ra, Ba
41	*Tetrachaetum elegans* Ingold	Staurosporae	+	+	Ps, Pr, Ra, Cd, Ao, Ba, Lo
42	*Tetracladium apiense* Sinclair and Eicker	Staurosporae	+	+	Pn
43	*T. marchalianum* de Wildeman	Staurosporae	+	+	Ps, Pr, Ra, Cd, Ao, Ba, Lo
44	*T. setigerum* (Grove) Ingold	Staurosporae	+	+	B, Pn
45	*Tetraploa aristata* Barkily and Broome	Staurosporae	+	+	B
46	*Tricladium aciculum* Nawawi	Staurosporae	+	+	N, Ql
47	*T. chaetocladium* Ingold	Staurosporae	+	+	Ps, Pr, Ra, Cd, Ao, Ba, Lo
48	*T. indicum* Sati and Tiwari	Staurosporae	+	+	Mi, N
49	*T. splendens* Ingold	Staurosporae	+	+	Ql
50	*Tripospermum cameleopardus* Ingold, Dann and Mc Dougall	Staurosporae	+	+	Ql
51	*Triscelophorus acuminatus* Nawawi	Staurosporae	–	+	Ps, Pr, Ra, Cd, Ao, Ba, Lo
52	*T. monosporus* Ingold	Staurosporae	+	+	Ps, Pr, Ra, Cd, Ao, Ba, Lo
53	*T. konajensis* Sridhar	Staurosporae	–	+	Ra, Lo
54	Volucrispora ornithomorpha *(Trotter) Haskins*	Staurosporae	+	–	–
	Total		44	48	

+ = Present, -= Absent
Ao = *Acer oblongum*, Ba = *Betula alanoides*, Cd = *Cedrus deodara*, Lo = *Lyonia ovalifolia*, Pn = *Polygonum nepalensis*, Pr = *Pinus roxburghii* Mi = *Myrica indica,* Ra = *Rhododendron arboreum*, Ql= *Quercus leucotrichophora,* Qf = *Quercus floribunda,* B = *Barberis* sp.

Lemonniera, Lunulospora, Lateriramulosa, Magdalannaea, Naidella, Pestalotiopsis, Setosynnema, Speiropsis, Tetrachaetum, Tetracladium, Tetraploa, Tricladium, Tripospermum, Triscelophorus, and *Volucrispora* were observed from submerged decaying leaf litters as well as water foam samples. Of these, 38 species were found to be common in their occurrence in foam samples as well as on submerged decaying leaf litter samples (Table 16.2).

Ten species viz *Acaulopage tetraceros, Alatospora flagellata, Articulospora tetracladia, Dimorphospora foliicola, Lemonniera alabamensis, L. cornuta, L. terrestris, Naidella fluitans, Triscelophorus acuminatus* and *T. konajensis* were recovered only from decaying submerged leaf litter samples, whereas six species viz *Diplocladiella scalaroides, Dwyaangam cornuta, Flabellocladia gigantea, Lateriramulosa uni-inflata, Magdalannaea monogramma* and *Volucrispora ornithomorpha* were observed from water foam samples only. Out of the total observed species of waterborne conidial fungi, 81.48% species were encountered in foam samples while 88.88% species were recorded from decaying submerged plant materials, especially leaf litter. On the basis of similarity index value calculated for leaf litter and foam occurring species, and 82.6% similarity was found among the species occurring in these two habitats.

DISCUSSION

The species composition of waterborne conidial fungi in two selected streams ie Jageshwar and Jeoli was studied. A total of fifty-seven species were recorded collectively from both the streams, 35 species were present in the Jeoli stream while fifty-four species were recorded in Jageshwar stream. Thus Jageshwar stream showed greater species richness than the Jeoli stream. The former is high altitude stream with low temperature, while the latter is a low altitude stream with high temperature. The higher and lower occurrence of these fungi in the stream may be attributed to their altitudinal differences and this view was found in favour of Chauvet (1991).

Twenty-two species viz, *Acaulopage tetraceros, Alatospora pulchella, Anguillospora filliformis, A. furtiva, Dimorphospora foliicola, Diplocladiella appendiculata, Dwyaangam cornuta, Flabellocladia gigantea, F. tetracladia, Heliscella campanulata, Isthmotricladia gombakensis, Lemonniera alabamensis, L. aquatica, Lateriramulosa uni inflata, Magdalannaea monogramma, Naidella fluitans, Speiropsis pedatospora, Tricladium aciculum, T. chaetocladium, T. indicum, T. splendens and Volucrispora ornithomorpha* were found as restricted to Jageshwar stream. The remaining species were found common in their occurrence as these occurred in both the streams, Jeoli and Jageshwar, except for two species *Clavariopsis azlanii and Clavatospora tentacula,* which were found to be restricted to Jeoli. Species similarity index value for both the studied streams was also made and it showed 74.15% similarity between the species collected from both the study sites, Jageshwar and Jeoli.

The similarity index value calculated for the species occurring on decaying submerged leaf litter samples and water foam samples of both the sites, indicates 88.88% similarity at the Jeoli site and 82.6% similarity at the Jageshwar site. These results indicate that regular foam observation may also be useful to study stream flora. This view is also in agreement with Nilsson (1964) and Ingold (1975) who suggested that foam collection alone can give more than 75% picture of stream mycoflora.

As is evident from Tables 16.1 and 16.2, staurosporus conidial forms were found most prevalent, whereas sigmoid forms ranked second, followed by helicoid forms in order of occurrence. Though these fungi were colonised on a variety of submerged leaf litters, including gymnospermus plants, however, leaf litter of *Acer oblongum, Betula alanoides, Lyonia ovalifolia, Mangifera indica, Nerium species, Pyrus pashia, Polygonum nepalensis, Quercus floribunda, Q leucotrichophora, Rhododendron arboreum* and *Syzigium cumuni* were found to be a more favourable substrate for them.

ACKNOWLEDGEMENTS

The authors are thankful to Dr S.C. Sati for his kind guidance and Prof S.P. Singh, Head, Department of Botany, Kumaun University, Nainital for providing lab facilities to work.

REFERENCES

Barlocher, F and Kendrick, B. 1974. Dynamics of fungal population on leaves in a stream. J. Ecol., 62: 761-791.

Barlocher, F and Kendrick, B. 1976. Hyphomycetes as intermediaries of energy flow in streams. In; "Recent Advances in Aquatic Mycology", (E.B. Gareth Jones Ed.) 435-446.

Barlocher, F. and Kendrick, B. 1981. Role of aquatic Hyphomycetes in the trophic structure of streams. In "The Fungal community Its organisation and role in the ecosystem," (Wicklow, D.T. and Carroll, G.G. Eds), Marcel Dekker: Newyork, pp. 743-760.

Chauvet, E. 1991. Aquatic Hyphomycetes distribution in south-western France. Journal of Biogeography, 18: 699-706.

Gareth Jones, E.B. 1976. Topics of further interest 707-724 In *"Recent advances in Aquatic mycology"* Edited by E.B. Gareth Jones, Elek Seicnce London.

Ingold,C.T.1975. *An illustrated guide to aquatic and water borne Hyphomycetes (Fungi imperfecti) with notes on their biology.* Freshwater Biol. Assoc. Scient. Publ. No. 30 England 96 pp.

Ingold, C.T. 1976. The morphology and biology of fresh water fungi excluding Phycomycetes. In "Recent advances in aquatic Mycology". (Ed. E.B. Gareth Jones) Publ. Paul Elek . Ltd. 335-357.

Iqbal, S.H. and Webster, J. 1973. The trapping of aquatic Hyphomycetes spores by air bubbles.Trans.Brit.Mycol.Soc., 60: 37-48.

Kaushik, N.K. and Hynes, H.B.N. 1971. The fate of dead laves that fall into streams. Arch Hydrobiol., 68 : 465 -515.

Madhusudan Rao, M. and Manoharachary, C. 1982. Processing of leaves by fungi in aquatic ecosystem. Indian Phytopathology, 35 : 654-656.

Manoharacharya, C. and Madhususudan Rao, M. 1983. Ecological studies on Hyphomycetes associated with submerged leaves from India. Indian Phytopathology, 36 : 62-65.

Mishra, R. 1968. Ecology workbook. Oxford IBH Publication Co. Calcutta 244 page.

Mer, G.S. and Khulbe, R.D. 1981. Aquatic Hyphomycetes of Kumaun Himalaya, India. Sydowia. Annales Mycologic., 34: 118-124.

Nilsson, S. 1964. Freshwater Hyphomycetes. Taxonomy, Morphology and Ecology. Symb. Bot. Upsal., 18:1-30.

Sati, S. C., Mer, G. S. and Tiwari, N. 1989. Occurrence of water-borne conidial fungi on *Pinus roxburghii* needles, Curr. Science, 58: 918-919.

Sati, S.C., Tiwari, N. and Belwal, M. 2002. Conidial aquatic fungi of Nainital, Kumaun Himalaya, India. Mycotaxon, 81: 445-455.

Sati, S. C. and Tiwari, N. 1990. Freshwater Hyphomycetes from Jageshwar stream, Kumaun Himalaya, India. Nat. Acad. Sci. Letters, 13: 7-9.

Sati, S. C. and Tiwari, N. 1992a. A new species of *Tricladium* from Kumaun Himalaya, India. Mycol. Research, 96: 229-232.

Sati, S. C. and Tiwari, N. 1992b. Colonisation, species composition and conidial production of aquatic Hyphomycetes on chir pine needle litter in a freshwater Kumaun Himalayan stream. Int. Revue. Ges. Hydrobiol., 77: 445-453.

Sati, S.C. and Tiwari, N. 1997. Glimpses of conidial aquatic fungi in Kumaun Himalaya. In *"Recent Researches in Ecology, Environment and pollution"* X: 17-37. Eds. S.C.Sati, J. Saxena and R.C. Dubey. *Today and Tomorrow printers and Publishers,* New Delhi.

Suberkropp. K. 1984. Effect of temperature on seasonal occurrence of aquatic Hyphomycetes. Trans. Brit. Mycol. Soc., 82: 53-62.

Suberkropp, K., and Klug, M.J. 1976. Fungi and bacteria associated with leaves during processing in a woodland stream. Ecology, 57: 707-719.

Subramanian, C.V. and Bhatt, D.J. 1981. Conidia from freshwater foam samples from the Western Ghats, Southern India. Kavaka, 9: 45-62.

Tiwari, N. 1992. Taxonomy and species composition of Hyphomycetes in forested streams and their colonisation pattern on tree leaves in Nainital, Central Himalaya. Ph.D. Thesis, Kumaun University, Nainital.

Triska, F. J. 1970. Seasonal distribution of aquatic Hyphomycetes in relation to the disappearance of leaf litter from a woodland stream. Ph. D. thesis, University of Pittsburgh. Pittsburgh. PA.

Webster, J. and Descals, E. 1981. Morphology, distribution and ecology of conidial fungi in freshwater habitats. In the *Biology of Conidial Fungi* (Cole. G. T. and Kendrick, B. eds) N.Y. Academic Press, 295-355.

Webster, J. Moran, S.T. and Davey, R.A. 1976. Growth and sporulation of *Tricladium chaetocladium* and *Lunulospora curvula* in relation to temperature. Trans. Brit. Mycol. Soc., 67: 491-499.

17

Dynamics of Litter Decomposition by Fungi in Oak Forests of Kumaun Himalaya

R.C. GUPTA, G.P. PANDEY AND R.L. FOTEDAR

Department of Botany, Kumaun University, SSJ Campus, Almora –263601, India

SUMMARY

The present investigation aims to determine the effects of the nature and heterogeniety of substratum on the dynamics of mycoflora in a temperate oak forest of Kumaun Himalaya. A protected area of about 500 x 500 sq m was selected in Mirtola forest of Almora, Kumaun Himalaya. Although there were marked differences in the amount of chemical constituents remaining during different stages of decomposition in the nylon bags and pots, but there was a good resemblance in the release pattern of these chemical components. The species composition of the mycoflora varied on leaf litter during different stages, exhibiting a defined fungal succession. There was a distinct succession of fungi, both on the leaf surface and inside the leaf. A large number of macromycetes or higher fungi were also found to be associated with decomposing leaf litter in the forest during the rainy season.

Keywords: Nutrient dynamics, Litter decomposition, Oak forest, Kumaun Himalaya, Microbial decomposition

Forest ecosystems cover most of the vegetative land area on earth (Whittaker and Likens,1973) and have a higher sustained productivity than other natural terrestrial ecosystems. Forest ecosystems exhibit mechanisms for capture, distribution and conservation, all of which are essential for system persistence (Witkamp, 1971; Reiners, 1968; Satchell, 1974).

In most of the terrestrial systems, a greater proportion of carbon fixed by autotrophs is returned to the soil in the form of dead organic debris, which is termed as 'Litter' by ecologists (Rochow, 1974; Satchell, 1974). All this organic carbon either moves through the food web to the higher trophic levels to appear in one of the several possible forms or is degraded by microbial action to become incorporated into microbial tissue, along with the formation of chemically simpler substances. In many nutrient-poor ecosystems, most of the nutrients that are absorbed by the vegetation, originate from the decomposition of plant litter and humus (Staaf and Burg, 1977; Berendse *et al.*, 1987).

An understanding of the nutrient cycle in the forest ecosystem is recognised as being critical for continued long term forest use and management. Accumulation of litter and its decay are essential processes of the overall energy flow and nutrient cycling phenomenon within an ecosystem. Decomposition is the major process involved in the dissipation of energy stored in organic matter in the forest ecosystem. Detritus contains energy and nutrients which can be exploited, particularly by microorganisms and animals in wooded ecosystems. In addition, a food chain was recognized which channelled dead organic material through bacteria, fungi and 'detritus feeders'. In fact, decomposition is a subtle process, which involves a complex set of interactions and a variety of time scales, and its intensities fluctuate in response to the production of organic matter and also the feed back on that production.

At elevations ranging from 1000m of timber line (about 3600m) in the Central Himalaya, one or other species of oak (*Quercus* spp.) forms climax vegetation (Champion and Seth, 1968 a,b). Oaks are the 'People's species', providing fodder for livestock, fire-wood for cooking, tools for agricultural activities and leaves for manuring crop fields (Singh and Singh, 1986). Oaks have been related to water conservation species and therefore, oak is also known as the 'Gold' of Uttarakhand (Joshi *et al.*, 1988).

Quercus leucotrichophora, a dominant species of oaks in Kumaun Himalaya, is gregarious and sometimes forms pure forests to a considerable extent. *Rhododendron arboreum, Pinus roxburghii, Lyonia* sp., *Myrica nagi* are some of its commoner companions. It grows on a variety of geological formations, including micaceous sandy soils. It attains its larger dimension on deep moist soil in cool northern aspects (Anonymous, 1969).

The present investigation aims at determining the effects of the nature and heterogeniety of the substratum on the dynamics of mycoflora in a temperate oak forest of Kumaun Himalaya.

MATERIALS AND METHODS

Litter Production

For studying litter production, a protected area of about 500 × 500 sq m was selected in Mirtola forest of Almora. Litter fall was trapped by placing six litter traps (50 × 50 cm each) on the forest floor randomly. Each trap was made of 2 mm mesh nylon, suspended from a galvanised wire supported by four sticks at a height of 50 cm, and allowed to drop to a depth of 25cm. Litter from these bags was collected separately in paper bags and brought to the laboratory, where the sample was sorted out into three main categories, (1) leaf litter (2) wood litter (3) miscellaneous litter.

Litter Decomposition

Litter decomposition study was carried out at the Mirtola forest, just after leaf fall. The freshly fallen leaves were stored and air dried to a constant weight. 10 gm of dried leaves were placed in a nylon mesh bag firmly closed by inserting metal pins (Crossley and Hoglund, 1962). A total of 75 bags were prepared and placed under dense oak trees in a specific trench. Five bags were picked randomly each month and brought to the laboratory for analysis. The litter of two bags was used for mycofloral analysis and the rest of the bags were weighed and dried at 60°C for 48h, and kept in butter paper bags for organo—chemical analysis.

Exponential decay model was used to estimate the rate of annual leaf litter weight loss. The half life time and turnover was calculated by standard equations (Jenny *et al.*,1949; Olson,1963).

Physico-chemical Analysis

Moisture content, H_2SO_4 fraction, NaOH soluble fraction, Acetone soluble fraction and crude fibre were analysed (Kanwar and Chopra, 1967) at four months' intervals starting from the freshly fallen leaves.

Bio-chemical Analysis

The total organic carbon, total organic matter, total nitrogen, C/N ratio, sodium and potassium were determined by standard techniques followed by various workers (Walkley and Black, 1934; Robinson, 1952; Jackson, 1958; Piper, 1944).

Estimation of Energy Content

The release of energy during litter decomposition was estimated by using a bomb calorimeter. The methodology given by Lieth (1968) was followed.

Isolation of Mycoflora

Mycoflora of decomposing oak leaf litter was screened, qualitatively as well as quantitatively at monthly intervals for confined litter, and bi-monthly for potted litter. Direct observation methods such as (a) leaf clearing (b) impression films were used. Humid chamber, material plating and dilution plate techniques were followed for detailed analysis of the litter decomposing fungi. The number of fungi per unit area of the decomposing litter was calculated by using a standard formula. Aeromycoflora was recorded at monthly intervals at Mirtola forest during the course of the investigation. Importance Value Index (IVI) of each fungal species inhabiting the litter in each population and sampling period was calculated following the method of Joshi and Chauhan (1982). Concentration of Dominance (Cd) value was calculated using the formula given by Simpson (1949). Similarity quotient was obtained by the formula of Sorenson (1948) and percent constancy was calculated by the method given by Hawksworth (1974).

RESULTS AND DISCUSSION

The Litter Fall

The leaf litter constituted 73.2% of the total litter fall in the present study and is roughly near the value of 70% as computed by Bray and Gorham (1964), and is within the range of 40% to 85% as reported for temperate forests by Rodin and Bazilevich (1967). Meentemeyer *et al.*, (1982) calculated 70% of leaf litter in the total litter fall in forests around the world. The estimate (3.65 t/ha/yr) of present leaf litter production falls within the range of mean litter fall at 3.5 t/ha/yr as reported by Rodin and Bazelevich (1968) for the oak forests of the world, although it is lower than the value of 4.0 to 4.5 t/ha/yr as reported for temperate regions of USSR by Rodin and Bazilevich (1967).

The wood constituting 22.5% total litter falls within the the range of 21% to 30% as published by Bray and Gorham (1964). However, it is higher than the value (0.94 t/ha/yr) reported by Mehra and Singh (1985) in mixed oak-pine woodlands of the Central Himalayan region.

The rate of leaf litter decomposition for Central Himalayan oak forests ranges between 1.50–1.70 mg/g/day (Singh and Singh, 1986) and matches with the rate of 1.68 and 1.54 mg/g/day, as estimated in the present study, though, it is higher than most of the values reported for temperate oak forests (Fig. 17.1). Actually, litter decomposition is a result of the complex interaction between biotic factors, such as soil fauna, microorganisms, nutrient and carbon compounds of litter, and of biotic factors like temperature and moisture (Witkamp, 1971). The trend of decomposition demonstrated a most rapid mass reduction during the rainy season and minimum reduction during the winter season. Thus, the present dry weight loss of confined leaf litter for summer, rainy, post-monsoon and winter season, respectively amounted to 11.01% 23.25%, 12.95% and 14.0% (Fig. 17.2). Relatively lower rates during post-monsoon and summer season may be due to the paucity of soil water, while slow humidification during the winter season, due to low temperature (Singh and Gupta, 1977).

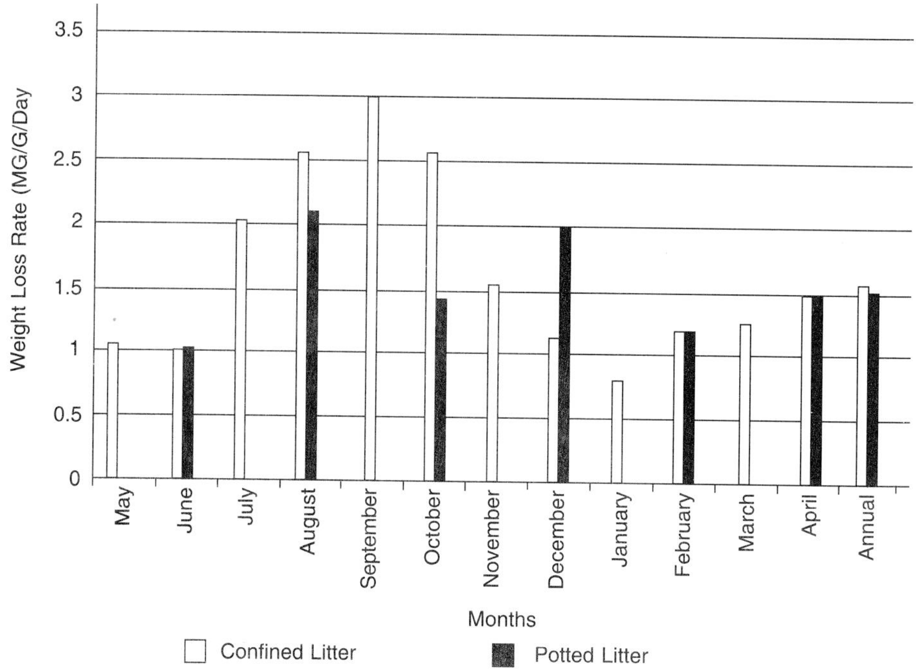

Fig. 17.1: Monthly Variation in Wt Loss Rate.

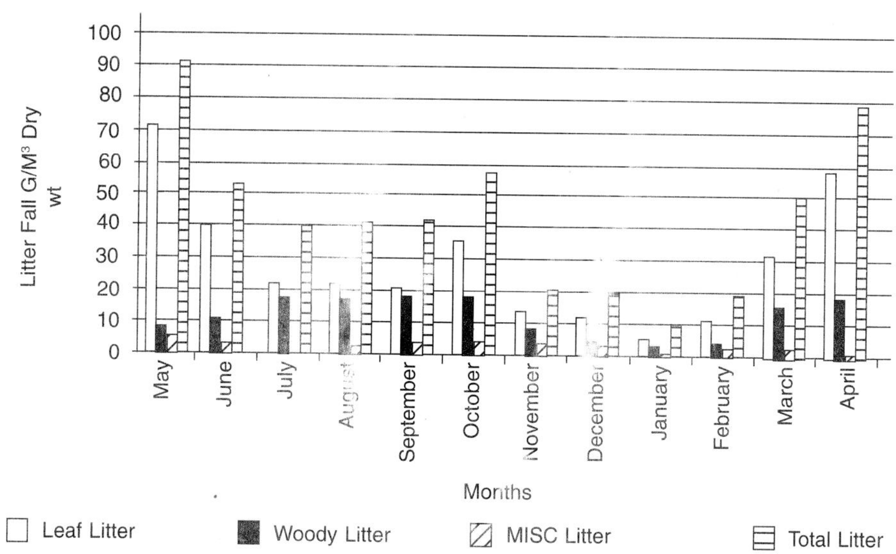

Fig. 17.2: Litter Fall Pattern in OAK Forest.

Physico-chemical properties

In the present study, although there were marked differences in the amount of chemical constituents remaining during different stages of decomposition in the nylon bags and pots, there was a resemblance in the release pattern of these chemical components. In both cases, H_2SO_4 soluble fraction disappeared rapidly, but crude fibre component disappeared the least. Results obtained by Pandey and Singh (1984) also show that maximum rapid loss of available carbohydrates from oak leaf litter occurred in between 29 to 121 days of decomposition.

The relative percentage of crude fibre content increased all the time. It thus appears that the decomposition rate of the crude fibre fraction becomes rate determining for the decomposition of the whole leaf litter. Berg *et al.*, (1982) observed that the degradation of lignin and growth of fungus only occurred after the addition of a carbohydrate (eg cellulose), that could be used as an energy source (Fig. 17.4). Lignin decomposes slowly and has been identified as a big factor in controlling litter decomposition in several different ecosystems (Shukla and Singh, 1984).

The nitrogen dynamics of the decomposing oak leaf litter were consistent with the nitrogen dynamics observed by various authors like Das and Ramakrishnan, 1985; Upadhyay and Singh,1989). In general, the loss of leaf litter was in the order of K > Na >Ca > N.

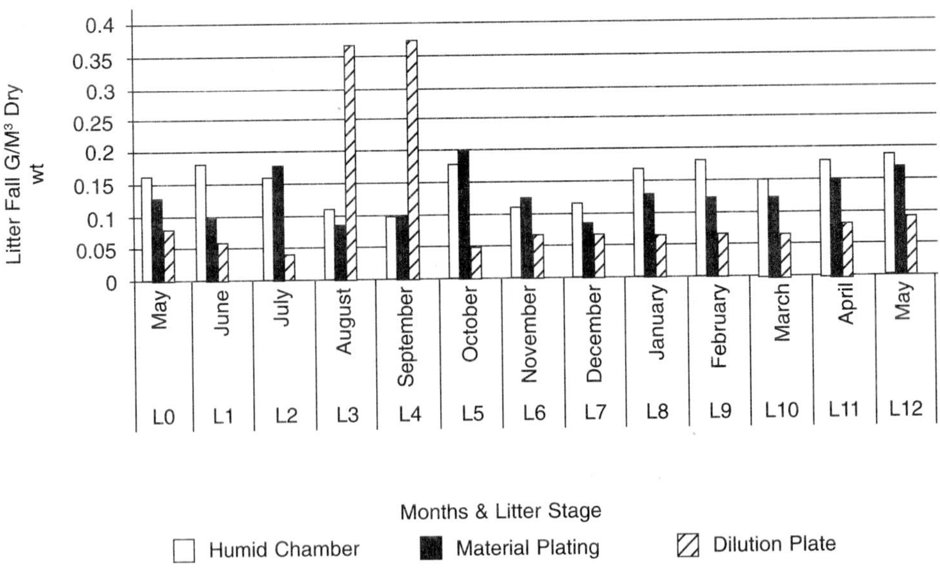

Fig. 17.3: Conc. of Dominance (CO) of Litter Mycoflora.

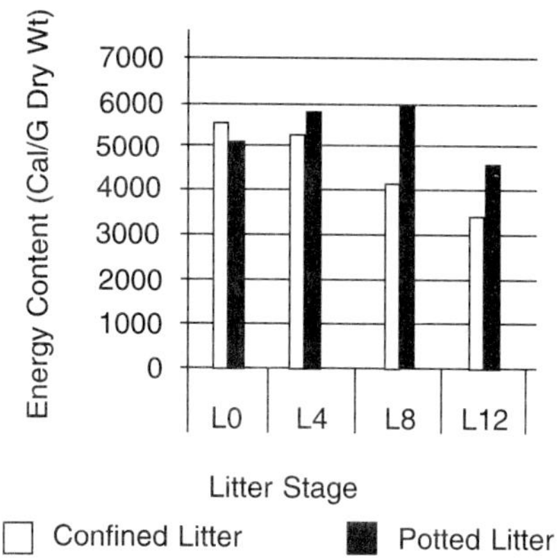

Fig. 17.4: Energy Cintent of Leaf Litter

Dynamics of Litter Mycoflora

The examination of litter mycoflora recorded by employing different techniques at several stages of decomposition has shown that by the time leaves reached the soil surface as litter they were substantially colonised by a variety of saprophytic fungi. A perusal of results obtained during this study revealed that the species composition of the mycoflora varied on the litter during different stages, exhibiting fungal succession. There was a distinct succession of fungi, both on the leaf surface and inside the leaf.

The mycoflora of oak litter altered quantitatively as well as qualitatively (Fig. 17.3) as the leaf material degraded. Garrett (1957) attributed changes in mycoflora to the progressive breakdown of the newly invaded substrate by a succession of heterotrophic microorganisms, and proposed that there was a sequence corresponding with the dominance of different 'substrate groups' of fungi. One of the features that has led to the derivation of the substrate group hypothesis, was the observation that the initial rate loss of different components could be clearly ranked; Sugar > hemicellulose > cellulose > Lignin.

Among the fungal group isolated in the present study, it seems that Deuteromycotina, which constituted about 79% of the total mycoflora, played an important role in the decomposition process. The initial colonisers of fresh fallen leaves were *Fusarium oxysporum, Phoma sorghina, Cladosporium cladosporioides, Coelophoma cylinderocarpon, Aspergillus luchuensis* and *Pestalitia* sp. It was noted that Ascomycotina and Sphaeriopsidale members were recorded frequently on blotter by Humid Chamber Incubation method and on Direct Observation method (Fig. 17.5). It was also noted that Basidiomycetous fungi were very rarely isolated in this study, however, their mycelia were quite often observed on direct observation and humid chambers. The abundant

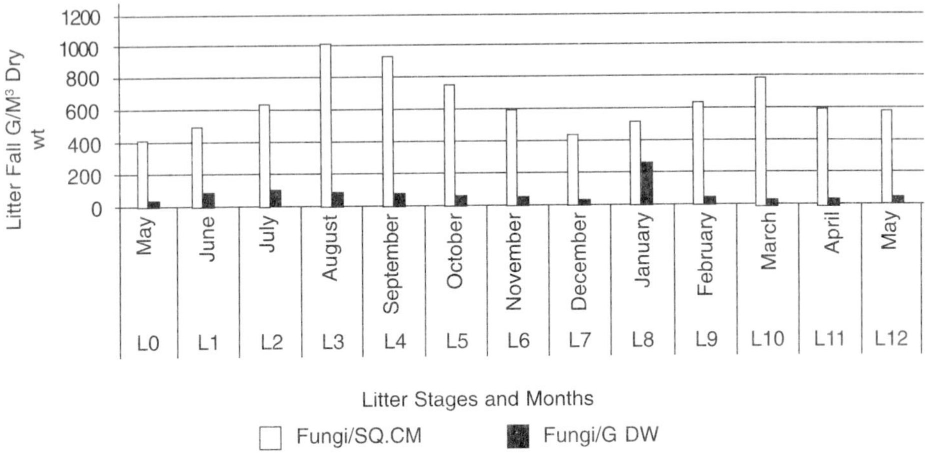

Fig. 17.5: Change in Mycoflora During Decomposition.

growth of Basidiomycetous hyphae was seen either in the rainy season or in the litter decomposing at very moist places. This frequent observation of basidiomycetous mycelia is most probably indicative of their importance in the utilisation of cellulose and lignin in the litter substrate.

According to Hudson's (1968) proposed general scheme for the successive colonisation of fungi on decaying leaves of plants, succession, begins with the colonisation of living leaves by parasites, followed by primary saprophytes. The primary saprophytes are then followed by secondary saprophytes, consisting of Ascomycetes, Basidiomycetes and Fungi Imperfecti. The final stage in Hudson's succession is the 'soil inhabitants' under which he included the Mucorales and species of *Trichoderma* and *Penicillium.*

The present study indicates that fungal succession is directly influenced by the chemical composition of substrates. The appearance of secondary saprophytes which utilised cellulose and sugars most, was common in potted litter up to L10 and L12 stage, where the loss of those chemicals from the litter was relatively slower than in the confined litter.

The second wave of secondary saprophytes, the Basidiomycetes, were not isolated during the study, as a result of the method used. However, the results obtained by direct observation of leaf litter clearly emphasise that several strains of Basidiomycetous hyphae were present on the leaf surface and a *Marasmius* sp. was collected frequently on parially decomposed litter. The secondary saprophytes on oak leaf litter were *Alternaria tenuis, A.alternata, Mortierella subtilissima, Chaetomium globossum, Rhizopus nigricans, Mucor hiemalis,* white sterile mycelium and *Fusarium* sp.

The increase in the total number of fungi on partially decayed leaves may be attributed to the increase litter surface area for colonisation, and the amount of available nutrients on the litter. The well moist substrate and favourable temperature might have favoured fungal growth. The reduction in fungal species in the later stages of decay may most probably be due to the increase in lignin content in the litter substrate, and during winter, unfavourable atmospheric conditions might have been responsible for a fall in the number of litter inhabiting fungi. Importance of environmental factors and moisture content of the substrate in fungal colonisation has been recognised by several workers (Webster,1956; Dix,1984; Rai and Kumar,1988). Dix (1985) observed that after decay, wood and leaves have a higher water potential as compared to undecayed material at the same water content, and concluded that improvement in water availability through decay may be one of the principal underlying forces which gives direction to fungal sucession on various substrates. The number of species in the rainy season increases on partially decomposed substrate, while it decreases on those substrates which were mostly decomposed, but remained aboveground. In spite of the most favourable environmental conditions in the rainy season the decline in the number of species in the mostly decomposed substrate is attributed to the depletion of organic nutrients, and competition amongst the microorganisms which colonise the substrate in the later part of decomposition. The accumulation of antagonistic substances in the substrates might play an important role in decreasing the number of fungal species.

Scopinella barbata which was isolated from the litter in Humid Chamber method and recorded frequently by direct examination of litter, has been reported from oak leaf litter for the first time from India (Gupta and Pandey, 1990). It's occurrence during advanced stages of decomposition and subsequent disappearance at later stages indicates that it falls in the category of secondary saprophytes, playing a very important role in the decomposition of oak litter in Kumaun Himalaya.

AIR MYCOFLORA

Mycofloral population dynamics of air, in relation to litter decomposition, reflect the fact that dominant fungal species in relative months of decomposition were also detected in decomposing leaf litter. Recently, Reddy and Reddy (1986) reported similar observations and concluded that the spores reached the decaying leaf litter through wind currents and got lodged in the decaying leaves, thus acting as a source for litter decomposing fungi. On the other hand, it might be possible that the litter decomposing fungi sporulated and released their spores into wind currents, which subsequently acted as a means of transport or dispersal agents.

ROLE OF MACROMYCETES

A large number of macromycetes or higher fungi which mainly appeared in the rainy season, might have utilised the lignin fraction of litter. The

Basidiomycetes are well known for their lignicolous property. Graiger (1946) studied the ecology of larger fungi and stated that species of higher fungi have definable preferences for particular ranges of pH of the substrate. The majority favoured acid conditions, but a minority preferred alkaline reactions. He also observed that the most humicolous Basidiomycetes require a high average water content in the substrate, a factor which is of greate importance. The acidic character, with high soil moisture, and high organic carbon and nitrogen in the present oak woodland might have favoured the growth of these fungi. Apart from the macromycetes a saprophytic angiosperm plant *Monotropa uniflora* was collected frequently from the Mirtola oak forest. Its appearance in abundance on the heap of litter indicates its importance in the decomposition process. Amongst the higher forms, the species of *Agaricus, Amanita, Clavaria, Ramaria, Boletus, Mycena, Marasmius, Laccaria* were the most common. Ascomycetous forms like *Leotia lubrica, Verpa* sp.,*Trichoglossum walteri, Helvella crispa, Peziza* sp. and *Bulgaria* were collected frequently.

ACKNOWLEDGEMENTS

The authors are thankful to the Prof G.C.Joshi, Head, Department of Botany, Kumaun University, SSJ Campus, Almora, for providing laboratory facilities. One of the authors (G.P.Pandey) is grateful to the U. G.C., New Delhi for providing JRF. We are also thankful to the Directors of Vivekanand Parvatiya Anusandhan Sansthan and. G. B. Pant Paryavaran Sansthan, Almora, for providing Library facilities.

REFERENCES

Anonymous 1969. The wealth of India.Vol. VIII. Publication and Information Directorate, CSIR, Hillside Road, New Delhi, India

Berendse, F. and Bosatta, E. 1987. The effect of lignin and nitrogen on the litter in nutrient-poor ecosystems-a theoretical approach. Can. J. Bot. 65:1116–1120.

Bray, J.R. and Gorham, E. 1964. Litter production in the forests of the world. Adv. Ecol. Res.2:101–157. Berg B., Hannus, K.P.T and Theander, O. 1982 Changes in organic chemical components of needle litter during decomposition: long term decomposition in a scot-pine forest I. Can. J. Bot. 60: 1310–1319

Champion, H.G. and Seth, S.K. 1968a. A revised summary of forests types in India. Delhi Manager of Publications. p 29

Crossley, D.A. Jr. and Hoglund, M.P. 1962. A litter bag method for the study of micro arthropods inhabiting the leaf litter. Ecology 43:571–573.1968b General siviculture for India Delhi Manager of Publications , p.329.

Dix, N.J. 1984. Moisture content and water potential of abscissed leaves in relation to decay. Soil Bio. Biochem. 16:367–370.

Dix, N.J. 1985. Changes in relationship between water potential after decay and its significance for fungal successions. Trans. Br. Mycol. 85:649–653.

Das, A.K. and Ramakrishnan, P.S. 1985. Litter dynamics in Khasi pine *Pinus kesia* Royle ex Gorden of North Eastern India. For. Ecol. Manage. 10:135–153.

Gochenaur, S.E. 1978. Fungi of a Long Island oak birch forest I. Community organisation and seasonal occurrence of the opportunistic decomposers of the A horizon. Mycologia 70:975–994.

Graiger, J. 1956. Ecology of the larger fungi. Trans. Br. Mycol. Soc. 29:52–63.

Gupta, R.C. and Pandey, G.P. *1990 Scopinella*: A new record from India. Ind. J. Mycol. Pl. Path.

Hawksworth, D.L. 1974 Mycologist's Hand Book. C.M.I., Kew, England.

Hudson, H.J. 1968. The ecology of fungi on plant remains above the soil. New Phytol. 67:837–874.

Jackson, M.L. 1958. Soil chemical analysis. 498 pp. Prentice Hall, New Jersey.

Jenny, H., Gessel, S.P. and Bingharn, T. 1949. Comparative study on decomposition rate of organic matter in temperate and tropical regions. Soil Sci. 68:419–432.

Joshi, I.J. and Chauhan, R.K.S. 1982 Investigations into the soil mycoecology of Chanmbal ravines of India.I.Fungal communities and seasonal successsion. Plant and Soil 66:329–338.

Joshi, S.C. Pangtty Y.S., Joshi, D.R, and Dani, D.D. 1988. "Uttarakhand ka Sona-Banj". pp.26. Published by Uttarakhand Seva Nidhi, Almora , India.

Kanwar, J.S. and Chopra, S.L. 1967. Analysis and practical chemical agriculture chemistry. S Chand and Co., New Delhi.

Lieth, H. 1968. In: Functioning of terrestrial ecosystems at the primary production level. Ed. E E Echardt, UNESCO pp179_186.

Mehra, M.S., Pathak, P.C. and Singh, J.S. 1985. Nutrient movement in litter fall and precipitation components of Central Himalayan Forests. Ann. Bot. 55:153–170.

Meentmeyer, V. 1978. Macroclimate and lignin control of litter-decomposition rates. Ecology 59:465–472.

Olson, J.C. 1963. Energy storage and the balance of products and decomposers in ecological systems. Ecology 44: 322–331.

Pandey, U. and Singh, J.S. 1984. Nutrient changes and release during the decomposition of leaf litter in a Himalayan oak-conifer forest. Can. J. Bot. 62:1824–1831.

Piper, C.S. 1944. Soil and Chemical analysis. The University of Adelaide, Australia.

Robinson, W.O. 1952. Soils: Their origin, constitution and classification. Thomas Murbi and Co., London.

Reddy, S.S. and Reddy, S.M. 1986. Aeromycoflora of Paskhal forest in relation to leaf litter decomposition. Proc. Nat. Acad. Sci. India 56(B) IV: 355–358.

Reiners, W.A. 1968. Carbon dioxide evolution from the floor of three Minnesota forests. Ecology 44: 471–483.

Rochow, J.J. 1974. Litter fall relations in a Missouri forest. Oikos 25: 80–85.

Rodin, L.E. and Bazilevich, N.I. 1967. Production and mineral cycling in terrestrial vegetation (English translation) ed by FOGG G E). Oliver and Boyd, Edinburgh, Paris p.45–50.

Rai, B. and Kumar, A. 1988. Microbial decomposition of forest leaf litter as influenced by certain edaphic factors. J.Ind. Bot. Soc. 67:18–26.

Shukla, A.N. and Singh, I.D. 1984. Biodegradation of *Shorea robusta* Gaertn. leaf litter and the cycling of minerals in a tropical sal forest. Plant and Soil 81:403–409.

Simpson, E.H. 1949. Measurement of diversity. Nature 163: 688.

Singh, J.S. and Singh, S.P. 1987. Forest vegetation of the Himalaya Bot. Rev. 53:80–192.

Sorenson, T. 1948. A method of establishing groups of equal amplitude in plant society based on similarity of species content Dansk. Vidensk. Biol. Skr. 5:1–34.

Staaf, H. and Berg, B. 1977. Mobilisation of plant nutrients in a Scots pine forest moor in Central Sweeden. Silv. Fenn. 11:210–217.

Upadhyay, V.P. and Singh, J.S. 1989. Nitrogen release pattern in decomposing oak and pine litter in Nainital Hill, India. Indian for. 115:320–326.

Walkley, A. and Black, I.A. 1934. An experiment of the Degtjaress method for determining soil organic matter and a proposed modification of the chromic acid titration method. Soil Sci. 37:29–38.

Webster, J. 1956. Succession of fungi on decaying cocksfoot culms. J. Ecol. 44:517–544.

Whittaker, R.H. and Likens, G.E. 1973. Carbon in biota. In: Carbon in the Biosphere. Eds. Woodwell and Pecan pp.281–302. US Dept. Commerce, Springfield, Virginia 22151.

Witkamp, M. 1971. Soil as components of ecosystems. In: Annual review of ecology and systematics. vol 2. Eds. Johnson *et al.*, Annual Rev., Inc. Palo Aceto, California. Pp.85–110.

18

Leaf Litter Decomposition in Kumaun Himalayan Forest: A Review

S.C. SATI AND NEETU GUPTA

Department of Botany, Kumaun University, Nainital -263002, (U.A.), India

SUMMARY

Leaf litter decomposition is one of the important processes in the biogeochemical cycle and nutrient dynamics of any ecosystem. A good amount of work on this line has also been done in the Kumaun Himalaya by some investigators. The present paper deals with a comprehensive review on leaf litter decomposition studies made by different workers in the forests of Kumaun Himalaya so far.

Keywords: Leaf litter decomposition, Kumaun Himalaya, Microbial decomposition

Plants, in the form of biomass, tap a larger part of solar energy and earth minerals. For the utilisation of this energy and nutrients, the decomposition of plant parts (litter) is a must. The process of decomposition plays a prime role in the functioning of the ecosystem. In all mature ecosystems the bulk of net primary productivity is shed as plant litter. The components of decomposition, in the form of dead organic matter are known as detritus. The detritus is broken down by the combined action of the decomposer community, which is predominantly composed of microorganisms (Bacteria and fungi) and invertebrates. These organisms feed on the detritus and utilise energy, carbon and other nutrients for their own growth.

The decomposition of plant litter is one of the most crucial processes in the biogeochemical cycle of the forest ecosystem (DeCantanzaro and Kimmins 1985). It is governed by an interaction of biotic, abiotic components and substrate quality (Singh and Gupta 1977; Meentmeyer 1978; Swift, Heal and Anderson 1979; Melillo, Aber and Muratore 1982; McClaugherty *et al.*, 1985).

Primarily, the rate of decomposition varies with different plants, depending upon the physiochemical properties of litter, environmental temperature, soil atmosphere, including moisture content and potentiality and quantity and quality of microorganisms (Daubenmire and Prusso 1963; Dickinson and Pugh 1974; Williams and Gray 1974). In coniferous forests, the soil fauna is less prominent than in the deciduous forests but it is probably responsible for a part of nitrogen mineralisation (Presson *et al.*, 1980).

According to Anderson (1973), decomposition refers to the catabolic degradation of the litter constituents by soil organisms and it is a highly complex and often prolonged process, influenced by several factors, such as the substrate quality, chemical composition and soil climate which are important factors influencing litter decomposition in the forest ecosystems. It best indicated for pine forests (Meentmeyer and Berg 1986).

Berg (1986) proposed a model for litter decomposition and suggested that decomposition of litter may be divided into at least two phases. In the first phase, soluble substances and non-lignified carbohydrates (cellulose and hemicellulose) are decomposed. The concentration of soluble material is easily degraded. In the other phase, primarily lignin and lignified cellulose remains are decomposed. Berg and McClaugherty (1989) suggested that lignin and humic compounds, the latter formed during decomposition, are slowly decomposed. Furthermore, nitrogen is incorporated into humic substances during decay (Berg and Theander 1984; Stevenson 1982).

McClaugherty and Berg (1987) reported that cellulose, lignin and nitrogen levels are the major regulatory factors in litter decomposition. Gilbert and Boccock (1960) found that rapid decrease in C:N ratio is associated with faster decomposition of plant litter. Singh (1969) reported the lowest decomposition rate and highest C:N ratio in litter which is relatively poor in ash and nitrogen content.

Parnas (1975) proposed a model of decomposition of organic material, and concluded that the decomposition of any substrate is directly proportionally to the growth rate of decomposers. The major litter decomposers are bacteria, including actinomycetes, fungi, protozoa, nematodes etc.

According to Berg and Soderstrom (1979), fungi appear to play a crucial role in the process of accumulation of nitrogen. A large number of investigators including Ramacle (1970 and 1971), Rusco (1971), Vissner and Parkinson (1975) and Shukla *et al.* (1977), Hudson (1962) and Sati *et al.* (1989) found that the decomposition rate was influenced by fungal population. Singh and Gupta (1977), Singh and Singh (1992), Singh *et al.* (1990) and Gassner and Chauvet (1994) highlighted the importance of fungi in decomposition.

Fungi have cellulolytic, proteolytic and keratinophilic properties. The microbial decomposition of cellulose has been studied by several workers (Pandey *et al.* 1988, Mehrotra and Aneja, 1979; Awasthi and Gupta 1983, Sharma 1980; Pandey 1985, Mitra and Bose 1978). Srivastava *et al.* (1981) made a comparative study on cellulolytic ability of mesothermic and thermophilic fungi isolated from manure and plant refuse.

Water is one of the most important climatic variable which effectively regulates the rate of decomposition process. Water not only accelerate biological activity but also leaches of soluble substances from the litter. Edwards and Heath (1963) observed the rapid rate of decomposition of leaves during the rainy season.

Micola (1960), Vancleave (1971), Weight and McGinnis (1975) found that temperature is another favourable and decisive factor for leaf litter decomposition. However in the tropical zone, moisture is found often a limiting factor (Madge 1965, Hopkins 1966). In the temperate zone, moisture or temperature may affect the decomposition rate, depending on the local conditions (Edwards and Heath 1963).

In the terrestrial ecosystem, the hydrogen ion concentration (pH) also plays an important role in litter decomposition, which influences the microflora (Gray and Williams 1971a) (Warcup, 1951) and the fauna (Kuhnelt, 1961) of acid and alkaline soils. The pH may affect the growth of various saprophytes.

Thus pH, moisture content, substrate quality and temperature ultimately govern the distribution of microflora and process of decomposition (Swift *et al.*, 1979, Howard and Howard 1980).

The Study Area

Kumaun Himalaya, the eastern part of Uttaranchal (UA), a young state created in November 2000 from Uttar Pradesh, is one of the northern states of India. It stretches between 29°5′ and 31°25′ N latitudes, and between 77°48′ and 81°0′ E longitudes in the western part of Central Himalaya and comprises mostly hilly temperate zones (Fig. 18.1).

The Kumaun Himalayan region is well known for its beautiful landscapes with many famous tourist and religious destinations. It has a varying topography and climatic setup over the Shiwaliks of the Central Himalaya. The Kumaun division administratively comprises six districts, Nainital, Almora, Bageshwar, Pithoragarh Champawat and Udham Singh Nagar (Fig. 18.1). Most part of this region are dominated by lofty mountains with deciduous and non-deciduous forest.

The different climatic conditions of Kumaun Himalaya have provided a unique opportunity for studying the leaf litter decomposition, both in natural and laboratory conditions. The forest area of this region is covered by several forest plants such as *Quercus, Rhododendron, Aesculus,* etc. Due to low temperatures and slow microbial activity, a large quantity of litter gets accumulated in the temperate forests (oak and pine) of Kumaun Himalaya.

Sometimes, this leads to disastrous conditions due to forest fires, particularly during summer months. The forests of Kumaun Himalaya also act as a natural sponge for absorbing and retaining the rain water, that would otherwise cause floods in the plains of North India during the monsoons.

Work Done

In the Kumaun Himalayan forests, the decomposition pattern of leaf litter have been extensively studied by many workers in different areas (Pandey and Singh 1981 a, b; Jha 1995; Lodhiyal 2000; Sati and Pant 2000). Pandey (1979) is one of the pioneers who studied leaf litter decomposition in Kumaun Himalaya. She investigated the rate of leaf litter decomposition for six tree species and one shrub species, using litter bag technique. She observed 100% decomposition for *Daphne cannabina, Aesculus indica, Ilex dipyrena, Quercus floribunda,* 94.56% for *Q. leucotrichophora,* 81.81% for *Cedrus deodara* and 72.22% for *Cupressus torulosa* by the end of her study. Pandey and Singh (1984) concluded that the

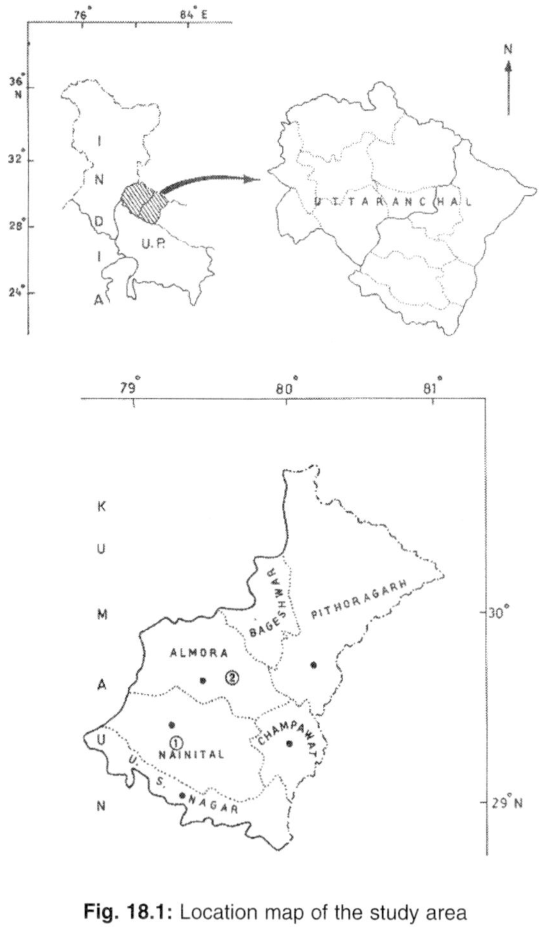

Fig. 18.1: Location map of the study area

chemical composition of litter material influenced decomposition rates. The leaf litter, characterised by high contents of nitrogen, water soluble compounds and low content of lignin and cellulose, decomposed more rapidly than the leaf litter having a converse composition. A linear combination of rainfall and temperature explained 80% of the variability in monthly weight loss (Pandey and Singh 1982).

Rawat (1983) studied three species of oak, namely, *Quercus leucotrichophora, Q. floribunda, Q. lanuginosa* and one species of gymnosperm tree, namely, *Pinus roxburghii*. He observed 62–100% weight loss per annum among the studied species. *P. roxburghii* showed a lower decomposition rate than the *Quercus* species. Singh and Gupta (1977) also concluded that in temperate forests, the rate of decomposition was relatively higher for leaf litter of broad-leaf species than for other tree species. The mean annual temperature, altitude and lignin content were the principal regulating factors for weight loss and nutrient cycle.

Leaf litter decomposition in *P. roxburghii* forest ecosystem was studied by Chaturvedi (1983). Over a period of one year, he recorded 88.03% decomposition for *P. roxburghii*, and 100%, 99.5% and 97.3% for *Q. floribunda, Q. leucotrichophora* and *Q. lanuginosa*, respectively. He also observed that in all decomposing leaf litter, the concentration of nitrogen and phosphorus increased, while the concentration of water soluble compounds, total available carbohydrate and potassium decreased. Later Upadhyay (1984) studied leaf litter decomposition of ten woody species viz, *Shorea robusta, Mallotus philippensis, Lyonia ovalifolia, Rhododendron arboreum, Pinus roxburghii, Myrica esculenta, Quercus glauca, Q. floribunda, Q. leucotrichophora* and *Q. lanuginosa* in five different forest types of Kumaun Himalaya. He found that the annual weight loss ranged from 47–100% for all types of leaf litter. Differences between site and species were also found to be significant (Upadhyay and Singh 1989).

Upadhyay and Singh (1985a, b) also observed that the leaf litter having high N-contents had higher weight loss rates, whereas species having low N-contents exhibited lower rates. The litter quality and abiotic components also influenced the rate of weight loss (Gupta and Singh 1981, Upadhyay *et al.* 1989). Upadhyay *et al.* (1985) showed that habitat also influenced the rate of decomposition in standard tree species.

One of the most interesting investigations on microbial decomposition of leaf litter was conducted by Pandey (1985). She took five tree species, viz *Shorea robusta, Lyonia ovalifolia, Pinus roxburghii, Quercus leucotrichophora* and *Q. lanuginosa* at different localities in Kumaun Himalaya. She observed a 73–98% weight loss in the leaf litter. The increase in temperature and decrease in altitude enhanced the rate of decomposition. However, the chemical composition of the litter also affects the rate of decomposition. Her work also emphasised the role of microorganisms on decomposing leaf litter. A total of 153 fungal species were isolated from different decomposing material, with most of them (97%) belonging to Deuteromycetes (Pandey and Khulbe 1984).

In a study on high altitudinal oak species (*Quercus semecarpifolia*), Rawat (1990) reported a 65.51% annual dry weight loss. She also observed that nitrogen concentration in decomposing leaf tissue shows remarkable variations in different durations. A 61.27% annual weight loss of *Q. leucotrichophora* leaf litter has also been reported by Pandey (1990a) in the oak forest of Kumaun Himalaya in Almora. He also emphasised the role of fungi in decomposition and reported 53 fungal species associated with oak leaf litter.

Pandey (1990b) studied the leaf litter decomposition of ringal (*Arundinaria falcata*) and compared the decomposition rates of senesced leaves and green leaves. She reported that green leaves decompose faster (99.37%) than senesced leaves (94.15%).

The decomposition of *Eucalyptus hybrid* has been studied by Bargali (1990) and Joshi (1991). *Q. semecarpifolia* and *C. torulosa* studied by Saxena (1990) were found to show 67.9% and 63.7% decomposition, respectively. She found that the maximum weight loss during the rainy season might be due to greater leaching and temperature.

Joshi (1991) worked on leaf litter of six tree species, viz, *P. roxburghii, Eucalyptus hybrid, Grevillea robusta, Lancaena leucocephala, Populus deltoids* and *Tectona grandis,* and observed 78–99% decomposition. Adhikari (1992) studied leaf litter decomposition in Kharsu oak and Silver fir forest, at the Pindari Glacier of Central Himalaya. He selected four tree species, *Q. semecarpifolia, R. arboreum, Abies pindrow* and *A. indica,* and reported a 66–97% weight loss during the study period. In his study the initial C:N ratio ranged from 24:1 to 42:1, which was lower than those reported by Upadhyay and Singh (1989) for leaf litter (41:1 to 82:1).

In the same year, Garkoti (1992) studied leaf litter decomposition of four plant species, *Acer cappadocicum, R. arboreum, R. campanulatum* and *Betula utilis,* which showed 74.4%, 41.5%, 42.3% and 52.6% weight loss, respectively, in a one year period. However, Khulbe (1992) observed 66% and 61% annual weight loss in *Q. leucotrichophora* and *P. roxburghii,* respectively.

While studying decomposition in the aquatic ecosystem, Tiwari (1992) carried out a comparative study of the decomposition rate of oak, pine and their mixed litter in a fresh water stream. She reported that the cumulative weight loss after 13 months was 89% in oak and pine, and 85% in the oak-pine mixed litter. Sati and Pant (2000) suggested that the higher weight loss in pine litter, which equalled that of oak litter, may be attributed to their needle-shaped leaves. They found that in aquatic ecosystems, there was no positive correlation between C:N ratio of substrate and weight loss, as it acts as a major determinant in terrestrial decomposition (Sati and Pant 2000).

Tiwari (1992), in her study observed a definite role of aquatic Hyphomycetes in aquatic decomposition. She identified a total of thirty species from decomposing leaf litter of oak, pine and their combination, and suggested that oak litter is a preferential substrate for aquatic fungi as it was colonised by twenty-six species, as compared to twenty-two and eighteen species that colonised oak-pine mixed and pine litter, respectively.

While working with the decomposition of forest leaf litter, Lodhiyal (2000) studied shisham (*Dalbergia sissoo* Roxb.) forests in Tarai and Bhabhar belts of Kumaun and found 95.2% decomposition in one year.

REFERENCES

Adhikari, B.S. 1992. Biomass, productivity and nutrient cycling of Kharsu oak and Silver fir forests in Central Himalaya. Ph.D. Thesis, Kumaun University, Nainital, India.

Anderson, J.M. 1973. The breakdown and decomposition of sweet chestnut (*Castanea sativa*) and beech (*Fagas sativa*) leaf litter in two deciduous woodland soils. I. Breakdown, Leaching and Decomposition. Oecologia (Berlin) 12:215-274.

Awasthi, P.B. and Gupta, S.C. 1983. Production of cellulytic enzymes by fungi growing on decaying plant. Geobios., 10(6):281–282.

Bargali, S.S. 1990. *Structure and functioning of eucalyptus plantations in Tarai belt of Kumaun Himalaya.* Ph.D. Thesis, Kumaun University. Nainital, India. 338 p.

Berg, B. 1986. Nutrient release from litter and humus in coniferous forest soils–a mini review. Scand. J. For. Res., 1:359–369.

Berg, B. and McClaugherty, C. 1989. Nitrogen and phosphorus release from decomposing litter in relation to the disappearance of lignin. Can. J. Bot., 67:1148–1156.

Berg, B. and Soderstrom, B. 1979. Fungal biomass and nitrogen in decomposing Scots Pine needle litter. *Soil Biol. Biochem.* 11: 339–341.

Berg, B. and Theander, O. 1984. The dynamics of some nitrogen fractions in decomposing Scot Pine needle litter. Pedabiologia, 27:161–167.

Chaturvedi, O.P. 1983. Biomass structure, productivity and nutrient cycling in *Pinus roxburghii* forest. Ph.D. Thesis, Kumaun University, Nainital, India.

Daubenmire, R. and Prusso, D.C. 1963. Studies on the decomposition rates of tree litter. Ecology, 44:589–592.

DeCatanzaro, J.B. and Kimmins, J.P. 1985. Changes in the weight and nutrient composition of litter fall in three forest ecosystem types in Coastal British Columbia. Can. J. Bot. 63:1046–1056.

Dickinson, C.H. and Pugh, G.J.F. (eds.) 1974. Biology of plant litter decomposition, Vols. I & II. Academic Press, London and New York.

Edwards, C.A. and Heath, G.W. 1963. The rate of soil animals in the breakdown of leaf material. In: Soil organisms (eds. J. Docksen and J. Vander Drift), North Holland Publication Company, Amsterdam, 76–84 pp.

Garkoti, S.C. 1992. High altitude forests of Central Himalaya : Productivity and nutrient cycling. Ph.D. Thesis, Kumaun University, Nainital, India.

Gessner, M.O. and Chauvet, E. 1994. Importance of stream microfungi in controlling breakdown rates of leaf litter. Ecology, 75:1807–1817.

Gilbert, O. and Bocock, K.L. 1960. Changes in leaf litter when placed on the surface of soils containing humus types. II Changes in the nitrogen content of oak and ash leaf litter. J. Soil Sci., 11:10–19.

Gupta, S.R. and Singh, J.S. 1981. The effect of plant species, weather variables and chemical composition of plant material on decomposition, in a tropical grassland. Plant and Soil, 59: 99–117.

Hopkins, B. 1966. Vegetation of the Olokemeje forest reserve, Nigeria. IV. Litter and soils with special reference to their seasonal changes. J. Ecol., 64: 687–703.

Howard, P.J.A. and Howard, D.M. 1974. Microbial decomposition of tree and shrub leaf litter. I. Weight loss and chemical decomposition of decomposing litter. Oikos, 25:341–352.

Hudson, H.J. 1962. Succession of microfungi on ageing leaves of *Saccharum officinarum*. Trans. Brit. Mycol. Soc., 49:185.

Jha, K.K. 1995. Structure and function of an age series of teak (*Tectona grandis* Linn.) plantations of Kumaun Himalaya. Ph.D. Thesis, Kumaun University, Nainital.

Joshi, C.S. 1991. Litter fall, decomposition and nutrient release in certain plantation of Kumaun Himalaya. Ph.D. Thesis, Kumaun University, Nainital, India.

Khulbe, A. 1992. Decomposition of oak and pine forest litter and associated mycoflora. 150 pp. Ph.D. Thesis, Kumaun University, Nainital, India.

Kuhnelt, W. 1961. "Soil Biology", Faber and Faber, London.

Lodhiyal, N. 2000. Structure and function of shisham (*Dalbergia sissoo* Roxb.) forests in Tarai and Bhabar belts of Kumaun Himalaya. Ph.D. Thesis, Kumaun University, Nainital.

Madge D.S. 1965. Leaf fall and litter disappearance in tropical forests. Pedobiologia 5: 273–288.

McClaugherty, C.A., Pastor, J., Aber, J.D. and Mellilo, J.M. 1985. Forest litter decomposition in relation to soil nitrogen dynamics and litter quality. Ecology, 66:266–275.

Meentemeyer, V. 1978. Macroclimate and lignin control of litter decomposition rates. Ecology, 59: 465–472.

Meentemeyer, V. and Berg, B. 1986. Regional variations in rate of mass loss of *Pinus sylvestris* needle litter in Swedish pine forests as influenced by climate and litter quality. Scandinavian Journal of Forest Research, 1: 167–180.

Mehrotra, R.S. and Aneja, K.R. 1979. Microbial decomposition of *Chenopodium album* litter. I. Succession of decomposers. J. Ind. Bot. Soc., 58:189–195.

Melillo, J.M., Aber, J.D. and Muratore, J.F. 1982. Nitrogen and lignin control of hardwood leaf litter decomposition dynamics. Can. J. Bot., 63:621–626.

Micola, P. 1960. Comparative experiment on decomposition rates of forest litter in Southern and Northern Finland. Oikos, 11:161–166.

Mitra, D.K. and Bose, S.K. 1978. Isolation of cellulolytic fungi. Indus Fern Symp. Proc. 3:133–134.

Pandey, G.P. 1990 a. Studies on litter decomposition by fungi in oak forests of Kumaun Himalaya. Ph.D. Thesis, Kumaun University, Nainital, India.

Pandey, K. 1985. Studies on microbial decomposition of cellulose and selected leaf litter in different forests of Kumaun Himalaya. Ph.D. Kumaun University, Nainital, India.

Pandey, K. and Khulbe, R.D. 1984. Microflora on decomposing organic matter. In: Integrated ecological study of Eastern Kumaun Himalaya with emphasis on natural resources. J.S. Singh and S.P. Singh (eds.), Final Report (Vol. 2) submitted to Dept. of Science and Technology, New Delhi.

Pandey, N., Pandey, H. and Saxena, J. (1988) Cellulolytic activity of oak litter (*Quercus leucotrichophora* Camus ex Bahadur) by colonising fungus. Nat. Acad. Sci. Letters. 11:11.

Pandey, P. 1990 b. Structure and function of Ringal (*Arundinaria falcata* Nees) community in the oak zone of Kumaun Himalaya. Ph.D. Thesis, Kumaun University, Nainital, India.

Pandey, U. 1979. Leaf-fall, leaf litter decomposition and nutrient release in a mixed oak forest. Ph.D. Thesis, Kumaun University, Nainital, India.

Pandey, U. and Singh, J.S. 1982. Leaf litter decomposition in an oak-conifer forest in Himalaya: The effects of climate and chemical composition. Forestry, 55:47–59.

Pandey, U. and Singh, J.S. 1984. Nutrient changes and release during decomposition of litter in an Himalayan oak-conifer forest. Can. J. Bot. 62:1829–1831.

Pandey, U., and Singh, J.S. 1981b. A quantitative study of forest floor, litter fall and nutrient return in an oak-conifer forest in Himalaya. II. Pattern of litter fall and nutrients. Oecol. Gener., 2(2):83-99.

Pandey, U., and Singh, J.S. 1981a. A quantitative study of forest floor: litter fall and nutrient return in an oak-conifer forest in Himalaya. I. Composition and dynamics of forest floor. Acta Oecologia, 2:49-61.

Parnas, H. 1975. Model for decomposition of organic material by microorganisms. Soil Biol. Biochem., 7:161–169.

Persson, T., Baath, E. Clarholm, M., Lundkvist, H., Soderstrom, B. and Sohlenius, B. 1980. Trophic structure, biomass, dynamics and carbon metabolism of soil organisms in a Scots Pine forest. Ecol. Bull., 32: 419–462.

Rawat, L. 1990. Forest floor dynamics, litter fall, litter decomposition and nutrient release in a *Quercus semecarpifolia* Smith forest of Kumaun Himalaya. Ph.D. Thesis, Kumaun University, Naini Tal, India.

Rawat, Y.S. 1983. Plant Biomass net primary production and nutrient cycling in oak forests. Ph.D. Thesis, Kumaun University, Nainital, India.

Remacle, J. 1970. La microflora des litterens. Bull. Soc. Roy. Bot. Belg., 103:83–96.

Remacle, J. 1971. Succession in the oak litter microflora in forests at Mealin-Eglise (Ferage), Belgium. Oikos, 22:411–413.

Ruscoe, Q.W. 1971. The mycoflora of living and dead leaves of *Nothofagus truncata*. Trans. Bri. Mycol. Soc., 56:463–474.

Sati, S.C. and Pant, N. 2000. Microbial decomposition of leaf litter in a fast flowing forest water stream of Kumaun Himalaya. In: Microbes: Agriculture, Industry and Environment, 197–209. 2003.

Sati, S.C. and Tiwari, N. 1992. Colonisation, species composition and conidial production of aquatic Hyphomycetes on chir-pine needles litter in a fresh water Kumaun Himalaya stream. Int. Rev. Ges. Hydrobiologia, 77:445–453.

Sati, S.C., Mer, G.S. and Tiwari, N. 1989. Occurrence of water-borne conidial fungi on *Pinus roxburghii* needles. Current Science, 58:918–919.

Saxena, A. 1990. Ecological studies on high elevation Cypress-surai and Kharsu oak forest ecosystems of Kumaun Himalaya. Ph.D. Thesis, Kumaun University, Nainital, India.

Sharma, P.D. 1980. Studies on cellulose decomposition by microorganisms. Ph.D. Thesis, University of Sagar, Sagar, M.P., India.

Shukla, A.N., Arora, D.K. and Dwivedi, R.S. 1977. Effects of microbial culture filtrates on sal (*Shorea robusta* Gaernt.) leaf litter fungi. Soil Biol. Biochem., 9:217–219.

Singh, J.S. and Gupta, S.R. 1977. Plant decomposition and soil respiration in terrestrial ecosystems. Bot. Rev., 43: 449–528.

Singh, J.S. and Singh, S.P. 1992. Forests of Himalaya. Gyanodaya Prakashan Nainital, India. pp. 294.

Singh, K.P. 1969 Studies in decomposition of litter of important trees of tropical deciduous forest at Varanasi. Trop. Ecol., 10:292–311.

Singh, S.P., Pande, K., Upadhayay, V.P. and Singh, J.S. 1990. Fungal communities associated with the decomposition of a common leaf litter (*Quercus leucotrichophora* A. Camus) along an elevational transect in the Central Himalaya. Biol. Fertil. Soil., 9:245–251.

Srivastava, R.B., Narain R. and Mehrotra, B.S. 1981. Comparative cellulolytic activity of mesophic and thermophilic fungi isolated in India from manure and plant refuse. Ind. J. Mycol. Plant Pathol., 11:66–72.

Stevenson, F.J. 1982. Humus chemistry genesis composition, reactions. J. Wiley and Sons, New York.

Swift, M.J., Heal, O.W. and Anderson, J.M. 1979. Decomposition in terrestrial ecosystems. Blackwell Scientific Publication. Oxford, p. 35–39.

Tiwari, N. 1992. Taxonomy and species composition of Hyphomycetes in forested streams and their colonisation pattern on tree leaves in Nainital, Central Himalaya. Ph.D. Thesis, Kumaun Himalaya, Nainital, India.

Upadhyay, V.P. 1984. Leaf litter decomposition in certain forest ecosystems of Kumaun Himalaya. Ph.D. Thesis, Kumaun University, Nainital, India.

Upadhyay, V.P. and Singh, J.S. 1985a. Nitrogen dynamics of decomposing hardwood leaf litter in a Central Himalayan forest. Soil Biol. Biochem., 17: 827-830.

Upadhyay, V.P. and Singh, J.S. 1985b. Decomposition of woody branch litter on an altitudinal transect in the Himalayas. Vegetatio, 69:44–54.

Upadhyay, V.P., Singh, J.S., Pandey, U and Singh, J.S. 1985. Effect of habitat on decomposition of standard leaf litter species. Biology and Fertility of Soils, 1:201–207.

Upadhyay, V.P., Singh, J.S. and Meentemeyer, V. 1989. Dynamics and weight loss of leaf litter in Central Himalayan Forests. Abiotic versus litter quality influences. J. Ecol., 77: 147–161.

Van Cleave, K. 1971. Energy and weight loss functions for decomposing foliage in birch and aspen forests in Interior Alaska. Ecology, 52:720–723.

Visser, S. and Parkinson, D. 1975. Fungal succession on aspen poplar leaf litter. Can. J. Bot., 53:1640–1651.

Warcup, J.H. 1951. The ecology of soil fungi. Trans. Brit. Mycol. Soc., 34: 379–399.

Wieght, R.E. and Mc Gimms J.T. 1975, Ecology. 56, 129–140

Williams, S.T. and Gray, T.R.G. 1974. Decomposition of litter on the soil surface. In: Biology of plant litter decomposition (eds. C.H. Dickinson and G.J.F. Pugh). Academic Press, London and New York, 2:611–632.

Section D

Fungal Biotechnology/Symbiosis

19

Biotechnological Approaches for Improvement of Mycorrhizal Fungi and Their Application in Nursery Practices

MANOJ BHATT, ANURAG YADAV AND R.C. DUBEY

Department of Botany & Microbiology, Gurukul Kangri University, Haridwar-249404 (U.A.) India.

SUMMARY

Nowadays, successful forest plantations requires deliberate introduction of mycorrhizal fungi and, hence, such fungi are termed inoculants. There are a variety of sources of mycorrhizal inocula. In the present paper, different aspects of biotechnological approaches for the improvement of mycorrhizal fungi and their applications in nursery practices has been reviewed.

Keywords: Mycorrhizal inoculants/Fungi, Microbial consortium, Biotechnological approaches

Mycorrhizal fungi are the integral components of the forest ecosystem showing mutual dependency between them and higher symbionts for natural functioning. It is now a well established fact that adequate mycorrhizal development on seedlings is critical for the success of plantations on harsh sites with low native populations of mycorrhizal fungi. In forestry, mycorrhizal inoculation has two purposes: (i) introduction of mycorrhizal fungi into soils where they are absent, and (ii) replacement of the naturally occurring fungi by more efficient species.

Genetic diversity among ectomycorrhizal fungi affects mycorrhizal development. Therefore, selection of suitable strains of mycorrhizal fungi are very important for out-planting success. It is quite possible that a strain which is effective in one region may fail to give similar results in another region. Marx *et al.* (1984) reported that an isolate of *Pisolithus tinctorius, which* originated from the South and Eastern regions, was very effective in those regions but did poorly in North-western nurseries in the United States. In contrast, the North-West regions, encouraging results have been reported with an ectomycorrhizal fungus, *Laccaria laccata,* which performed well on container grown seedlings (Molina and Chamard, 1983) and in bare-root nurseries (Molina and Trappe, 1984).

These findings lend support to the contention that fungal species adapted to particular regions and habitats, should be selected for use in those regions for desired results. There exists considerable symbiotic variations in *P. tinctorius* and the other ectomycorrhizal fungal species. Marx (1981) has suggested that several isolates representing different sources of potential variation should be examined prior to evaluation of the symbiotic potential of any fungal species. Besides being symbiotically efficient, the ectomycorrhizal (ECM) fungi should be a good competitor so that it may survive and dominate the out-planting site.

Marx (1980) and Garbaye (1982) have suggested the following criteria evaluate for the competitiveness of an ECM fungus for its field applications: (i) it should prevent the development of poor competitors on inoculated seedlings in the nursery, (ii) it should be able to survive on inoculated seedlings when out-planted, (iii) it should act as a pioneer or dominant fungus and progressively colonise the newly formed root tips on all parts of the root system, and lastly, (iv) it should be more diversified and have a greater colonising ability than the native mycobionts.

One of the most outstanding achievements of scientists of the Forest Research Institute (Dehradun) is the establishment of mycorrhizal banks in different parts of the country like Burnihat (Assam), Yachuli (Arunachal Pradesh), Palni Hills (Tamil Nadu), Vishakhapattnam and East Godavari District (Andhra Pradesh) and Bastar (M.P). The mycorrhizal banks which serve as store houses of mycorrhizal fungi were established by introducing mycorrhizae in nurseries. The purpose of these banks is to cater to the needs of plantation forestry with regards to mycorrhizal inoculum for tropical pines in the respective regions. These banks are, in fact, instrumental in the success of large-scale pine plantations raised in these states.

Improvement in performance of tree seedlings due to *P. tinctorius* ectomycorrhiza has been reported on routine reforestation sites in many states in the U.S.A. on a boreal forest site in Canada (Navratil *et al.*, 1981), on afforestation sites in Nigeria (Ekwebelam and Odeyinde, 1985), Congo

(Delwaulle *et al.*, 1982), Brazil (Krugner and Filho, 1930), Mexico (Valdes, 1986), Liberia (Marx *et al.*, 1985) and Philippines (De La Cruz, 1987).

In the light of tremendous success achieved in re-vegetating wastelands in the U.S.A., there is no doubt that *P. tinctorius* is a candidate fungus for practical use in the aforestation of adverse sites. The *P. tinctorius* ectomycorrhizal inoculation can be obtained from Mycorr. Tech., Inc., University of Pittsburgh, Pennsylvania (U.S.A). In addition to this fungus, species of *Hebeloma* sp., *Laccaria laccata, Scleroderma aurantium* and *S. bovista* are also currently available for inoculation purposes. Spore pellets, spore-encapsulated seeds and bulk spores of *P. tinctorius* are available from the International Forest Tree Seed Company, Odenville (Alabama) and South Pine, Inc., Birminghal (Alabama).

Successful forest plantation sometimes requires deliberate introduction of ectomycorrhizal fungi (Smith and Read, 1997). Hence, such fungi are used as inoculants (Le Tacon *et al.*, 1992). A research programme has been developed in France over the last 15 years for improving the growth of Douglas fir (*Pseudotsuga menziesii*) by inoculation of strains of *Laccaria bicolor* (Le Tacon *et al.*, 1992). This fungus is distributed worldwide and is common in both forest nurseries (Henrion *et al.*, 1994) and mature forest strands. Soil growing dikaryotic mycelia are regarded as the main vegetative stage, forming mycorrhiza with tree roots and giving rise to fruit-bodies.

MYCORRHIZAL INOCULANTS

There are a variety of sources of mycorrhizal inoculum. They vary from commercial suppliers to semi-commercial producers or research organisations that can supply limited amounts of fungus from experimental stocks. In 1982, Marx and his colleagues summarised some of the factors to be considered in producing an inoculum of ectomycorrhizal fungi or VAM fungi, respectively. An inoculum can come in a range of forms, such as soil, spore preparations, dried mycelium, infected roots or more sophisticated products, such as alginate beads with incorporated mycelial homogenates. Some ectomycorrhizal fungi also produce small sclerotia (Grenville *et al.*, 1985), but the usefulness of these as inocula has not been assessed in detail. Most of the commercial interest in mycorrhizal inoculants which has developed over the past 20 years, mostly involves adaptation of existing techniques for large-scale production. Relatively few patent applications have been made, most of them involving the particular growth substrates, modification of established methodology or seed pelletisation. Besides the direct marketing of inoculum, an aspect that has not been well developed, is the use of mycorrhizal fungi in industries that produce plants for resale.

The plants would then be marketed as having improved survival characteristics and being superior (and possibly more natural) to their non-mycorrhizal equivalents. Various factors would have to be considered when determining whether the production of such plants is economically viable, including the cost of initial research and development, the cost of inoculation, the size of the market, the post sale performance and of course the increased revenue that could be recovered from such products. A good marketing strategy would be needed, as the potential for this area is large.

Commercial Preparation of Ectomycorrhizal Inoculants

Introduction of large quantities of soil from foreign sources can offer the cheapest form of introducing non-indigenous species to some ecosystems and it is widely practiced at present for the introduction of ectomycorrhizal species (Mikola, 1980). A purer and more consistent form of inoculum is the introduction of macerated mycelium of laboratory grown cultures. This approach is time-consuming and requires considerable space.

The final preparation of the inoculum from flask-grown batch cultures is inconvenient and labour-intensive. If the expertise and facilities are available for this approach, it is more feasible to opt for batch fermentation of larger quantities of submerged mycelium under optimised conditions. This can be followed by a systematic and partially mechanised method of product extraction and packaging into an appropriate delivery medium. Providing that the costs can be kept low, this could be done in a large commercial operation, as an alternative means of using the established fermentation facilities.

The most ambitious attempt for the production of commercial mycorrhizal inoculum arose with the involvement of the Abbott Laboratories with Marx of the USDA Forest Service after 5 years of intensive research that produced a commercial product of *Pisolithus tinctorius* (Pt) known as **MycoRhiz**. Marx has been instrumental in showing the potential benefits to be derived in commercial forestry from the inoculation of pines with the ectomycorrhizal fungus Pt. A Biotech company, Mycorr Tech. Inc. (MTI) was formed by one of the researchers from Sylan, which concentrates on the Pt product. MTI produced vegetative mycorrhizal inoculum of Marx's isolate of Pt in March 1988.

In the U.K. Biotal limited of Cardiff was ectomycorrhizal products. They supply mixed inocula of four mycorrhizal fungi: *Thelephora terrestris, Laccaria laccata, Paxillus involutus* and *Hebeloma crustuliniforme*. The inoculum is formulated in alginate beads, approximately 1 mm in diameter, each containing a total of 10 colony forming units of the four fungi. The inoculum has a shelf-life of 2 months, if stored at 4°C.

In Europe, the much progress has been made in France through the work of the Instiut National de la Recherche Agronomique (INRA). A conventional peat-vermiculite inoculum of mycorrhizal fungi, *L. laccata, P. involutus* and *H. crustuliniforme*, grown in plastic bags has been developed by a firm called, Somycel. Secondly, INRA has collaborated with Rhöne-Pöulenæ in the production of an inoculum of *Hebeloma* species, and has formulated the mycelium in alginate beads for delivery to the plant.

In Australia, the development of an alginate-based inoculum is currently underway in a joint project between the University of Western Australia and CSIRO. Batch fermented starter cultures are homogenised and incorporated into hydrogel beads before being incorporated into an air-lifted fermenter for final production by Interbac Aust. Pvt. Ltd that produces **MycoBeads** (fungal mycelia immobilised in hydrogel beads) of *L. laccata, Descolea maculata, Hebeloma* species etc. Pallet production of fungal inoculum is done by Philippine Agricultural Biotechnology Company (PABCO) in collaboration with the Biotech Centre of the University of the Phillipines at Los Banos, marketed under the trade name, **MycoGroe**. Some additional suppliers also produce spore-based inoculum of Pt. The International Tree Seed Company and South Pine Inc, both from Alabama, produce spore pellets, spore encapsulated seeds or bulk spores. Several other commercial mycorrhizal preparations such as **EndoNet, MycorisePro, BioterraPlus** etc. are available in the market.

In India, the synthesis of mycorrhizae with pure cultures of fungi has been attempted, and success has been achieved with *Scleroderma geaster, Corticium bicolor, Scleroderma verrucosum* (Bakshi, 1974). Sharma and Mishra (1985) have also successfully synthesised mycorrhizae in Khasi pine with pure cultures of *Boletus edulis, H. crustioides, L. laccata, Lactarius* species, *P. tinctorius, S. citrinum, Russula* species, under aseptic conditions. They reported that the seedlings inoculated with the mycelium of *S. citrinum* and with natural inocula showed an earlier development of mycorrhizae than the seedlings receiving the inocula of other forms. Further, they observed smaller length of mycorrhizae on the plants inoculated with *S. citrinum* than with other fungi. The total number of mycorrhizal roots was greater than the others. This led to an increased absorptive root surface area and provided a greater nutrient exploitation zone to the mycelium of the fungus, resulting in better growth of seedlings than the other fungi.

Commercial Preparation of Vesicular Arbuscular Mycorrhizal Inoculants

Progress in the development of an inoculum of VA-mycorrhizal fungi has been severely hampered by the inability to grow any of the mycobiants in axenic culture. All culture systems used so far, involve the use of a dual culture of suitable host plant roots and the appropriate fungus.

In 1988, an Asian Conference on Mycorrhiza (ACOM) was held in India which was useful in initiating the exchange of information in East, but far more needs to be done in Africa and Latin America. Much of the impetus for the use of VA-mycorrhizal inoculants in agriculture has come from the work of Menge *et al.* (1978) at the University of California in relation to citriculture. Since citrus trees are obligatorily dependent on good VAM development for growth, a solution was needed to alleviate the symptoms of nutrient-deficiency, when citrus trees were reintroduced into sterilised soils. The addition of VAM inoculum resulted in low infection, so the problem was wide spread in the stock. The operation has now been taken over by Brokaw Nursery, Inc., which has been supplying inoculum of *Glomus deserticola* since 1980. Its inoculum is produced by using conventional dual pot cultures (Menge *et al.*, 1978).

This approach has been taken up by Native Plants Inc. (NPI) located in Utah. They market VAM inoculum for use in horticultural industry under the trade, name **NutriLink**. NPI also sells single species material containing other VAM strains for research purposes. These include *G. deserticola, G. etunicatum, G. mosseae* and *Gigaspora margarita.*

Although not directly aimed at a commercial market, an important development has been the setting up of an International Culture Collection of VA-Mycorrhizal Fungi (INVAM) at the University of Florida. INVAM has been set up on an international basis with the following objectives: (i) to preserve germplasm of VA-mycorrhizal fungi, (ii) to develop improved procedures for germplasm of VA mycorrhizal fungi, (iii) to serve as a centre for personnel training on the culture and identification of VA-mycorrhizal fungi, and (iv) to provide scientists working on mycorrhiza with a reliable source of starter cultures of these fungi for research purposes. INVAM began its operation in the late 1985, and currently it holds stocks of over 600 isolates, representing 60 of the 120 described species of VAM fungi.

Standard pot culture methods were first used and modified, but more recently Nutrient Film Techniques (NFT) have been developed and patented by Mosse and Thompson (1983). The construction and development of this system, using *Phaseolus vulgaris* or *Zea mays* as the main plant hosts, has also been described by Elmes *et al.* (1984). Inoculum of *Glomus mosseae, G. fasciculatum, G. caledonium* and *Acaulospora laevis* has been successfully produced by this approach. Bean roots grown by this technique and infected with both *Glomus fasciculatum* and *Rhizobium phaseoli* were successfully used as an inoculum for red clover growing in non-sterile field soil.

A novel and interesting use of VAM in the biotechnological industry was developed by Twyford Plant Laboratories (TPL) in the U.K. This company specialises in the production of cloned and tissue cultured plantlets under strictly controlled and clean conditions. TPL has been producing cloned plantlets of date-palm for several years and have now devised appropriate inoculation protocols to establish VAM on their products. Four isolates of VA-mycorrhizal fungi *Glomus mosseae, G. monosporum, G. intraradices* and *G. aggregatum* were obtained from soils of neutral pH in the Middle East Asia, and California, which had similar characteristics to those soils into which the date-palm would eventually be out-planted. In glasshouse experiments, each of the isolates significantly increased the growth of the date-palm plantlets after 8 months in a sandy substrate, as compared with non-inoculated control plants.

Commercial Preparation of Ericaeous Mycorrhizal Inoculants

Although VA-mycorrhiza and ectomycorrhiza have contributed to most of the commercial interest in inoculum production, some progress has also been made in production of ericaceous mycorrhiza in New Zealand for the inoculation of established plants or rooted cuttings of blueberry (*Vaccinium corymbosum*) to increase fruit yield. The inoculum is produced by MiniTech Limited (NZ) and marketed under the name, **MycoAid.** It comprises homogenised and dried mycelial preparation of *Pezizella ericae*. Initial field trials carried out in mycorrhizal deficient soils from 1971 to 1981 showed a significant increase (25% to 40%) in fruit yield.

MOLECULAR APPROACHES USED FOR MONITORING OF INTRODUCED MYCORRHIZAL FUNGI

In France, strain S238N of *L. bicolor* is used commercially to inoculate Douglas-fir in nurseries and forest plantations. This strain increased the total volume of wood produced by 60% (Selosse *et al.*, 2000). Unfortunately, morphological identification of fruiting bodies and mycorrhiza is not sufficient to distinguish *Laccaria* genets (genetic individuals); hence molecular tools are required for such differentiation. Survival of S239N strain has been studied in nurseries and in Douglas fir plantations, using PCR-based techniques (Henrion *et al.*, 1994). Molecular typing of *Laccaria* sporophores with Random Amplified Polymorphic DNA (RAPD) markers confirmed that this inoculant strain was still present and dominant for 14 years after inoculation (Selosse *et al.*, 1998).

The electrophoresis analysis of isozymes, extensively used in studies of plant genetics, has been of great importance in fungal taxonomy and population genetics. This approach is now being applied for similar studies of mycorrhizal fungi. Variations in isozymes mobilities of six enzymes from the resting spores of vesicular arbuscular mycorrhizal fungi has enabled the characterisation of six species of *Glomus* (Sen and Hepper, 1986) and the biochemical detection of intraspecific variations of similar isolates of various species of *Glomus*. In ectomycorrhizal studies, Ho and Trappe (1987) showed that isolates representing six species and three host-related sections of *Rhizopogon* sp. had acid phosphatase isozyme mobilities that differ greatly between sections, but were more conserved within species. A similar analysis of *L. laccata* enabled the division of isolates into three host-related groups (Ho, 1987).

Maintenance and stress resistance of beech and oak plantlets is considerably improved when seedlings are inoculated in the nursery with *Paxillus involutes,* but the success of mycorrhization is strongly related to fungal isolates. To determine the success of individual *P. involutes* isolates used in inoculation, molecular characterisation was carried out and used to survey the mycorrhizae and fruit bodies obtained in the nursery. Persistency of isolates was also surveyed on mycorrhizae in roots, 1 and 2 years after planting the seedlings to the field. Molecular characters for particular surveys of the isolates were the length of the PCR product using primers CLN 12/5SA for amplification and the restriction fragment length polymorphism of the intergenic spacer (IGS1) digested by *Hae*III. Sequencing of the IGSI confirmed the length polymorphism, but did not allow further discrimination of the isolates. Differences in IGSI sequences between isolates increased with larger geographical distance of isolate origin, but genetically closely related isolates could not be separated. The molecular survey indicated that the *P. involutes* inoculants were not necessarily those on the mycorrhizae, but some were more competitive than the others.

Only the most efficient isolates were found to form mycorrhizae with seedlings in the nursery when several isolates were mixed for inoculation. An Australian *P. involutes* isolate failed in mycorrhization of beech and oak in the nursery and the seedlings became mycorrhizal by another isolate. Fruit-body production in the nursery could be linked to the inoculated isolates. The isolate remained associated with the inoculated beech and oak for 2 years after out-planting to a forest area and arable land. The molecular survey thus helped to select efficient isolates of *P. involutes* for forestry application.

USDA (Beltsville) has also been successful in raising a monoclonal antibody that distinguishes a *G. mosseae* isolate from other species and is now testing cross-reactivity within the species complex (P.D. Millner, Personal Communication). These techniques will have value in the selection of the most appropriate fungi as inoculants, especially in those cases where successful studies for competition between different fungi on the same root systems have been accomplished. It is clear that the major mycorrhizal fungus in terms of root occupancy does not necessarily have a major effect on plant growth. Thus, an inoculant fungus established at the nursery stage or placed beneath the seed in direct drilled crops, can have determining effects on plant growth which are not necessarily related to its subsequent spread or its ability to become the main mycorrhizal symbiont as the plant matures.

DNA-mediated Transformation for the Improvement of Isolates of Mycorrhizal Fungi

The genetically altered fungi provide more benefits to the plant and offer a great potential to improve the establishment and yield of many crop plants. Certainly, genetic manipulation of mycorrhizal fungi, by involving genes for agronomically important traits or for improved symbiosis, has the potential to improve agricultural and forest production worldwide. DNA-mediated transformation extends the potential for genetic manipulation, even among established genetic systems, and provides access to genetic experimentation in fungi less amenable to study by conventional genetic analysis. Transgenic introduction of genes and/or their control mechanisms has proven to be important in the current research, involving several filamentous fungi (Leong and Berka, 1991). There are presently a variety of transformation/selection systems available and in theory; such systems are applicable to any fungus that can be grown *in vitro*.

Fungal transformation typically involves the use of regenerative protoplasts, identification of selectable marker genes, and stable integration of the transforming DNA into the chromosome of the organism (Ballance, 1991). Vectors developed for fungal transformation are typically unable to replicate autonomously in the target fungus, since the transformation event is expected to be integrative. Once integrated, this transforming DNA is stably inherited as a chromosomal or genomic insert.

Various fungal species have been transformed, following incubation of protoplasts of desired fungus with a DNA insert that encodes genes for selected traits, such as required metabolite synthesis (Yelton *et at*, 1984), antibiotic or pesticide resistance (Henson *et at*, 1989), or utilisation of some novel nitrogen source eg acetamide (Geisen and Leistner, 1989). A number of basidiomycetes have been transformed using genes that complement auxotrophic mutations. Tryptophan-requiring auxotrophs, of both *Coprinus cinereus* and *Schizophyllum commune*, were transformed to prototrophy using homologous *trpC* genes and regulator signals (Munroz-Rivas *et at*, 1986). An adenine auxotroph of *Phanerochaete chrysosporium* was transformed to an adenine biosynthetic enzyme gene from *S. commune* (Alic *et al.*, 1989).

Some progress has been made recently in ectomycorrhizal fungi, with the development of protoplast fusion techniques for the isolation of mutants (Hebraud and Fevre, 1988). This method has also been used for the characterisation of plasmids of *Morchella* species as a prerequisite for vector development (Meninhart and Essar, 1987). There is considerable scope in the future for the development of super-strains as inoculants in the field, provided the current difficulties in releasing genetically altered organisms into the environment are overcome and the influence of the host genotype on the development of mycorrhizae within the root system. Different cultivars of pearl millets have to become colonised to differing degrees when faced with a similar mycorrhizal challenge. Genetic differences between cultivars of leafless peas have also been shown by Estuan *et al.* (1987) to markedly affect the development of the mycorrhizal symbiosis. There is a consequent need to consider host genotype-dependent differences in the mycorrhizal symbiosis in current programmes of plant breeding.

Transformants of the basidiomycetes yeast, *Rhodosporidium torutoides*, were selected for the ability to use phenylalanine as the sole nitrogen source by virtue of the expression of the homologous phenylalanine ammonia-lyase gene (Tully and Gilbert, 1985). More recently, an ectomycorrhizal basidiomycete, *L. laccata* has been transformed to the aminoglycoside antibiotic hygromycin B with a bacterial-derived gene conferring resistance (Barrett *et al.*, 1990). This result provides the first evidence for transformation in a mycorrhizal fungus.

These studies on the transformation of basidiomycetes illustrate several features typical of fungal transformation systems.The transforming DNA is integrated into high molecular weight chromosomal DNA and maintained for stability in the fungal genome. There is some inter-specific expression of genes, as *P. chrysosporium* will express the *S. commune ade* gene (Alic *et al.*, 1989). This represents a conservative example of heterologous gene expression involving fungi that belong to the same taxonomic class. The successful transformation of *L. laccata* was done by employing promoter and terminator signals derived from an unrelated fungus, *Aspergillus nidulans.* This result is especially fortuitous as it indicates that control signals of ascomycetous origin, and perhaps other eukaryotic sources are able to function in a basidiomycete.

Current research with ectomycorrhizal fungi involves the use of protoplasts as a starting material for genetic manipulation through DNA-mediated transformation (Barrett *et al.*, 1990). Transformation procedures are expected to extend more traditional approaches to genetic study and thus, enhance the potential for selective breeding and strain improvement of these fungi. Through the use of protoplasts and procedures of molecular genetics, it should be possible to decipher the genetic behaviour of ectomycorrhiza-formation, and to identify traits which affect symbiosis. The goal of such research is to develop improved fungal strains for use as inocula in sylvicultural practices related to tree seedling production, forest productivity and afforestation.

Introduction of novel genes or the elimination of unwanted genes in the ectomycorrhizal fungi, by transformation, is an area of research that promises to provide genetically superior fungal inocula for improved practices, for forest nursery production and / or tree seedling out-planting. To accomplish this goal it will be necessary to obtain fundamental information on relevant genes and their expression, in order to carry out genetic manipulation leading to improved ectomycorrhizal fungi. The success of this approach will depend on interdisciplinary research efforts and not just upon molecular genetic manipulations in the laboratory.

In considering control strategies to improve symbiosis, it will be important to avoid non-target effects that could disrupt the balance among rhizosphere components. The overall intention of the research is to optimise symbiosis through genetic manipulation of the fungal component, without impairment of the components of rhizosphere ecology, that may be supportive of that symbiosis. Regardless of the specific gene(s) introduced and perhaps in some cases, impaired by transformation, transformed phenotypes of ectomycorrhizal fungi should be stable due to integration of the foreign DNA into the fungal chromosome. This stability adds to the desirability of developing transgenic fungi as inocula to improve symbiosis in order to ameliorate stress factors that adversely affect plants.

The preparation of transgenic inocula in sufficient quantity and of high quality, and a careful assessment of ectomycorrhizal formation following inoculation, should also be observed as part of an evaluation for overall benefits and aspects of risk assessments, as related to the release of genetically engineered organisms to the environment.

Limitations in Conventional Genetic Study

The extent of mutualistic interdependence between a mycorrhizal fungus and plant root system, on overall benefit to both partners, limits the ability to study the fungus as an experimental or genetic system. The endomycorrhiza or vesicular-arbuscular mycorrhiza involves a limited number of species of zygomycetous fungi. These fungi do sporulate during vegetative or mitotic growth, often in profusion, but they are developmentally dependent biotrophs with no known sexual or mitotic spore states. The vesicular arbuscular mycorrhizal fungi, as obligate symbionts with no known sexual phase, thus offer little or no opportunity for conventional or other genetic investigation. By contrast, the ectomycorrhizal fungi collectively involve more than 1,000 species, of mainly basidiomycetous and ascomycetous fungi, mostly with an ability to grow saprophytically.

Many ectomycorrhizal fungi can thus grow vegetatively, often quite well, in pure culture, but in general they do not sporulate in the vegetative state and are rarely able to complete a sexual cycle under standard laboratory conditions. Their sexual or meiotic spores, when produced, germinate asynchronously and often at low frequency (Kropp, 1988). Such intrinsic limitations for sexual competence are the principal reasons for ectomycorrhizal fungi not being examined extensively through conventional genetic methods. To date, only very few ectomycorrhizal basidiomycetes species, principally from the genera *Laccaria* and *Hebeloma,* have been examined critically through breeding experiments.

Various Methods Employed in Genetic Fields

Protoplast Fusion

Protoplasts have been liberated *in vitro* from a wide variety of fungi, including several ectomycorrhizal basidiomycetes (Barrett *et al.,* 1989). The absence of mitotic spores from ectomycorrhizal species requires mycelial fragments that serve as a starting material for protoplast formation with these fungi.

Commercially available enzyme mixtures, such as Novozyme 234 or Caylase C3, are generally effective for protoplast generation of fungi. These lytic enzyme preparations principally include β-1,3-and β-1,4-glucanases, α-1,3-and α-1,4-glucanases, as well as chitinase and glucuronidase activities. The preparations may also contain proteinase, detrimental to the stability of liberated protoplasts.

Optimum conditions for protoplast formation and maintenance involve enzyme concentration/exposure time, culture age, pH conditions of buffers/media, all of which vary with individual species. Once the cell wall is removed, an osmotic stabiliser such as 0.6 M KCl or 1.2 M sorbitol or 0.5 M mannitol, must be present in maintenance regeneration media, at least until some re-synthesis of cell wall material has occurred (Peberdy and Ferenczy, 1985). Regeneration rates also vary and among ectomycorrhizal fungi they are low, often only 1–5%, with yields in the order of 10^7 viable protoplast regeneration per gram fresh weight of mycelia (Barrett *et al.*, 1989).

The ability to form and regenerate large number (less than 10^6) of protoplasts is an incentive for attempting transformation with any fungus, provided a suitable vector employing a selectable marker gene is available. A few species of the genera *Laccaria* and *Hebeloma* are the only ectomycorrhizal fungi among several evaluated for protoplast yield/regeneration with a potential for such experimentation (Kropp and Fortin, 1986; Barrett *et al.*, 1990). The only vectors presently available for such experimentation, however, are those developed for studies involving other filamentous fungi, primarly ascomycetous species.

Representative Vectors

The vector pAN7 -1, developed originally for studies with *Aspergillus nidulans* is representative of the type of vector used in fungal transformation studies (Punt *et al.*, 1987). This vector was adopted by Barrett and co-workers (1990) for the successful transformation of *L. laccata.* Transformation was based on positive selection for resistance to Hygromycin B (HmB), using the *Escherichia coli* aminocyclitol phosphotransferase *hpt* gene bracketed by an *Aspergillus nidulans* glyceraldehydes-3-phosphate dehydrogenase *gpd* promoter and the transcription terminator region of *A. nidulans* tryptophan synthetase *trpC* gene.

The pAN7-1 is 6.7 kb in size, and in addition to the *hpt* gene and adjacent control signals, it contains sequences derived from the bacterial plasmid pBR322. The latter sequences are required for replication and recovery of pAN7-1 DNA from *E. coli* clones and includes a replication *ori* of bacterial origin, and a selectable marker gene for ampicillin amp^r resistance (Maniatis *et al.*, 1982). The pAN7-1 DNA obtained from *E. coli* clones is normally purified by caesium chloride/ethidium bromide density gradient centrifugation for use in transformation experiments.

The successful application of pAN7-1 in *L. laccata* transformation encourages the development of comparable vectors, either with selectable marker genes or with reporter genes or genes for still other traits. Reporter genes that can be easily and quantitatively assayed for function could be effectively adopted to study promoter specificity/activity in transformed *L. laccata.* Two reporter genes of promise in this regard are the *E. coli uidA* gene, coding for β-glucuronidase (GUS) activity (Roberts *et al.*, 1989), and the bacterial-derived A-B *lux* fusion gene, coding for luciferase activity. Such genes might be introduced onto a vector in tandem with a selectable marker genes or placed upon a separate vector, the latter suitable for a co-transformation experiment involving a second vector carrying a dominant selectable marker gene.

Other Procedures

One strategy used recently for transformation with fungal systems is 'electroporation' (Goldman *et al.*, 1990). Electroporation is the non-destructive permeabilisation of biological membranes for short-duration in high-amplitude electric fields. This technique may provide a method of introducing DNA into species of fungi not amenable to, or inefficient for transformation, by the traditional calcium ion and polyethylene glycol (PEG) treatments. Yet another technique is a ballistic process involving DNA bound to a micro-projectile (Sanford, 1988), which can introduce DNA into cells by high velocity propulsion in a partial vacuum. This procedure has been effective in transformation studies involving mammalian cells, as well as intact plant cells, and might be especially useful if adapted to fungal systems, where cell wall removal and efficient protoplast generation have not been successful.

Prospects of Molecular Genetic Manipulation

Results with the *L. laccata* system for transformation encourages research to transform this fungus, as well as other ectomycorrhizal fungi with specific useful genes, for improving their symbiotic benefits to plants. The intention of such research would be to introduce useful genes into the root-associated fungus, rather than into the host plant genome. It is anticipated that relevant gene expression at the level of the fungus, on the surface of the root, will exert more direct influence on rhizoplane activities. Since the growth of the mycorrhizal fungus extends beyond the immediate root zone, it may be possible to influence the total rhizosphere and surrounding soil, generally through transgenic functions introduced into the fungus. Candidate genes from plant or microbial sources for these transformation studies may involve genes encoding production of phytohormones that stimulate root and mycorrhizal development, genes encoding metabolites that foster osmo-tolerance or disease resistance, genes for substances repellent to insects or nematode predation, genes for resistance to pesticides, or genes encoding enzymes such as phosphatase or nitrate reductase that influence nutrient availability or alleviate various edaphic stress factors.

Transgenic fungal inocula may effectively substitute for fertilisers and/or pesticides during forest seedling-production, providing a biotechnological alternative to expensive chemical treatment in the control of biotic and abiotic stress phenomenon.

In addition to engineering novel gene activities into an ectomycorrhizal fungus, it should be possible to disrupt undesirable genes resident in the fungal genome by transformation. Procedurally, this has been accomplished with *A. nidulans* by homologous recombination, using a defective gene or one that is interrupted by some selectable marker gene insert (Miller *et al.*, 1985).

Field performance of seedlings tailored with specific ectomycorrhizae, other than Pt, has been reported during the last one decade and the results show that these fungi are promising for eventual practical applications. *Scleroderma* sp. improved growth of *Pinus* and *Eucalyptus* in mixture with Pt in Philippines (De La Cruz, 1987), *S. aurantium* improved growth of *Quercus* (Beckjord and McIntosh, 1984). *Scleroderma bovista* improved growth of *Q. leucotricophora* (Yadav, 2002) and *Cedrus deodara* seedlings (Bhatt, 2002). *Rhizopogon vinicolor* triggered growth of Douglas-fir seedlings (Castellano and Trappe, 1985), *Suillus granulatus* stimulated growth of *Pinus* and *Picea* spp. (Grossnickle and Reid, 1982), *Cenococcum geophilum* improved growth of *Tsuga heterophylla* (Kropp *et al.*, 1985) and *Quercus rubra* (Beckjord and McIntosh, 1984), *Thelophora terrestris* and *Laccaria laccata* improved growth of *Picea sitchensis* (Thomas and Jackson, 1983) and *T. terrestris* improved growth of *Pinus taeda* (Ruehle, 1982).

In France, ectomycorrhizal inoculation is aimed at improving field performance in re-forestation, and increasing production of edible mycorrhizal fungi. Seedlings of Douglas-fir, Norway-spruce and Scot-pine form abundant mycorrhizae when inoculated with vegetative inocula of *L. laccata, L. bicolor* and *Hebeloma crustuliniforme.* About 95% of the seedlings in France are grown in bare root nurseries, where soil fumigation is common.

Douglas-fir seedlings usually take 3–4 years to attains transplantable size. The plantable Douglas-fir seedlings can be produced in fumigated soil after 2 years, following inoculation with *L. laccata.* Truffle production in France has been increased considerably through artificial inoculation of *Quercus* species with *Tuber melanosporum* or *T. uncinatun* vegetative inocula (Le Tacon *et al.*, 1988). Similarly, the production of another edible mushroom *Suillus granulatus* has been increased by artificial inoculation of *Pinus pinaster* seedlings and subsequent out-planting. A comparable difference in root and shoot length, and collar diameter was observed in comparison to control seedlings.

CONCEPT OF MICROBIAL CONSORTIUM

The recent findings of Duponnois and Garbaye (1991) on Mycorrhization Helper Bacteria (MHB) are of considerable interest and have brightened the prospects of utilising MHB as a tool in conjunction with mycobionts to enhance mycorrhization of seedlings in the nursery. They have reported the stimulatory effect of dual inoculation of Douglas-fir seedlings on mycorrhization with *L. laccata* and MHB *Bacillus amyloliquefaciens* (MB3), *Pseudomonas* sp. (BBcl), *Pseudomonas fluorescens* (BBc6), *Pseudomonas* sp. (BBc5) and *Bacillus* sp. (SHB1). Out of 5 MHBs, the isolate MB3 was most efficient, as the mycorrhization rate at the end of the growing season increased from 60% to 88% in Peyrat-lé-Chateau and from 83% to 99% in Champenoux.

Bhatt (2002) and Yadav (2002) isolated an ectomycorrhizal fungus *S. bovista* associated with deodar and oak forests and used its vegetative inoculum to observe the effect of *S. bovista,* co-inoculated with mycorhizosphere bacteria, on deodar and oak seedlings, which are predominant canopy of the hilly area of Uttranchal.

S. bovista showed better results when it was mixed with mycorrhizosphere isolated bacteria as it resulted in increasing seedling growth and biomass yield. The infection rate in the roots of deodar and oak seedlings with *S. bovista,* showed the helping nature of these two microorganisms. Although the initial stages showed some fluctuations, after some time, astonishing results in plant growth were observed as they posed a synergistic effect on growth of both plants (Figure 19.1 and 19.2, Table 19.1).

Table 19.1: Growth of Oak (*Q. leucotrichophra*) Seedlings under Different Inoculation Treatments[a]

Treatment	Growth of Seedlings (cm)								
	Three Months			Six Months			Nine Months		
	Shoot Length	Root Length	Collar Diameter	Shoot Length	Root Length	Collar Diameter	Shoot Length	Root Length	Collar Diameter
Control	6.26	9.56	0.156	8.04	15.72	0.168	11.42	16.50	0.246
EMF	8.12	12.50	0.138	11.16	24.82	0.234	13.92	27.66	0.278
B_1	6.60	11.94	0.150	8.68	16.38	0.174	10.22	13.80	0.228
B_2	6.38	10.88	0.146	8.50	15.08	0.166	12.18	17.38	0.206
EMF+B_1	8.76	14.20	0.146	12.50	25.86	0.248	13.12	29.64	0.296
EMF+B_2	8.58	14.68	0.148	12.10	24.08	0.262	15.26	32.94	0.320
EMF+B_1+B_2	8.28	15.24	0.148	12.02	23.10	0.250	14.74	36.86	0.304
S.E.	0.83	1.45	0.04	1.32	2.15	0.044	1.78	2.60	0.027
C.D. at 5%	1.09	1.90	0.0027	1.73	2.81	0.057	2.33	3.39	0.035
F	8.70[**]	10.43[**]	0.327[ns]	10.78[**]	24.41[**]	4.879[**]	5.21[**]	60.34[**]	12.464[**]

[a]Values are mean of five replicates, EMF=*S. bovista*, B1=*P. fluorescens*
B2= *B. subtilis*
[**]Significant at 1% A NOVA, [ns]not significant

Alternatively, MHB inoculation could be used to obtain the same rate of mycorrhiza formation with a lower dose of fungal inoculum. Moreover, during out-planting, it is an advantage for the introduced fungus to be massively associated with the plant in order to be in a strong competitive position against the other symbiotic fungi at the planting site (Villeneuve *et al.,* 1990). Hence, inoculation with MHB is clearly beneficial to control mycorrhization practice and is likely to contribute greatly to re-forestation and afforestation sites.

CONSTRAINTS ON MYCORRHIZAL RESEARCH

The research in the field of mycorrhiza has led to remarkable achievements, but still there are many problems to be dealt with. It has generally been observed that after transplanting, the seedlings into natural forest soil, the nursery

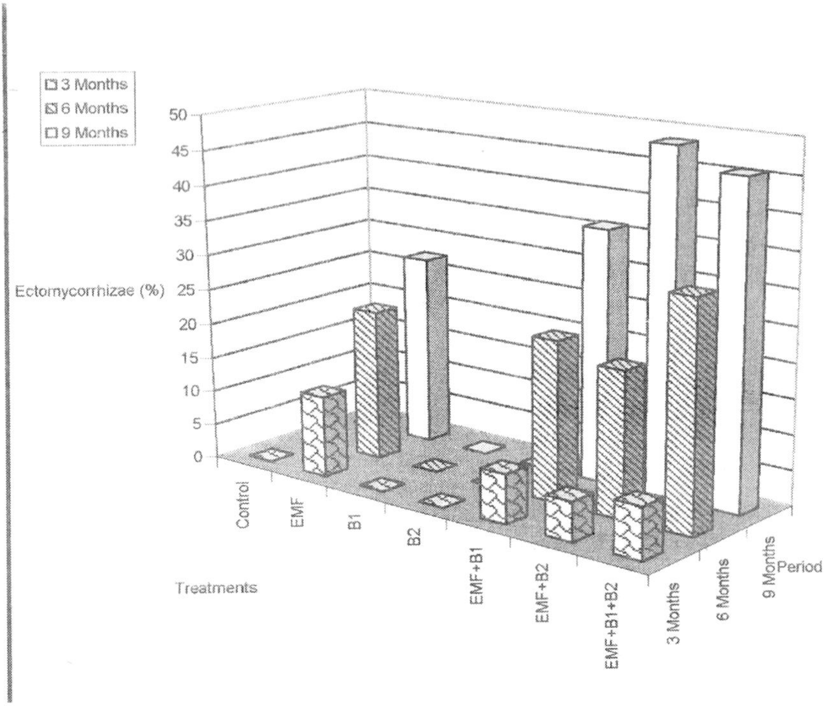

Fig. 19.1. Occrrence of Ectomycorrhizae (%) under Different Treatments on *C. deodara*. EMF=*Scleroderma bovista*, B_1=*Pseudomonas florescens*, B_2= *Bacillus subtilis*

mycorrhiza are replaced by indigenous species of the site. Obviously, this replacement may cause stagnation of the seedlings for some time after transplanting. Introduction of suitable fungi in nurseries is often difficult, as the unfavourable ecological conditions may pose a serious problem for introduced species to compete and survive. Screening the symbiotic efficiency of different species in natural conditions is as important for the success of mycorrhizal inoculation, as developing suitable inoculation techniques.

Another major threat is that the inoculum formulated on application may show good results in one region and may not prove to be effective at another site. Marx and his colleagues (1984) reported that *P. tinctorios*, which originated in South and South-Eastern regions, was very effective in these regions, but showed poor results in the Western nurseries in the United States. So, the fungal strains adapted to particular regions and habitats, should be replaced for use only in those regions so as to get the desired results.

In nurseries, management problems arise and when steps are taken to alleviate those problems, the mycorrhizal populations are most often disrupted. For example, spots of root rot may develop in a nursery because of poor localised drainage. If fungicides are applied to control the root rot, mycorrhizal fungi also may be decimated. The resulting mycorrhizal deficiency is then

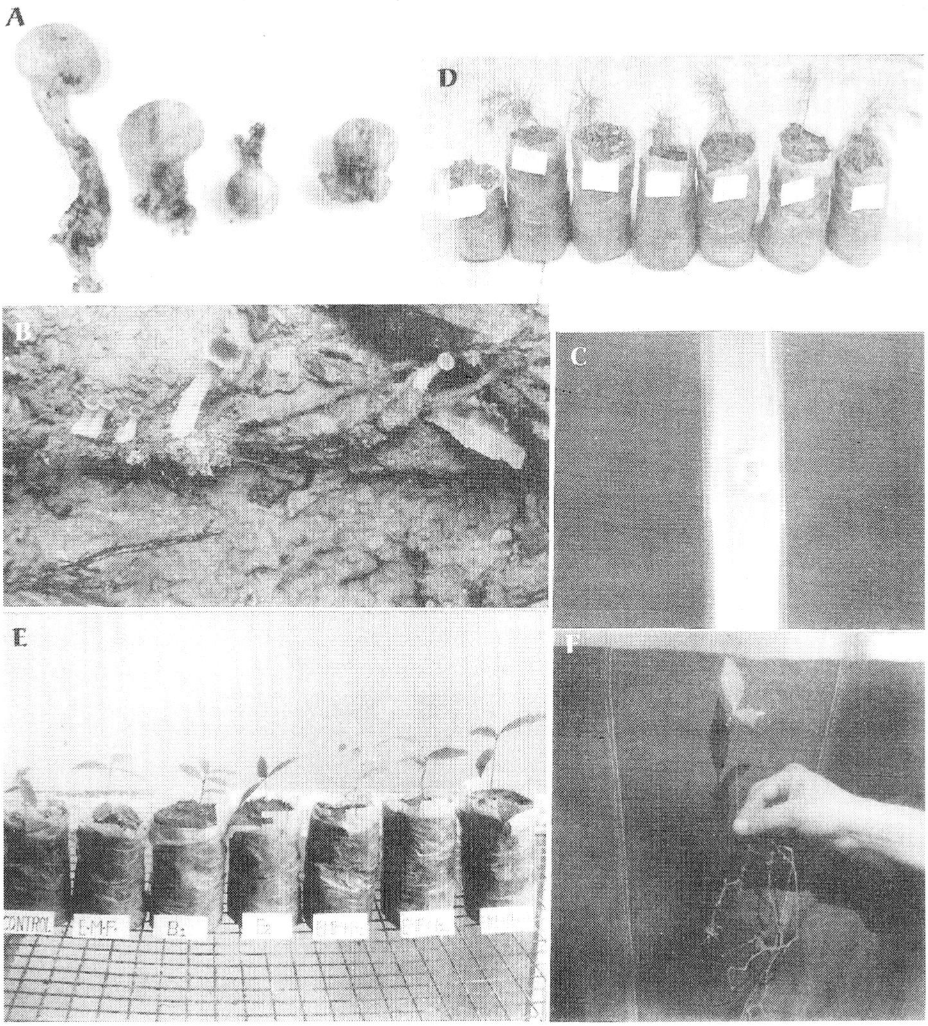

Figure 19.2. A. Sporocarps of *Scleroderma bovista*; B. Fruit-bodies of *Suillus* sp. developing on bare roots; C. Growth of mycelia of *S. bovista* on MMN medium; D. Nursery experiments showing synergistic effect of *S. bovista* and mycorrhizosphere bacteria (*Pseudomonas fluroescens* and *Bacillus subtilis*) on deodar seedlings; E. Same effect on oak seedlings; and F. Uprooted oak seedling showing mycorrhizal roots.

reflected by nutrient deficiency. If that symptom is treated by extra fertilisation, mycorrhiza formation may even further be depressed. Once the nutrient-starved-seedlings stop growing, their root systems are open to attack by other pathogens for lack of protection by mycorrhizal fungi.

As far as VAM fungi are concerned, they are cultured in bulk on the roots of living plants grown either in soil or in a nutrient solution. These systems are crude and expensive, requiring a lot of money and space. Generation periods are long, averaging two to six months. Adequate field testing of VAM inoculation, and commercial exploration of the potential benefits of VAM fungi still rests on the development of a low cost technology for raising and multiplying VAM inoculum.

But what are the criteria to be followed in the selection of the most efficient mycorrhizal fungi? What physiological criteria have to be employed? Could these criteria be achieved by gene transfer techniques, which are now available for application to mycorrhizal fungi?

CONCLUSION

There is great potential for increasing the efficiency of growth of the agricultural crops by mycorrhizal inoculation and management, particularly in semi-arid and tropical zones. Although the role of mycorrhiza is principally the same in these zones, as in temperate climates, these regions present different problems, and management strategies may need alternative approaches to those adopted in developed countries. This concept is not new, but there is a tendency for pessimism or frustration in certain quarters, because increased crop yields have not been rapidly or universally achieved or inoculation with mycorrhizal fungi does not have direct application in well-developed areas of agriculture. The usefulness of mycorrhiza is especially appropriate to the development of sustainable systems of agriculture (Mosse, 1980).

Stress should be laid on the continual need for extensive field trials over a series of seasons before definitive statements can be made about the benefits that mycorrhizal inoculation can bring about. These trials should be targeted into those regions and crops where improvements in the efficiency of agriculture might be expected, irrespective of the likely monetary gains. Mycorrhizal inoculants will never be financially lucrative, but their tremendous value lies in area of low soil fertility. Introduction of effective fungi into such soil must remain a priority, especially in relation to their ability to use limited amounts of phosphorus fertilisers (Gianinazi, 1986).

REFERENCES

Alic, M., Kornegay, J.R., Pribnow, D. and Gold, M.H. 1989 *Appl. Environ. Microbiol.* **55**:406–411.

Backjord, P.R. and McIntosh, M.S. 1984 *Can. J. Bot.* **2**:1571–1574.

Bakshi, B.K. 1974 *Mycorrhiza and its role in Forestry.* P.L.480 Project Report. Forest Research Institute, Dehradun.

Ballance, D.J. 1991 *Molecular Industrial Mycology* (S.A. Leong and M. Berka) 1–29. Marcel Dekker, New York.

Barrett, V., Dixon, R.K. and Lemke, P.A. 1990 *Appl. Microbiol. Biotechnol.,* **33:** 313–316.

Barrett, V., Lemke, P.A. and Dixon, R.K. 1989 *Appl. Microbiol. Biotechnol.* **30**:381–387.

Bhatt, M. 2002 *Ph.D. Thesis,* Gurukul Kangri University, Haridwar.

Castellano, M.A. and Trappe, J.M. 1985 *Can. J. For. Res.,* **15**:613–617.

De La Cruz, R.E. 1987 Terminal Report, National Institute of Biotechcnology. Applied Microbiol. National Oil Co. Philippine.

Delwaulle, J.C., Garbaye, J. and Okombi, G. 1982 *Bios. For. Trop.*, **196:**25–32.

Duponnois, R. and Garbaye, J. 1991 *Plant Soil*, **138:**169–176.

Ekiwebelam, S.A. and Odeyinde, M.A. 1985 In: Proc. 6th NACOM, Oregon, USA.

Elmes, R.P. and Mosse, B. 1984 *Can. J. Bot.*, **62:**1531–1536.

Elmes, R.P., Happer, C.M., Hayman, D.S. and O'Shea, J. 1983 *Ann. Appl. Biol.*, **104:**437–441.

Estaun, V., Calvet, C. and Hayman, D.S. 1987 *Plant Soil*, **103:**295–298.

Garbaye, J. 1982 *Les Mycorrhizae, Partie integrante de la plante; Bilogie et Perspectives d'Utilization.* S. Gianinazzi, V. Gianinazzi Pearson and A. Touvelot (eds) pp 303–312.

Gianinazzi-Pearson, V. 1986. *Fertilisers Agriculture* **92:**3–12.

Goldman, G.H., Van Montagu, M. and Herrera-Estrella, A. 1990 *Curr. Genet.*, **17:**169–174.

Grenville, D.J., Piche, Y. and Peterson, R.L. *Can. J. Microbio1*, **31:**1085–1088.

Grossnickle, S., C. and Reid, C.P.P. 1982 *Can. For. Sci.* **12:**354–361.

Hebraud, M. and Fevre, M. 1988 *Can. J. Microbiol.*, **34:**157–161.

Henrion, B., Di, Battista, C. Bouchard, D., Vairelles, D., Thompson, B., D., Le Tacon, F. and Martin, F. 1994 *Mol. Ecol.*, **3:**571–580.

Henson, J.M., Blake, N.K. and Filgeram, A.L. 1989 *Curr. Genet.*, **14:** 113–118.

Ho, I. 1987 *Can. J. For. Res.*, **17:**855–858.

Ho, I. and Trappe, J.E. 1987 *Mycologia*, **79:**553–558.

Kropp, B.R. 1988 *Canadian Workshop on Mycorrhizae in Forestry.* (eds M. Lalonde and Y. Piche,), 131–133. Universite Laval, Quebec, Canada.

Kropp, B.R. and Fortin, J. A. 1986 *Can. J. Bot.*, **64:**1224–1226.

Kropp, B.R., Castellano, M.A. and Trappe, J.M. 1985 *Tree Planters notes*, **36:**13–16.

Krugner, T.L. and Filho, M.T. 1980 *Piraracicaba*, **21:**41–51.

Le Tacon, F., Alvarez, I.F., Bouchard, D., Henrion, B., Jackson, R.M., Luff, S., Parlde, J.I., Pera, J., Stranstrom, E., Villeneuve, N. and Walker, C. 1992 *Mycorrhizas in ecosystems* (Eds Read D. J., Lewis, D. H., Fitter, A. H. and Alexander, I. J.). CAB Wallingford, pp 119–134.

Le Tacon, F., Garbaye, J. and Bouchard, D. 1988 *Canadian Workshop on Mycorrhizae in Forestry.*, pp. 51–74.

Leong, S.A. and Berka, R.M. (Eds) 1991 *Molecular Industrial Mycology.* Marcel Dekker, New York.

Maniatis, T., Fritsch, E.F. and Sambrook, J. 1982 *Molecular cloning: A Laboratory Manual.* Cold Sring Harbour, New York.

Marx, D.H. 1980 USDA Forest Services, pp 27–28.

Marx, D.H. 1981 *Can. J. Fores. Res.*, **11:**168–174.

Marx, D.H., Cordell, C.E. and Kenny, D.S. 1984 *For. Sci. Monograph.*, **76:**1–101.

Marx. D.H., Ruehle, J.L., Kenney, D.S., Cordell, C.E., Riffle, J.W., Molina, R.J. Pawuk, W.H., Navratil, S., Tinus, R.W. and Goodwin, O.C. 1982 *Forest Sci.*, **28:**373–400.

Meinharat, F. and Eager, K. 1987 *Appl. Microbiol. Biotechnol.*, **27:**276–282.

Menge, J.A., Lambanuska, C.A., Johnson, E.L.V. and Platt, R.G. 1978 *Soil Sci. Soc.*

Mikola, P. 1970 *Int. Rev. For. Res.*, **3:**123–196.

Miler, B.L., Miller, K.Y. and Timberlake, W.E. 1985 *Mol. Cell. Biol.*, **5:**1714–1721.

Molina, R. and Charmard, J. 1983 *Can. J. For. Res.*, **13:**89–95.

Molina, R. and Trappe, J. M. 1982 *For. Sci.*, **28:**425–457.

Mosse, B. 1986 *Biol. Agri. Hort.*, **3:**143–152.

Mosse, B. and Thompson, J. P. 1984 *Can. J. Bot.*, **62:**1523–1530.

Mosse, B. and Thompson, J. P. 1983 UK Patent GB 2043688B.

Munroz-Rivas, A.M., Specht, C.A., Drummond, B.J., Froelinger, E., Novotny, C.P. and Ullrich, R.C. 1986 *Mol. Gen. Genet.*, **205:**103–106.

Navratil, S., Phillips, N.J. and Waynia, A. 1981 *For. Chron.*, **57**:212–217.
Pebery, J.F. and Ferenczy, L. 1985 *Fungal Protoplasts: Applications in Biochemistry and Genetics.* Marcel Dekker, New York.
Punt, P.J., Oliver, R.P., Dangemanase, M.A., Powells, P.H. and Van Hondel, C.A.M.J.J. 1987 *Gene,* **56**:117–124.
Roberts, L.N., Oliver, R.P., Punt, P.J. and Van Hondel, C.A.M.J.J. 1989 *Curr. Genet.,* **15**:177–180.
Ruchle, J.L. 1982 USDA *For. Serv. Gen. Tech. Rep.*, pp 133–137.
Sandford, J.C. 1988 *Trends Biotechnol.,* **6**:299–309.
Selosse, M.A., Bouchard, D., Martin, F. and Le Tacon, F. 2000 *Can. J. For. Res.,* **30**:360–371.
Selosse, M.A., Jacquot, D., Bouchard, D., Martin, F. and Le Tacon, F. 1998 *Mycol. Ecol.* **7**:561–573.
Sen, R. and Hepper, C. M. 1986 *Soil Biol. Biochem.,* **18**:29–34.
Sharma, G.D. and Mishra, R.R. 1988 *Mycorrhiza for Green Asia,* ACOM, Madras, Deptt of botany, NEHU, Shillong., pp 319–331.
Smith, S.E. and Read, D.J. 1997 *Mycorrhizal Symbiosis.* 2d ed Academic, New York.
Thomas, G.W. and Jackson, R.M. 1983 *New Phytol.,* **95**:223–229.
Tully, M. and Gilbert, H. 1985 *Gene,* **36**:235–240.
Valdes, M. 1986 *Can. J. Bot.,* **64**:885–888
Villeneuve, N.F., Le Tacon and Bouchard, 1990 *Plant Soil,* **135**:95–107.
Yadav, A. 2002 *Ph.D. Thesis* Gurukul Kangri University, Haridwar.
Yelton, M.M., Hamer, J.E. and Timberlake, W.E. 1984 In: *Proc. Natl. Acad. Sci.* USA **81**:1470–1474.

20

Mycorrhizae as Biocontrol Agents Against Plant Pathogens: A Review

MANOJ TIWARI, A. TIWARI AND S.C. SATI

Department of Botany, D. S. B. Campus, Kumaun University, Nainital-263 002, Uttaranchal, India

SUMMARY

With a few exceptions, crop plants have mycorrhizae, but the degree of root colonisation by mycorrhizal fungi and the effects of symbiosis may vary, depending on the total interaction between the host, symbionts and the environment. In most cases, mycorrhizae significantly change the physiology and chemical constituents of the host, the pattern of root exudation, and the microbial composition of mycorrhizosphere soil. These changes can greatly influence the growth and health of plants, in part due to the biological suppression of plant diseases. The mechanisms involved are probably multiple, and will depend on the conditions of the test. The major effect of mycorrhiza may be improved nutrition, but secondary effects induced thereafter may contribute significantly to observed effects on diseases.

At present, it is not possible to predict the usefulness of mycorrhizal fungi in disease control at the operational level. However, this is a relatively new field of research as compared to the biological control of diseases on crops. Despite the difficulties associated with mycorrhizal fungi, their large scale use shows potential both as a component of integrated pest management programmes and for their environmental friendliness.

Keywords: Biocontrol, Mycorrhizal symbiosis, Plant pathogens

Plant pathologists have been taking adequate protection measures to combat pests, and to minimise the losses caused to agricultural production due to various pathogens. During the past few decades, emphasis has been placed on the application of chemical control measures. During this period, fungicides have provided an important and economic means of crop protection. Various fungicides have successfully controlled seed rot, root rot and damping-off of many plants. Biological methods could provide an alternative to reduce the use of fungicides. Addition of biological control agents to the growth medium may provide economical protection to plants against diseases, without the use of fungicides. Various microbial antagonists, including strains of the bacterial genera *Bacillus, Pseudomonas* and *Enterobacter* and strains of the fungal genera *Gliocladium, Pythium* and *Trichoderma* have shown biocontrol activity (Sreenivasaprasai and Maniphushanrao, 1990; Howell, 1991; Wolfhechel and Jensen, 1992). Some of these biocontrol agents have been developed into commercial products (eg a *Bacillus subtilis* strain for cotton and peanut [Quantum 4000] and a *Gliocladium* strain for glasshouse soil mixes [SoilGard]). However, the available products cover a small range of crops and have a very small share in the crop-protection market. Several investigators have also paid attention towards the interaction of mycorrhizal fungi with pathogenic agents and its possible role as the biocontrol agent.

MYCORRHIZA

Frank (1885) described mycorrhizae (*myco* for 'fungus' and *rhiza* for 'root'), as a symbiosis and reported considerable growth enhancement of mycorrhizal seedlings as compared to non-mycorrhizal ones. Ever since, this root-fungus symbiosis, called mycorrhiza, has been known and extensively investigated. The fungus receives food from the host plant in the form of sugars and other organic molecules, and performs numerous services in turn such as: enhancing water and nutrient uptake, particularly nutrients such as phosphorus that do not readily move in soil (Read, 1991); decomposing organic matter and cycling the nutrients contained therein, directly back to their host plants (Smith and Read, 1997); protecting plants against root pathogens (Novak, 1998); extending the lifetime of small roots (Majdi *et al.*, 2001); binding soil particles together into large aggregates, and thereby, producing a favourable structure for water retention and gas exchange (Finley and Söderström, 1992); linking individual plants of the same or different species, together into a common hyphal network, through which carbon and nutrients pass (Tiwari *et al.*, 2003; Robinson and Fitter, 1999) and; helping in reducing competition between plants of different species (Tiwari *et al.*, 2004; Sanders *et al.*, 1995).

Almost every plant species has mycorrhizal associations with one or more species of fungi (Molina *et al.*, 1992). All conifer species have proved to be mycorrhizal in their natural habitat (Okland and Eilertsen, 1993). Of over 6500 species of angiosperms that have been studied so far, 70% are obligatory mycorrhizal and the other 12% are facultatively mycorrhizal in that they may or may not form mycorrhizae (Trappe, 1987). It is considered that nearly 95% of the world's flora is characteristically mycorrhizal (Barker *et al.*, 1998), although mycorrhizal status has been examined for only about 3% of the total flora (Smith and Read, 1997). While the majority exhibits VAM symbiosis, the temperate timber plants are predominantly ectomycorrhizal (Smith and Read, 1997). Ectomycorrhizal symbiosis occurs mainly between woody plants and filamentous fungi. This interaction involves only 5% of the seed plants but the wide geographic range of trees makes the ectomycorrhizal symbiosis important for plant biomass.

Biocontrol

Microbial populations in soil concentrate around plant roots stimulated by root exudates, sloughed-off cells and tissue fragments, cell lysates, secretions, etc, supplied by the plant. This region is defined as the rhizosphere (Curl and Truelove, 1986). This increase in microbial activity, the 'rhizosphere effect', in turn, affects the plant because the stimulated microorganisms, saprophytes, parasitic symbionts ('pathogens'), and mutualistic symbionts ('symbionts') carry out a range of activities important to plant growth. The development of the rhizosphere is a dynamic process, incorporating physical, chemical, and biological modifications occurring at the root-soil interface. In this context, events occurring at the root-soil interface like microbe to microbe interactions are crucial to understand the dynamic processes characteristic of rhizosphere establishments and maintenance (Suslow, 1982). This particularly applies to interactions taking place in the soil of the rhizosphere which influence the microbial colonisation of the root surface, the rhizosphere, and, ultimately, root infection by pathogens and mutalists. These facets must be taken into consideration when trying to manage rhizosphere microorganisms, as, for example, the biological control of plant pathogens, to improve plant nutrition, or to establish a microbial population on the root of axenically propagated plants (Linderman, 1986).

Mycorrhiza formation induces changes in the mineral composition and physiology of plant tissue, and, as a consequence, in root exudation. This fact, together with the development of the fungal mycelium around roots, produces some chemical and physical modifications of the environment surrounding the mycorrhizae. These changes affect the microbial populations. Therefore, Oswall and Ferchau (1968) coined the term 'mycorrhizosphere" to describe the zone of influence of the mycorrhiza, and consequently, the so-called 'mycorrhizosphere effect' (Linderman, 1988). For both VAM and ectomycorrhizae, reports are available to show selective changes, which affect, both qualitatively and quantitatively, the microbial groups developing in the rhizosphere (mycorrhizosphere + mycosphere) ecosystem. Thus, microbial interactions in the mycorrhizosphere are of increasing interest (Rambelli, 1973).

As stated by Harley and Smith (1983), "mycorrhizas are active, living components of the soil population, having some properties like those of roots and some like those of microorganisms". Since mycorrhizae are key components in influencing microbial equilibrium in the soil, they can affect microorganisms causing diseases of plant roots. As stated by Marx (1973), both mycorrhizal fungi and pathogens are associated in the same substrate, the fine feeder roots of their hosts. Thus, it was postulated that both types of symbionts, the mutalist mycorrhizal fungus (eusymbiont) and the parasitic microorganisms (dissymbiont), could interact as common components of the rhizosphere soil or when colonising the root. Evidence overwhelmingly shows that both ecto-and VA-mycorrhizae may act in this way. Thus, they could be considered as biological control agents to soil-borne diseases of plants (Bagyaraj, 1984; Tello *et al.*, 1987).

Fungal Root Pathogens

Since mycorrhizal fungi are major components of the rhizosphere, it is logical that they could affect the incidence and severity of root diseases. Most of the literature suggests that mycorrhizal fungi reduce soil borne diseases or the effects of disease caused by fungal pathogens (Bhatt, 2000). However, reports are still confusing, with some indicating no effect of the mycorrhizal fungus on diseases (Baath and Hayman, 1984; Zambolim and Schenck, 1984), and others indicating increased disease severity (Davis and Menge, 1981). Drawing a conclusion is difficult, partly because so many different pathogens and diseases are involved, and partly because of the experimental conditions of each study. Clearly, one should expect varied results, even if the mycorrhizal fungi used had been the same (Schenck, 1987).

Pathogenic Root Nematodes

Root infections by pathogenic nematodes are generally less severe on mycorrhizal plants than on non-mycorrhizal plants, but the responses may vary, and the mechanisms involved are controversial (Hussey and Roncadori, 1982; Ingham, 1988). Symptoms of nematode infection are generally reduced, and often, but not always, nematode populations are reduced (as indicated by the number of galls, juveniles or eggs per unit root length) (Hussey and Roncadori, 1978). Atilano *et al.* (1981), however, showed an increase in nematode populations in mycorrhizal plants.

All the proposed mechanisms of interaction between the mycorrhizal fungi and the nematodes depend on mycorrhiza-mediated changes in host physiology. Changes in root exudation by mycorrhizal plants may change the attractiveness of roots to nematode pathogens. Mycorrhizal association may improve host plant vigour by increased nutrient supply, and thus reduce yield losses caused by nematode infection (Cooper and Grandison, 1987). Many studies suggest that mycorrhiza reduced nematode effects must also involve some other physiological changes in the roots. Studies by split root techniques (Strobel *et al.*, 1982; Oliveira and Zambolim, 1987) indicated that reduced nematode infection on VAM plants only occurred if the two were together on the same roots. These results suggest that competition for food or space was involved (Suresh *et al.*, 1985; Sitaramaiah and Sikora, 1982). Physiological changes in mycorrhizal roots could also change resistance to nematodes by increased production of inhibitory substances (Suresh *et al.*, 1985), or by changes in root exudation, which could alter mycorrhizosphere populations and affect nematode populations and survival. While the mechanisms are still controversial, the evidence strongly indicates that mycorrhiza suppress nematode infections of roots. Undoubtedly, the effects and mechanisms involved depend on the conditions of the test, the host plant, edaphic conditions, and the species of mycorrhizal fungus involved.

Bacterial Diseases

Not many studies have explored the effects of mycorrhiza on bacterial diseases. However, Garcia-Garrido and Ocampo (1989) showed that VAM tomato (*Lycopersicon* Miller) plants exhibited greater growth than non-VAM plants, inoculated with *Pseudomonas syringae* pv. *syringae* van Hall, when the pathogen was added to the foliage three weeks after the VAM fungus. Populations of the pathogen were lower in VAM than in non-VAM plants.

Viral and Foliage Diseases

While soil-borne diseases caused by fungal, nematode, and bacterial pathogens are most often reduced by mycorrhizal fungi, those caused by viral and other foliage pathogens are generally increased in mcorrhizal plants (Dehne, 1982; Jalali and Jalali, 1991; Schonbeck, 1979). Reports indicate that disease incidence, but not necessarily severity of effect on the plant, is increased in mycorrhizal plants as compared to non-mycorrhizal plants. Viruses apparently multiply faster in VAM than non-VAM plants. One might suspect that VAM fungi could acquire vector viruses from root to root between plants, but that apparently does not happen (Jabaji-Hare and Stobbs, 1984). VAM effects on viruses occur throughout the plant due to changes in host physiology. Foliage diseases caused by both obligate and non-obligate leaf pathogens can be increased on VAM plants, as compared to non-VAM plants, due to enhanced development of the pathogens, rather than to increased incidence or frequency of infections it could also be correlated with improved nutrition and higher physiological activities in VAM than in non-VAM plants (Dehne, 1982; Schonbeck, 1979).

MECHANISMS OF DISEASE CONTROL

Since mycorrhiza have such a significant effect on host physiology and rhizospheric biological interactions, they can affect the incidence and severity of plant diseases. The mixed responses and interpretations of the role played by mycorrhizal fungi in the biological control of plants have precluded any clear conclusion that mycorrhizal fungi always suppress plant diseases. The controversy revolves around the mechanisms of interaction between mycorrhizal fungi and plant pathogens. Disease protection by mycorrhizal fungi may involve several active and passive mechanisms that may act singly, simultaneously or synergistically (Harley and Smith, 1983; Schisler and Linderman, 1989; Zak, 1964). These mechanisms affect pathogens directly through inhibition, competition, mechanical or chemical exclusion; or indirectly by increased host vigour and modification of the basic physiology of plant roots. Mycorrhizae are thought to contribute to biological control of plant diseases, primarily by means of stress reduction (Baker, 1986, 1987; Cook and Baker, 1983). The biological control of plant diseases may be strongly influenced by mycorrhizal fungi, by one or more of the following mechanisms.

Enhanced Nutrition

The most obvious mycorrhizal contribution to reduced root disease is to increase nutrient uptake, particularly P, N and other minerals, resulting in more vigorous plants that are better able to resist or tolerate root diseases (Manion, 1981). Davis (1980) and Graham and Menge (1982) showed this type of response in their studies, where mycorrhiza or added P decreased root exudates used by the pathogen for spore germination and infection.

Competition for Nutrients and Infection Site

Even though mycorrhizal fungi depend on the plant for carbohydrates from photosynthesis, it is not clear whether they compete with root pathogens for nutrients. Dehne (1982) indicated that fungal root pathogens could occupy root cortical cells adjacent to those colonised by mycorrhizal fungi, indicating a lack of competition. Nematode pathogens, on the other hand, require host nutrients for reproduction and development, and direct competition with mycorrhizal fungi has been hypothesised as a mechanism of their inhibition (Dehne, 1982; Smith, 1988).

Morphological Changes Including Physical Barrier

Localised morphological effects, like increased lignification of host root cells of the endodermis have been shown to occur in VAM roots, which is responsible for reduced *Fusarium* wilt on tomato and cucumber (Dehne and Schonbeck, 1979). Wick and Moore (1984) also showed increased wound-barrier formation that inhibited *Thielaviopsis* black root rot of VAM *Ilex crenata* Thundb. Similarly, disease protection by ectomycorrhizal fungi has been associated with a physical barrier effect caused by the fungal mettle around ectomycorrhizal roots. These few examples indicate that mycorrhiza may induce morphological changes in root tissues.

Changes in Chemical Constituents of Plant Tissues

Many physiological changes, including increased concentrations of anti-fungal chitinase (Dehne *et al.*, 1978), increased arginine accumulation (Baltruschat and Schonbeck, 1975), increased concentrations of phytoalexin-like isoflavonoid compounds (Morandi *et al.*, 1984), synthesis of plant growth regulators such as Auxins and ethylene (Harley and Smith, 1983; Nylund, 1988; Rupp *et al.*, 1989) have also been reported to be involved in localised effects on root pathogens. It is postulated that such materials account for increased resistance to fungal and nematode root pathogens, as compared to non-mycorrhizal plants. Just what role mycorrhiza play in these processes remains unknown also direct evidence is lacking.

Alleviation of Abiotic Stress

Environmental stresses, such as nutritional stress (deficiency or excess), pH, soil drought, soil toxicity, etc., influence the metabolism of the host plant and mycorrhizal fungi, and thus, the incidence and severity of biotic plant diseases and cause some abiotic diseases (Brundrett, 1991; Strobel and Sinclair, 1991; Elad, *et al.*, 1980). Mycorrhizal association increases plant tolerance to such stresses by various mechanisms. Mycorrhizae contribute greatly to the general health of plants by helping to avoid nutrient stress and associated disease-predisposing effects. Extraradical hyphae of mycorrhizal fungi may absorb water under soil drought conditions, and thus, help plants to tolerate drought (Hardie, 1985). Others suggest that mycorrhizae change the physiology of plants in ways that make them more drought-tolerant than non-mycorrhizal plants (Davies *et al.*, 1992, 1993). Mycorrhizal plants are also more tolerant to soil toxicities and thus maintain a higher level of growth and health than non-mycorrhizal plants, and therefore they are less susceptible to diseases.

Microbial Changes in the Mycorrhizosphere

The concept of the 'mycorrhizosphere' implies that mycorrhizae significantly influence the microflora of the rhizosphere by altering root physiology and exudation. Several studies have indicated, both quantitative and qualitative changes in the mycorrhizosphere soil of mycorrhizal plants, which is time-dependent and dynamic with plant development (Ames *et al.*, 1984; Bagyaraj and Menge, 1978; Secilia and Bagyaraj, 1987; Vancura *et al.*, 1989). The microbial population also varied with different VAM fungal species on the host (Meyer and Linderman (1986). These studies demonstrate that mycorrhizae influence the microbial populations in the mycorrhizosphere soil; many of those microbial shifts could influence the growth and health of plants. Mycorrhizal formation alters the selective pressure on populations of soil microorganisms, some of which can antagonise root pathogens. For example, sporangium and zoospore production by the root pathogen *Phytophthora cinnamomi* Rand was reduced in the presence of rhizosphere leachates from VAM plants when compared to non-VAM plants. Similarly, more actinomycetes antagonistic to selected fungal and bacterial pathogens were isolated from VAM pot culture plants than from non-VAM plants. Mycorrhizal fungi also suppressed disease by mycorrhizosphere microbial populations. A reduction in *Fusarium* populations in the mycorrhizosphere soil of tomato, and a corresponding reduction in root rot in VAM plants was recorded by Caron *et al.* (1985, 1986), in comparison to non-VAM plants. Other studies, by Bartschi *et al.*, 1981 and Bhatt, 2000, indicated protection of the host against root pathogens when plants were pre-inoculated with a mixture of VAM pot culture inocula. These results indicate that VAM fungi are relatively tolerant of antagonists and are highly rhizosphere-competent. Changes in mycorrhizosphere populations of antagonists to pathogens are a likely explanation for many of the reported effects of mycorrhizae on diseases. Selective increase in the number of antagonists in the mycorrhizosphere is possible only if the antagonists are present in the background soil or growth medium. Thus, if potentially effective antagonists are present and are increased by mycorrhizae, then disease suppression could be expected.

FACTORS INFLUENCING BIOCONTROL BY MYCORRHIZAL FUNGI

Timing and Extent of Mycorrhiza Formation

Considering both the faster infection rate of most fungal root pathogens, as compared to mycorrhizal fungi, and the time needed for mycorrhizal effects on the host physiology to occur, studies by Bhatt (2000) on *Pythium* root rot of Solanaceous vegetables, by Stewart and Pfleger (1977) on *Pythium* and *Rhizoctonia* root rot of poinsettia (*Euphorbia pulcherrhima* Willd. Ex Klotzsch), by Bartschi *et al.*, (1981) on *Phytophthora* root rot of Lawson cypress (*Chamaecyparis lawsoniana* (A. Murr.)Parl.), and by Rosendahl (1985) with *Aphanomyces* root rot of pea (*Pisum sativum* L.), demonstrated that mycorrhizae must be established and functioning before invasion by the pathogens for effective suppression of root diseases. Physiological effects could be localised or systemic. Aphanomyces root rot of peas (Rosendahl, 1985) and Phytophthora root rot of citrus (Davis and Menge, 1980) were only suppressed by mycorrhiza when the two organisms were present on the same roots. However, Rosendahl (1985) showed that oospore production was reduced on non-VAM pea roots split from VAM roots, as compared to plants with no VAM on either root system. Similarly Davis and Menge (1980) showed that citrus roots opposite the split from VAM roots had less *Phytophthora* root rot than if split from non-VAM roots. These two studies suggest that VAM effects could be both localised and systemic, probably involving two separate mechanisms. The systemic effect could have been a P effect, while the localised effect was due to some morphological or physiological changes in the root tissues, in the immediate vicinity.

Inoculum Level of the Pathogen

The potential for biological control to occur in any production system is directly related to the inoculum potential of the pathogen. A high pathogen inoculum density can overwhelm biocontrol agents (Cook and Baker, 1983; Schenck, 1987).

Mycorrhiza Management and Strategies

It is biologically fundamental to expect different interactions to occur between different mycorrhizal fungi, host plants and plant pathogens. Considering the complex host-mycorrhiza-pathogen interaction and variations in the chemical and microbial composition of soil, interpretations and comparisons between different studies are difficult if not impossible. If mycorrhizae contribute to disease suppression, then agricultural practices that reduce populations of mycorrhizal fungi and associated antagonists could result in increased disease incidence. Thus, soils have to be managed under conditions that optimise both disease suppression and plant development. In those systems where mycorrhizal fungi and associated antagonists could be co-inoculated, plants can be benefited from optimum combinations of mycorrhizae and antagonists for protecting against invasion by pathogens.

REFERENCES

Ames, R.N.; Reid, C.P.P. and Ingam, E.R. 1984. Rhizosphere bacterial population responses to root colonisation by a vesicular-arbuscular mycorrhizal fungi. New Phytol., 96:555–563.

Atilano, R.A.; Menge, J.A. and Van Gundy, S.D. 1981. Interaction between *Meloidogne arenaria* and *Glomus fasciculatus* in grape. J. Nematol., 13:52–57.

Baath, E. and Hayman, D.S. 1984. No effect of VA mycorrhiza on red core disease of strawberry. Trans. Br. Mycol. Soc., 82:534–536.

Bagyaraj D.J. and Menge, J.A. 1978. Interactions between a VA mycorrhiza and Azotobacter and their effects on rhizosphere microflora and plant growth. New Phytol., 80:567–573.

Bagyaraj, D.J. 1984. Biological interactions with VA mycorrhizal fungi. In: Powell, C. L. and Bagyaraj, D. J. (Eds.) VA Mycorrhiza. CRC Press Inc., Boca Raton, FL. Pp. 131–153.

Baker, K.F. 1987. Evolving concepts of biological control of plant pathogens. Annu. Rev. Phytopathol., 25:67–85.

Baker, R. 1986. Biological control: An overview. Can. J. Plant Pathol., 8:218–221.

Baltruschat, H. and Schoenbeck, F. 1975. Studies on the influence of endotrophic mycorrhiza on the infection of tobacco by *Thielaviopsis basicola*. Phytopath. Z., 84:172–188.

Barker, S.J.; Tagu, D. and Delp, G. 1998. Regulation of root and fungal morphogenesis in mycorrhizal symbiosis. Plant Physiology, 116:1201–1207.

Bartschi, H.; Gianinazzi-Pearson, V. and Vegh, I. 1981. Vesicular-arbuscular mycorrhiza formation and root rot disease (*Phytophthora cinnamomi*) development in *Chamaecyparis lawsoniana*. Phytopath. Z., 102:213–218.

Bhatt, A. 2000. Studies on *Pythium* species in relation to their parasitic association with Solanaceous vegetables of Haldwani and Nainital (Kumaun Himalaya). Ph. D. Thesis, Kumaun University, Nainital, Uttaranchal, India.

Brundrett, M. 1991. Mycorrhizas in natural ecosystems. Adv. Ecol. Res., 21:271–315.

Caron, M.; Fortin, J.A. and Richard, C. 1985. Influence of substrate on the interaction of *Glomus intraradices* and *Fusarium oxysporum* f. sp. radicis-lycopersici on tomatoes. Plant Soil, 87:233–239.

Caron, M.; Fortin, J.A. and Richard, C. 1986. Effect of phosphorus concentration and *Glomus intraradices* on Fusarium crown and root rot of tomatoes. Phytopathology, 76:942–946.

Cook, R.J. and Baker, K.F. 1983. The nature and practice of biological control of plant pathogens, APA Press, St. Paul, MN.

Cooper, K.M. and Grandison, G.S. 1987. Effects of vesicular-arbusclar mycorrhizal fungi on the infection of tamarillo (*Cyphomandra betacea*) by *Meloidogyne incognita* in fumigated soil. Plant Dis., 71:1101–1106.

Curl, E.A. and Truelove, B. 1986. The Rhizosphere. Springer-Verlag, New York. Pp 288.

Davies, F.T.; Potter, J.R. and Linderman, R.G. 1992. Mycorrhiza and repeated drought exposure effect, drought resistance and extraradical hyphae development of pepper plants independent of plant, size and nutrient content. J. Plant physiol., 139:289–294.

Davies, F.T.; Potter, J.R. and Linderman, R.G. 1993. Drought resistance of mycorrhizal pepper plants independent of leaf P concentration-response in gas exchange and water relations. Physiol. Plant., 87:45–53.

Davis, R.M. 1980. Influence of *Glomus fasciculatus* on *Thielaviopsis basicola* root rot of citrus. Plant Dis., 64:839–840.

Davis, R.M. and Menge, J.A. 1980. Influence of *Glomus fasciculatus* and soil phosphorus on Phytophthora root rot of citrus. Phytopathology, 70:447–452.

Davis, R.M. and Menge, J.A. 1981. *Phytophthora parasitica* inoculation and intensity of vesicular-arbuscular mycorrhizae in citrus. New Phytol., 87:705–715.

Dehne, H.W. 1982. Interactions between vesicular-arbuscular mycorrhizal fungi and plant pathogens. Phytopathology, 72:1115–1119.

Dehne, H.W. and Schonbeck, F. 1979. The influence of endotrophic mycorrhiza on plant diseases. II. Phenolmetabolism and lignification. Phytopath. Z., 95:210–216.

Dehne, H.W.; Schonbeck, F. and Baltruschat, H. 1978. The influence of endotrophic mycorrhiza on plant diseases. 3. Chitinase-activity and ornithine-cycle. Z. Pflkrankh., 85:666–678.

Elad, Y.; Chet, I. and Katan, J. 1980. *Trichoderma harzianum*: A biocontrol agent effective against *Sclerotium rolfsii* and *Rhizoctonia solani*. Phytopathology, 70:119–121.

Finlay, R. and Söderström, B. 1992. Mycorrhiza and carbon flow to the soil. In: Allen, M. F. (ed) Mycorrhizal Functioning: An integrative plant-fungal process, Chapman and Hall, The Netherlands. Pp 134–160.

Frank, B. 1885. On root-symbiosis-depending nutrition through hypogeous fungi of certain trees, as translated by J. M. Trappe, 1985. In: Molina, R. Proceedings of the 6th North American Conference on Mycorrhizae. Pp 18–25.

Garcia-Garrido, J.M. and Ocampo, J.A. 1989. Effect of VA mycorrhizal infection of tomato on damage caused by *Pseudomonas syringae*. Soil Biol. Biochem., 21:165–167.

Graham, J.H. and Menge, J.A. 1982. Influence of vesicular-arbuscular mycorrhizae and soil phosphorus on take-all disease of wheat. Phytopathology, 72; 95–98.

Hardie, K. 1985. The effect of removal of extraradical hyphae on water uptake by vesicular-arbuscular mycorrhizal plants. New Phytol., 101:677–684.

Harley, J.L. and Smith, S.E. 1983. Mycorrhizal symbiosis. Academic Press, New York. 483 pp.

Howell, C.R. 1991. Biological control of *Pythium* damping-off of cotton with seed-coating preparations of *Gliocladium virens*. Phytopathology, 81:739–741.

Hussey, R.S. and Roncadori, R.W. 1978. Interaction of *Pratylenchus brachyurus* and *Gigaspora margarita* on cotton. J. Nematol., 10:18–20.

Hussey, R.S. and Roncadori, R.W. 1982. Vesicular-arbuscular mycorrhizae may limit nematode activity and improve plant growth. Plant Dis., 66:9–14.

Ingham, R.E. 1988. Interactions between nematodes and VA mycorrhizae. Agriculture Ecosystems and Environment, 24:169–182.

Jabaji-Hare, S.H. and Stobbs, L.W. 1984. Electron microscopic examination of tomato root co-infected with *Glomus* sp. and tobacco mosaic virus. Phytopathology, 74:277–279.

Jalali, B.L. and Jalali, I. 1991. Mycorrhiza in plant disease control. In: Arora, D.K.; Rai, B.; Mukerji, K.G. and Knudsen, G.R. (Eds.). Handbook of applied mycology, Soil and Plants, Vol. 1. Marcel Dekker, New York, NY. Pp 131–154.

Linderman, R.G. 1986. Managing rhizosphere microorganisms in the production of horticultural crops. HortScience, 21:1299–1302.

Linderman, R.G. 1988. Mycorrhizal interactions with the rhizosphere microflora: The mycorrhizosphere effect. Phytopathology, 78:366–371.

Majde, H.; Damm, E. and Nylund, J.E. 2001. Longevity of mycorrhizal roots depends on branching order and nutrient availability. New Phytol., 150:195–202.

Manion, P.D. 1981. Tree Disease Concepts. Prentice Hall, NJ.

Marx, D.H. 1973. Mycorrhizae and feeder root diseases. In: Ectomycorrhizae: Their Ecology and Physiology. Marks, G.C. and Kozlowski, T.T. (Eds.) Academic Press, New York. Pp 351–382.

Meyer, J.R. and Linderman, R.G. 1986. Selective influence on populations of rhizosphere or rhizoplane bacteria and actinomycetes by mycorrhizas formed by *Glomus fasciculatum*. Soil Biol. Biochem., 18:191–196.

Molina, R.; Massicotte, H. and Trappe, J.M. 1992. Specificity phenomena in mycorrhizal symbiosis: Community-ecological and practical implications. In: Allen, M. F. (ed) Mycorrhizal functioning: An integrative plant-fungal process. Chapman and Hall, New York. Pp 357–423.

Morandi, D.; Bailey, J.R. and Gianinazzi-Pearson, V. 1984. Isoflavonoid accumulation in soybean roots infected with vesicular-arbuscular mycorrhizal fungi. Physiol. Plant Pathol., 24:357–364.

Novak, B. 1998. Effect of endomycorrhizas on some vegetable crops. Poljoprivrendna Znanstvena Smotra, 63:187–198.

Nylund, J.A. 1988. The regulation of mycorrhiza formation-carbohydrates and hormone theories reviewed. Scand. J. For. Res., 3:465–479.

Okland, R.H. and Eilertsen, O. 1993. Vegetation-environment relationship of boreal coniferous forests in the Sdhomfjell area, Gjerstad, S. Norway, Sommerfeltia, 16:1–254.

Oliveira, A.A.R. and Zambolim, L. 1987. Interacao entro o fungo endomicorrhizico *Gloms etunicatum* e o nematoide de galhas *Meloidogyne javanica* em feijoeiro com raiz partida. Bras., 12:222–225.

Rambelli, A. 1973. The rhizosphere of mycorrhizae. In: Marks, G.L. and Kozlowski, T.T. (Eds.). Ectomycorrhizae. Academic Press, New York, NY. Pp 299–343.

Read, D.J. 1991. Mycorrhiza in ecosystems. Experientia, 47:376–391.

Robinson, D. and Fitter, A. 1999. The magnitude and control of carbon transfer between plants linked by a common mycorrhizal network. Journal of Experimental Botany, 50:9–13.

Rosendal, S. 1985. Interactions between the vesicular-arbuscular mycorrhizal fungus *Gloms fasculatum* and *Aphanomyces euteiches* root rot of peas. Phytopath. Z., 114:31–40.

Rupp, L.A.; Mudge, K.W. and Negm, F.B. 1989. Involvement of ethylene in ectomycorrhiza formation and dichotomous branching of roots of Mugo Pine seedlings. Can. J. Bot., 67:477–482.

Sanders, I.R.; Koide, R.T.; Shumway, D.L. 1995. Diversity and structure in Natural Communities: The role of mycorrhizal symbiosis. In: Verma, A. and Hock, B. (eds) Mycorrhiza: structure, function, molecular biology and biotechnology (2nd edition), Springer-Verlag, Heidelberg. Pp 571–593.

Schenck, N.C. 1987. Vesicular-arbuscular mycorrhizal fungi and the control of fungal root diseases. In: Chet, I. (Ed.) Innovative Approaches to Plant Disease Control. John Wiley and Sons Inc., New York, NY. Pp179–191.

Schisler, D.A. and Linderman, R.G. 1989. Selective influence of volatiles purged from coniferous forests and nursery soils on microbes of a nursery soil. Soil Biol. Biochem., 21:389–396.

Schonbeck, F. 1979. Endomycorrhiza in relation to plant diseases. In: Schippers, B. and Gams, W. (Eds.) Soil-borne plant pathogens, Academic Press, New York, NY. Pp 271–280.

Secilia, J. and Bagyaraj, D.J. 1987. Bacteria and actinomycetes associated with pot cultures of vesicular-arbuscular mycorrhiza. Can. J. Microbiol., 33:1069–1073.

Sitaramaiah, K. and Sikora, R.A. 1982. Effect of the mycorrhizal fungus *Glomus fasciculatus* on the host-parasite relationship of *Rotylenchulus reniformis* in tomato Nematologica, 28:412–419.

Smith, G.S. 1988. The role of phosphorus nutrition in interactions of vesicular-arbuscular mycorrhizal fungi with soil-borne nematodes and fungi. Phytopathology, 78:371–374.

Smith, S.E. and Read, D.J. 1997. Mycorrhizal Symbiosis (2nd edition), Academic Press, London.

Sreenivasaprasai, S. and Manibhushanrao, K. 1990. Biocontrol potential of fungal antagonists *Gliocladium virens* and *Trichoderma longibrachiatum*. J. Plant Dis. Prot., 97:570–579.

Stewart, E.L. and Pfleger, F.L. 1977. Development of poinsettia as influenced by endomycorrhizae, fertilisers and root rot pathogens *Pythium ultimum* and *Rhizoctonia solani*. Florist's Rev., 159:37, 79–80.

Strobel, N.E. and Sinclair, W.A. 1991. Role of flavanolic wall infusions in the resistance induced by *Laccaria bicolour* to *Fusarium osysporm* in primary roots of Douglas-fir. Phytopathology, 81:420–425.

Strobel, N.E.; Hussey, R.S. and Roncadori, R.W. 1982. Interactions of vesicular-arbuscular mycorrhizal fungi, *Meloidogyne incognita* and soil fertility on peach. Phytopathology, 72:690–694.

Suresh, C.K.; Bagyaraj, D.J. and Reddy, D.D.R. 1985. Effect of vesicular-arbuscular mycorrhiza on survival, penetration and development of root-knot nematode in tomato. Plant Soil, 87:305–308.

Suslov, T.V. 1982. Role of root-colonising bacteria in plant growth. In: Mount, R. and Lacey, C. (eds.) Phytopathogenic Prokaryotes, vol. 1. Academic Press, New York. Pp 187–223.

Tello, J.C.; Vares, F.; Notario, A. and Lacassa, A. 1987. Las micorrizas:un potencial a explotar en la lcha contra las enfermedades de las plantas. ITEA, 73:40–64.

Tiwari, M.; Singh, S.P.; Pandey, N. and Singh, R.P. 2004. Carbon transfer between *Quercus leucotrichophora* and *Pinus roxburghii* through ectomycorrhizal mycelial connections. Current Science, 87:674–680.

Tiwari, M.; Singh, S.P.; Tiwari, A. and Sundriyal, R.C. 2003. Effect of symbiotic associations on the growth of the host *Coriaria nepalensis* and its facilitative impact on oak and pine seedlings in the Central Himalayas. Forest Ecology Management, 184:141–147.

Trappe, J.M. 1987. Phylogenetic and ecologic aspects of mycotrophy in the angiosperms from an evolutionary standpoint. In: Safir, G.R. (ed) Ecophysiology of VA mycorrhizal plants, Boca Raton, CRC Press. Pp 5–25.

Vancora, V.; Orozco, M.O.; Grauova, O. and Prikryl, Z. 1989. Properties of bacteria in the hyphosphere of a vesicular-arbuscular mycorrhizal fungus. Agriculture Ecosystem and Environment, 29:421–427.

Wick, R.L. and Moore, L.D. 1984. Histology of mycorrhizal and non-mycorrhizal *Ilex crenata* 'Helleri' challenged by *Thielaviopsis basicola*. Can. J. Plant Pathol., 6:146–150.

Wolfhechel, H. and Jensen, D.F. 1992. Use of *Trichoderma harzianum* and *Gliocladium virens* for the biological control of post-emergence damping-off and root rot of cucumber caused by *Pythium ultimum*. J. Phytopathol., 136:221–230.

Zak, B. 1964. Role of mycorrhizae in root disease. Annu. Rev. Phytopathol., 2:377–392.

Zambolim, L. and Schenck, N.C. 1984. Effect of *Macrophomina, Rhizoctonia, Fusarium* and the mycorrhizal fungus *Glomus mosseae* on nodulated and non-nodulated soybeans. Fitopatologia Brasileira, 9:129–138.

21

Role of Fungi in Biodeterioration of Museum Objects

ARUN ARYA

Department of Botany, Faculty of Science Maharaja Sayajirao University of Baroda, Vadodara-390 002, Gujarat, India.

SUMMARY

Preservation of artefacts is essential as these objects serve as tangible links to our past and constitute our bequest to the future. Man's struggle to save his cultural heritage is not new. Museums are repositories of important assets, which need to be conserved. We should be in a position to preserve them for long time. Vadodara (a recent name of the old city Baroda) has two important museums th Baroda Museum and Fatesing Museum. The Baroda Museum and Picture gallery is situated in the famous Sayaji gardens, on the banks of river Vishwamitri. It houses a collection of European oil paintings and various other artefacts. Maharaja Fatesingh Museum is situated in Lakshmi Vilas Palace compound. The collection consists of objects collected by Sir Sayajirao Gaekwad III, the maker of Baroda city. Occurrence of various fungal biodeteriogens in indoor air and the threats posed by them on different objects are described in detail. Some preventive measures are also suggested to protect them from damage due to fungi.

Keywords : Fungi, Bio-deterioration, Baroda Museum and Picture gallery, Fatehsingh Museum

Museums are not merely storehouses but they are centres of education in the service of Society. Museums have collections of cultural, artistic, ethnological archeological, historical and scientific significance. Today museums have also become centres of non-formal education. Acording to Foster (1986)," A good museum attracts, entertains, arouses curiosity, leads to questioning and thus promotes learning. It is an educational institution that is set up and kept in motion, that it may help the members of the community to become happier, wiser and more effective human beings." (Dana1909).

The word MUSEUM is derived from the Greek word, *Mouseion*, which was a place of contemplation, a philosophical institution or a temple of the Muses. ICOM (1990) defines a museum as a non-profit making, permanent institution in the service of society and its developmenting, communicates and exhibiting for purposes of study, education and enjoyment, material evidence of people and their environment.

The Baroda museum and picture gallery is famous for its collection of European oil paintings all over Asia. The buildings of Baroda museum are situated in pastoral environment of Sayaji Park. This Museum was built in the Indo-Saracenic style of architecture in 1894. The department of Museology also has a good museum. This is a pioneer institution in the field of Museum training at the postgraduate level, associated with the M. S. University. The department was established fifty years back in the Baroda Museum and picture gallery. The department is presently housed in a new three storeyed building in the campus of the Faculty of Fine Arts, opposite Sayajibaug, near the river Vishwamitri. Although the department started its new building in 1976, the Museum was thrown open to the public in the year 1989 only. Museum of the Department possesses different types of objects, such as tribal musical instruments, woods carving, objects made of paper mache, paintings and paper manuscripts, etc.

The Museology and Baroda Museum are situated within a radius of 1 km in Sayajigunj area of Baroda City. Two other spots, Oriental Institute and Fatehsingh Museum are situated 7 km apart. The Oriental Institute is a part of the M.S. University of Baroda. It has a collection of a large number of books and old Journals, paper and palm leaf manuscripts. It is situated on Palace road. Earlier it was housed in Smt. Hansa Mehta Library.

Biodeterioration can be defined as any undesirable change in materials of economic importance, brought about by the activities of living organisms. Biodeteriogens may be plants, microbes and animals including, human beings. Biodeteriogens like fungi, bacteria and actinomycetes pose a severe threat to museum objects. Microbes gain entry into the indoor atmosphere through wind currents and settle on various objects by impaction (Gregory, 1973). Sudden Jumps-introductory phases required Aerospora form a normal component of the earth's atmosphere. Aerobiologists have correlated such data with clinical allergy. These studies are helpful in assessing the deterioration and management of the environment. The aerobiological data may be used for the formation of disease forecasting system of plants (Mishra 1992) and human beings (Davis and Smith 1973; Spieksma 1980). Role of certain fungi present in indoor air and biodeterioration caused by them is discussed in the present paper.

METHODS USED DURING INVESTIGATION

Various methods have been suggested for isolating indoor aeromycoflora. In the present study, use of culture plate exposure method has been used, as gravity slide exposure method is found to have certain drawbacks, (Nigam and Pathak 1991) like:

(i) Similar spores are not identified upto generic level (eg *Aspergillus*, *Penicillium* and *Mucor* etc.)
(ii) Some spores are too small or too transparent or do not have sufficient individual characteristics eg yeast and *Phoma* etc.
(iii) Fungal hyphae are not identified.
(iv) Many spores cannot be trapped due to strong wind.

During the isolation of fungi from the Egyptian mummy cellotape technique was adopted. This technique is also suitable for to studying surface fungi or dust etc. Care should be taken to use new cellotape and cut small pieces before plating them on PDA plates.

STUDY OF INDOOR AEROMYCOFLORA

It is noteworthy that twenty-four spp. of fungi belonging to fifteen genera were isolated from Baroda Museum in 1999 and twenty-two spp. belonging to seventeen genera were isolated in 2000 (Table 21.1). The presence of viable spores of *Aspergillus japonicus*, *A. wentii*, *Chaetomium atrobrunneum*, *C. ganglegarum*, *Lewia infectoria*, *Monascus rubur* and *Phoma nebulosa* is noteworthy in indoor air. These biodeteriogens have the potential causing plant as well a animal diseases, and biodeteriorating cultural property at the same time. According to Nair (1972), fungi that can grow on any suitable substrata and play a prominent role in the deterioration of museum materials. Tilak and Kulkarni (1972) and Tilak *et al.* (1970–1972) failed to detect such fungi during their surveys in Ajanta caves. Conidia of Aspergilli and Fungi imperfecti dominated in their slides. They concluded that the excreta of bats and birds residing in these caves serve as feeding source for the microorganisms and this acts as a secondary source of aerospora. Air in the Mummy chamber harboured different mycoflora (Table 21.2), out of which *Aspergillus* species dominated; in all twenty-eight colonies were produced in 5 petriplates.

Table 21.1: Indoor Aeromycoflora of Egypto-Babylonian Gallery in Baroda Museum and Picture Gallery in 1999–2000

Fungi Isolated	Sep-99		May-00	
	Colonies*	Percentage Occurrence	Colonies*	Percentage Occurrence
Zygomycotina				
Absidia repens			5	3.4
Mucor hiemalis	8	4.8	5	3.4
Rhizopus stolonifer	10	6.1	5	3.4
Syncephalastrum racemosum			2	1.4
Ascomycotina				
Aspergillus flavus	5	3	10	6.8
A. japonicus	7	4.2	5	3.4
A. niger	17	10.3	20	12.2
A. parasiticus			7	4.8
A. ustus	5	3		
A. wentii	7	4.2		
Cheatomium atrobrunneum	6	3.6		
C. gangligerum	10	6.1	5	3.4
C. globosum	5	3	7	4.8
Emericella nisulans	5	3	5	3.4
E. variecolor	2	1.2	2	1.4
Lewia infectoria	5	3	5	3.4
Monascus rubur			3	3.4
Penicillium citrinum	7	4.2	5	3.4
P. nigricans	5	3		
Fungi imperfecti				
Alternaria alternata	4	2.4	5	3.4
Cladosporium herbarum	2	1.2	5	3.4
Curvularia lunata	10	6	5	3.4
C. verruculosa	2	1.2		
Fusarium oxysporum	8	4.8	10	6.8
Monilia sitophila	5	3	8	5.5
Phoma nebulosa	8	4.8	2	1.4
Trichoderma lignorum	2	1.2	5	3.4
T. viride	5	3		
Sterile white mycelium	10	6	8	5.5
Sterile black mycelium	5	3	5	3.4
Total fungal colonies	**165**		**146**	

*Data represent the total number of colonies that appeared on 5 P D A plates after 7 days of incubation; -, absence of colony

Table 21.2: Percentage Occurrence of Different Fungi in Air and Dust of Mummy Chamber on 16 April 1999

Fungi Isolated	Colonies*	Mummy Chamber Air Percentage Occurrence	Dust Colonies*	Percentage Occurrence
Aspergillus flavus	10	35.7	8	29.6
A. fumigatus	5	17.8	2	7.4
A. niger	5	17.8	5	18.5
Emericella nivea	2	7.1	5	18.5
Penicillium citrinum	2	7.1	2	7.4
Mucor hiemalis	2	7.1	5	18.5
Unidentified white mycelium	2	7.1		
Total no. of colonies	28		27	

*Data represent the total number of colonies that appeared on 5 P D A plates after 7 days of incubation; -, absence of colony

Table 21.3: Potential Allergens found in Aeromycoflora of Baroda Museum and Department of Museology

Sr. No.	Baroda Museum	Department of Museology
1.	*Aspergillus flavus*	*Aspergillus fumigatus*
2.	*A. sydowii*	*A. niger*
3.	*A. versicolor*	*Aspergillus sp.*
4.	*Cladosporium herbarum*	*Penicillium sp.*
5.	*Emericella nidulans*	*Rhizopus stolonifer*
6.	*Monilia sitophila*	
7.	*Penicillium sp.*	
8.	*Rhizopus stolonifer*	

Table 21.4: Symptoms of Biodeterioration of Different Kinds of Museum Materials (Nair, 1980)

Sl. No.	Symptoms	Example
1.	Stains and Discolouration	Caused by surface growth of fungi on paper, leather, textile etc.
2.	Disfigurement	Positive changes in the surface of a painting due to fungus
3.	Fibrilization	Loss of coherence in fibrous material, as in wool attacked by microorganisms
4.	Impenetrations	Invasion of the lumen of certain natural fibres by fungi.
5.	Changes in mechanical properties	Loss of weight and strength of wood on reduction or breaking
6.	Changes in chemical properties	Breakdown of cellulose by microorganisms, foxing on paper etc.
7.	Development of odour	Mildew on paper, cloth etc.

PROBLEMS OF BIODETERIORATION

Microorganisms produce different types of effects on objects discolouration, changes in shape or disfigurement, loss in firmness, and, if not cared for they can spoil the whole aesthetic look within a short time. These changes may be further enhanced if environmental conditions are unfavourable.

Following is a short list of various fungi causing damage to museum objects in Baroda during 1988–2001 (Table 21.5). Different types of objects were studied, like miniature paintings, manuscripts and herbarium sheets on paper, musical instruments made of bamboo and bottle-gourd, leather strings of Tambura,

Table 21.5: Fungi Isolated from Different Museum Objects.

Objects	Place	Organisms
Rajasthani miniature painting	Baroda Museum and picture gallery	*Chaetomium gangligerum* Aspergillus versicolor *A. terreus, Penicillium* sp.
Miniature painting	Museology Dept.	*Aspergillus tamarii* *Cladosporium herbarum*
Wooden plaque Pavri A. (Musical instrument)	Museology Dept. Museology Dept.	*Aspergillus niger* *Aspergillus japonicus* A. flavus *Penicillium citrinum*
Pavri B (Musical Instrument)	Museology Dept.	*Cochliobolus pallescens* *Phoma* sp. *Fusarium semitectum*
Parsi Manuscript	Museology Dept.	*Cochleobolus lunatus* *Chaetomium cymbiforme* *Aspergillus niger*
Tambura	Museology Dept.	*Emericella nidulance*
Chinese wooden screen	Museum and picture gallery Baroda	*Aspergillus ustus* *A. nidulans*
Pastel colour Painting	Fatehsingh Museum	*Aspergillus niger* *A. flavus*
Palm leaf Manuscript	Oriental Institute	*A.niger* *Nigrospora sphaerica*
Egyptian Mummy	Baroda Museum and picture gallery	*Aspergillus state of* *Emericella nivea*
Paper Manuscript	Oriental Institute	*Chaetomium globosum* Chaetomium fusisporale *Aspergillus niger*
Herbarium sheet	Museum of Botany Department	*Cladosporium herbarum* *Alternaria alternata*
Herbarium sheet	Museum of Botany Department	*Curvularia clavata*

panther skin, even human skin in an Egyptian mummy kept in Baroda Museum, and a wooden plaque with decorative work.

Isolations made during March–April 1987, from two post-medieval period miniature paintings (Rajasthani) revealed the presence of *Chaetomium gangligerum* L. Ames. Here the painting surface turned black, obliterating the figures and the colourful aspect of the painting. Greenish black velvety coloured growth was observed. From another miniature painting kept in the Museology Department two more fungi *Aspergillus tamarii* Kita and *Cladosporium herbarum* Link. were observed

Wood has been used by man since palaeolithic times for varied purposes. Wood is a very stable, ligno-cellulosic polymer; most wood species deteriorate due to biodegradation. Being of organic origin, it is metabolised by many fungi for food. Spores and growing mycelia are mostly responsible for altering the aesthetic appearance of an art object, by creating residues. On the surface of the object, mycelia can form dry and brittle crusts, wrinkled and corrugated colonies and some times, loose elastic, fluffy structures. Walston (1976) has worked on the problems of deterioration of ethnographical objects in an Australian museum. In the Tibetan Gallery of Baroda Museum, a Chinese wooden screen developed a deterioration problem on one of its panels. The panel was placed near a seepage in the wall; due to fault in the outlet of the wash basin the lacquer got deformed and the precious stones got separated from the screen. Another effect was the flaking of the paint due to loss of adhesion in the binding medium. On isolation, *Aspergillus ustus* and *A. nidulans* were found.

A wooden musical instrument pavri, used by Kunkuna tribes of Dangs area in Gujarat was found affected by fungi. It is an air bellowing instrument. On isolation from the joints, growth of *Aspergillus japonicus* and *A. flavus* was observed, along with *P. citrinum* Thom.. From another Pavri instrument the presence of *Curvularia* state of *Cochliobolus pallescence* (Tsuda and Ueyama) Sivan along with *Fusarium semitectum* Berk. and Rav. and *Phoma* species was observed. To preserve these instruments Shah and Arya (2001) suggested some curative treatment like cleaning with absolute alcohol, favoured by fumigation with 5% solution of Thymol. A regular cleaning of show-cases and a periodical spray of 0.25% Pentachlorophenol is suggested.

An experiment was conducted (Shah and Arya 2001) in which biodegradation of wood was observed by four fungi. The rate of degradation differed in all the cases. Maximum biodegradation was observed in Deodar wood, while the lowest biodegradation was observed in Neem wood. *Chaetomium cymbiforme* caused equal degradation of neem and mango wood. The fungi studied included *Curvularia verruculosus, Aspergillus ustus, Penicillium citrinum* and *C. cymbiforme*. Seifort (1983) reported decay of wood by members of Decrymycetales (Basidiomycotina). He found greater percentage of moisture content in blocks of wood showing more weight loss. The percentage weight loss was directly correlated to the percentage change in lignin and cellulose. Lakshmikant and Mathur (1991) reported a number of fungi from Gorakhpur causing damage to wooden materials, including *Penicillium citrinum*, one of the four fungi included in the present study.

The coffin box (as seen in Plate I fig A) is at present removed for treatment against insect borers. White coloured powdery growth was observed on the fingers and toe of the right leg (Plate I fig B). After identification, the fungus was confirmed as *Emericella nivea* (Wiley and Simmons Arya *et al.*, 2001). Its conidial stage *Aspergillus* was also observed in the culture, along with the perfect stage. Fig B shows the cut ends of the linen bandage on the mummy.

Two spp. of *Chaetomium C. globosum* (Plate H) and *C. fusisporale* were present on the paper manuscript. Ricelli (1999) studied fungal growth on paper samples. Fungi like *Aspergillus terreus, Chaetomium eslatum, Stacybotrys atra* and *Penicillium chrysogenum* grew on paper stored in high relative humidity. The addition of some new synthesised compounds to the different samples of paper inhibited fungal growth.

The deterioration of herbarium sheets was studied. Some of the sheets were as old as 1908. These were procured from the Cryptogemic Plant collection from Pusa. A sheet mounted with *Marchantia polymorpha* showed the presence of *Curvularia pallesance* and *Aspergillus niger*. Herbarium sheet of *Cycas* shows fungal growth of *Cladosporium herbarum* (Plate 1 fig G). On the paper document, *A. niger* was isolated on the cloth document *Chaetomium globosum* was isolated.

Maharaja Fatehsingh Museum was opened to the public in April 1961. It houses world famous oil paintings by Raja Ravi Varma. Many objects of bronze, marble and oils by the Venetian artist Felici are displayed here. The first floor is a ceremonial room of an 18th century French Palace with four pastel paintings being displayed. The pastel painting in coloured chalk depict the Crown Princess of France. The engravings are after paintings by Angelica Kauffmann. During the rainy season in 2004, a severe fungal attack was observed on these paintings (Plate II fig C and D). On isolation, *Aspergillus flavus* and *A. niger* were identified.

From an Orrian Palm Leaf manuscript, isolations were made in Jan–Feb 2004. Fungus *Nigrospora spherica* Penz. was isolated from black coloured spots of the manuscript (Plate I Fig. E and F). The conidia are dark black in colour and are this is a leaf infecting fungi producing dark pigment in culture, thus damaging the beauty of museum objects.

CONTROL MEASURES

Chemicals such as Para dichloro benzene, Thymol, Methyl bromide, Ethylene oxide are used to control fungal infections. Ethylene oxide has been banned in many countries due to its carcinogenic nature. Thymol, is used even today to control fungal growth in museums, thereby causing artwork framed in them

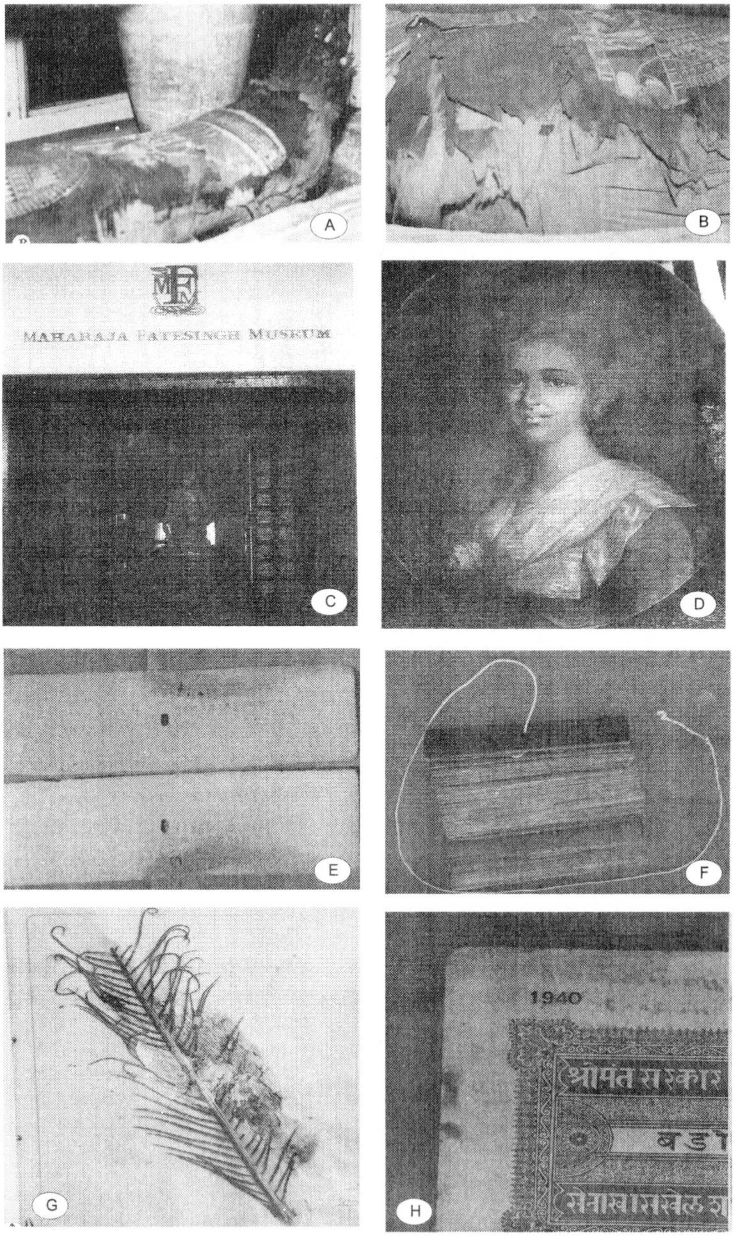

Plate I A. Lower part of the Egyptian mummy showing white coloured fungal growth on toes. Also visible is the big white canopic jar. **B**. Body of mummy showing cut ends of linen bandage and painting done on upper part of the body. **C**. Entrance of Maharaja Fatesingh Museum, Vadodara. Placed in front of gate is marble structure of Ruth. **D**. Pastel colour painting showing growth of *Aspergillus niger* and *A. flavus* on right hand side of painted lady. **E**. Two damaged leaves showing black colour marks produced by fungus *Nigrospora* **F**. Complete manuscript after fumigation treatment. **G**. Herbarium sheet showing deteriorated *Cycas* leaf. **H**. A paper related to Gaekwad court document showing fungal growth.

to adhere to the glass. Also reported is the softening of oil paints and yellowing of paper in numerous cases.

Some of the natural plant products screened for their fungicidal and fungistatic properties are *Allium satiuvm* (Kshemakalyani *et al.*, 1990; Arya *et al.*,1995; Abdul Hafez and El-said 1997), *Curcuma longa* (Saju *et al.*, 1998), *Eucalyptus citrodora* (Baruah *et al.*, 1996) and *Eugenia caryophyllus* (Chingduang *et al.*,1995), etc.

Botanical pesticides should be looked at as long term and effective alternatives for chemical pesticides. Shah and Arya (2001) found cashewnut oil, neem oil, eucalyptus oil and Godavaj powder (Sweet flag) effective against four fungal biodeteriogens. Aqueous leaf extracts of a large number of plants has been tried (Sekhawat and Prasad 1971; Khanna and Chandra 1972; Dixit *et al.*, 1983; Arya and Mathew 1990; Arya *et al.*, 1995). Dry rhizome powder has volatile oil with insecticidal properties. Its use is common in the Oriental Institute of Baroda as has been found to be antifungal, too. *Eucalyptus* leaves may be effective due to the presence of Cineole in them (Arya 1988). It has been observed that the activity of extracts may be lost due to autoclaving(Lokesha *et al.*, 1986).

CONCLUSION

The need of the hour is to make periodic surveys of museum objects to detect damages caused by microbes, to isolate the fungi, study the detailed behaviour of fungi and suggest suitable control methods. These methods need not be just effective but should meet the other criteria like preserving the aesthetic quality, be economically viable and eco-friendly, with no side effects, so as to preserve the works of art for posterity.

ACKNOWLEDGEMENTS

I thank Prof. G.P. Senan, Head, Department of Botany, M.S. University of Baroda, Vadodara for providing the necessary laboratory facilities. I am grateful to the Director, Agharkar Research Institute, Pune, and the Head, Division of Plant Pathology, IARI, New Delhi, for confirming the identity of the fungi. Thanks are due to U.G.C. for awarding a Project on 'Eco-friendly approach to control fungal biodeteriogens in Museums of Gujarat'.

REFERENCES

Abdul Hafez, S.I.I. and El-Said, A.H.M. (1997) Effect of garlic, onion and sodium benzoate on the mycoflora of pepper, cinnamon and rosemary in Egypt. *Int. Biodet. and Biod.* **39**:100–1001.

Arya, A. (1988) Control of Phomopsis fruit rots by leaf extract of certain medicinal plants. In: *Indigenous Medicinal Plant Symp.*(Kaushik M.P. ed.) Pub. By Today and Tomorrow's Printers and Publishers New Delhi, 41–46 pp.

Arya, A., Chauhan, R. and Arya, C. (1995) Effect of allicin and extract of garlic and *Bignonia* on two fungi. *Indian J. Mycol. Pathol.* **25 (3)** 316–318.

Arya, A. and Mathew, D.S. (1990) Control of chiku fruit rots by leaf extracts of certain medicinal plants. *Res. J. Pl. Env.*, **6 (1):**31–33.

Baruah, P., Sharma R.K., Singh R.S. and Ghosh, A.C. (1996) Fungicidal activity of some naturally occurring essential oils against *Fusarium moniliforme*. *J. of Essential oil Research*, **8 (4)**:411–412.

Chile, S. (1993) Study of airborne fungal spores at Jabalpur with special reference to allergenicity. Ph. D. Thesis R. D. University, Jabalpur.

Chingduang, S., Manoch, L., Tantayaporn, S., Vong Kaluang, C. and Siriacha, P. (1995) Fungal deterioration of book in libraries. In Preprints of: *International Conference on Biodeterioration of Cultural Property*, Bangkok, pp. 472–480.

Davis, R.R. and Smith, L.P. (1973) Aspects of human allergy and fungal spores. *Clinical Allergy*, **3**:23–267.

Dixit, S.N., Dubey, N.K. and Tripathi, N.N. (1983) Fungitoxic essential oils vis-à-vis disease control. In: *Recent Advances in Plant Pathol.* (Eds. Husain, A., Singh, K., Singh, B.P. and Agnihotri, V.P.) Print House (India) Lucknow, 248–252.

Foster, R.A. (1986) The larger provincial museums and art galleries. In: *Manual of Curatorship.* Butterworths, London, 66–75.

Gregory, P.H. (1973) *The Microbiology of the Atmosphere,* Leonord Hill Books, Aylesbury, England 1973 2nd ed. pp. 377.

ICOM (1990) *ICOM Statutes code of Professional ethics*, Paris pp. 3.

Kshemakalyani, S.B., Telore, R., Madhavi, B. and Patel, G.S. (1990) The effect of Allicin and extracts of garlic on *A. flavus* and *A. parasiticus. Indian J.of Mycol. and Pl. Pathol.*, **20 (5)**:247–248.

Khanna, K.K. and Chandra S. (1972) Antifungal activities in some plant extracts. *Proc. Nat. Acad. Sci. India,* **42 (13):** 300–302.

Lakshmikant and Mathur S.N. (1991) Biodeterioration of cultural property by cellulolytic fungi. In: *Biodeterioration of Cultural Property* (eds. Agrawal O. P. and Dhawan S.) MacMillan India Ltd. New Delhi 168–172.

Lokesha, S., Vasantha, K. and Shetty, H.S. (1986) Effect of plant extracts on growth and sporulation of *Aspergillus flavus. Plant Ds. Res.* **1 (1–2)**: 7981c. f.: *Rev. Pl. Path.,* **67** (19880:4864

Mishra, G. and Mishra, K.B. (1992) Aeromycoflora of a mica mine. In : *National symposium on Management of microbes in Service of Mankind,* Allahabad Univ., 19–21.

Nair, S.M. (1972) In: *Conservation in Tropics*. Proc. of AST-Pacific Seminars on con. of cul. Properties (Ed. Agrawal, O. P.) Rome 150–158.

Nair, S.M. (1980) Insect pesticides of textile materials in museum. In: *Conservation in tropics* ed. O. P. Agrawal, pp. 150

Nigam, R.K. and Pathak, N.C. (1991) Environmental mycology of a tannery and a textile unit at Kanpur: Allergical aspects, In: *Biodet. of Cultural Prop.* (Eds. Agrawal O. P. and Dhawan S.) Macmillan India Ltd. New Delhi, 375–386.

Riecelli, A. (1999) Fungal growth on samples of paper: Inhibition by new antifungals, *Restaurator,* **20 (2):** 97–107.

Saju, K.A.,Venugopal, M.N. and Mathew, M.J. (1998) Antifungal and insect-repellant activities of essential oil of turmeric (*Curcuma longa L.*) *Curr. Sci.*, **75** 660–662.

Sekhawat, P.S. and Prasad, R. (1971) Antifungal activities of some plant extracts on Inhibition of spore germination. *Indian Phytopath.*, **24**: 800–802.

Shah, N.R. and Arya, A. (2001) *Studies on some fungal biodeteriogens.* Pub by Bhartiya Kala Prakashan, New Delhi, 212.

Spieksma, F.T.M. (1980) Importance of aerobiological studies for the prediction of human diseases. In: *Aerobiological Studies for the prediction of human diseases* Aerobiol. Symp. Munich, 307–315.

Tilak, S.T., Sharma, B.R.N., Sengupta, S.R. and Kulkarni, R.L. (1970) The deterioration of Ajanta and Ellora paintings. In *Con. of Cul. Prop.* In India Proceedings of V. Seminar, New Delhi, 77–82.

Verma, K.S. and George, A.M. (1997) Aeromycoflora of Jabalpur. *Res. J. Chem. Environ.*, **1**: 29–31.

Wadhwani, K., Srivastava, A.K., Jamil Z., Rizvi, H.I., Mishra, A.K. and Negi, H. (1986) *Jour. Rec. Adv. Appl. Sci.*, **1**, 8–12,

Walston, S. (1976) *Proceedings of the ICCM National conference,* Canberra, pp. 145.

22

Mushroom Research and Development: A Perspective

R.P. SINGH AND PRATIBHA BHATT

Mushroom Research and Training Centre, G.B. Pant Univ. of Agriculture and Technology, Pantnagar-263 145, (U.P.), India

SUMMARY

Mushrooms have been picked up and eaten by men since times immemorial. They are considered to be a good source of proteins, vitamins with low fat. In this article, different aspects of mushroom research and development, including its importance have been highlighted. The Strength, Weakness, Opportunities and Threat (SWOT) analysis of mushroom production along with future strategies have been pointed out.

Keywords: Mushrooms, Mushroom Production, Importance

IMPORTANCE OF MUSHROOMS

Nutritional

Today, the popularity of mushrooms is not only due to their culinary value but also to their potential as a source of proteins that can enrich human diet, especially in some developing countries where meat may be rare and expensive. Gram for gram, mushrooms contain more protein than either fruits or vegetables. They can be eaten, as they are, cooked or raw, unlike other protein sources such as soya. Mushrooms are also low in cholesterol. Besides their protein content, mushrooms are also high in certain vitamins such as B, C and D.

Although devoid of Vitamin A, mushrooms contain high levels of riboflavin, thiamine and nicotinic acid. Their content of the antipellagra vitamin, niacin, is nearly equivalent to the levels found in pork or beef, considered to be the richest source of this vitamin. Mushrooms are also a good source of iron, potassium and phosphorus, in addition to folic acid, an ingredient known for enriching the bloodstream and preventing deficiencies. Mushrooms are also low in kidney ailments.

Medicinal

Several edible fungi like *L. edodes, F. velutipes, P. ostreatus, A. bisporus, P. nameko, Tricholoma matsutake* and *A. auricula* (Ikekawa *et al.*, 1969), have been reported to exhibit anti-tumour activity. Aqueous extracts of *A. auricula, Flammulina velutipes, L.edodes, Pholiota namiko, Pleurotus ostreatus, P. spodoeucus* and *tricholoma matsutake* had some inhibitory effect on the growth of tumours by 43–92% as compared to the control groups (Ikekawa *et al.*, 1969). Some mushroom extracts have also induced the formation of interferon, a defence mechanism against some virus infections and have displayed hypochlesteroemic activity or an ability to lower cholesterol levels (Cochran, 1978).

Use of Agricultural Wastes

Mushrooms are grown on organic substrates, either raw or composted. Examples of such materials are straw, corncobs, sawdust, bagasse, wood pulp, cotton waste, oil palm waste, banana leaves, poultry waste, coconut husks, tree bark and leaves. Used compost from mushroom growing may also be recycled for use as animal feed, soil conditioning and fertilisers.

Labour Intensive

Mushroom cultivation involves various technologies. The choice of these technologies depends partly upon the available substrates, but most of all upon the provision of capital for the equipment required to facilitate and simplify the different steps involved. In the instances where no or limited capital is available, methods that require simple equipment can be used. Most of these low-cost methods, suitable for rural projects, are labour-intensive and can provide employment in both the semi-urban and rural areas. In fact, some methods may use family labour, thus providing all members of the family with employment.

Absence of Crop Competition

While availability of land may be a limiting factor in most crop production operations, it is not so for mushroom culture. Mushrooms require little space because they can be grown under cover, inside mushroom houses, in basements, garages or other suitable areas within the house. They may be located on any unused land or available space between trees or besides buildings or houses, so that they do not compete with other crops for planting space.

INTERNATIONAL SCENARIO

Mushrooms are cultivated in more than hundred countries, with an estimated total production of 6.16 million tonnes (1997). Overall, world mushroom production increased more than 12% annually from 1981 to 1997. However, the total world production of *Agaricus* mushrooms decreased from 71.6% in 1981, to 31.8% in 1997), even though the actual production of this mushroom increased from 0.9 million tons in 1981 to 1.95 million tons in 1997, an increase of 2.2 times. The reason is that other mushrooms gradually became popular, particularly *Lentinula edodes,* which increased both in percentage of world production, from 14.3 to 25.2%, and in output, from 0.18 million tons to 1.56 million tons, and *Pleurotus* from 2.8 to 14.2% and from 0.035 million tons to 0.87 million tons, respectively. China has become an enormous producer and consumer of cultivated edible mushrooms and also a major producer of medicinal mushrooms. Total production in China in 1997 was 3.91 million tons, which amounted to 63.6% of the total world output. *Lentinula edodes* is a leading cultivated mushroom in China. Its production was 1.39 million tons in 1997, which is equal to 35.6% of the total Chinese mushroom production.

The largest importer of pre-prepared and preserved mushrooms is Northern America, with about 81,000 tons in 1996 followed by EU, importing about 62,000 tons. Most of these supplies come from China and East European countries, while Indian mushrooms have an insignificant presence. It is rather interesting to note that USA, despite producing about 3,57,000 tons of mushrooms, further imports 1,00,000 tones, thus becoming the largest consumer of mushrooms in the world. For India too, USA is the major market for its mushrooms to concentrate upon. Though China has been the largest exporter to USA, their presence is likely to suffer due to antidumping duty imposed on China, which could be an emerging opportunity for India to increase its exports to USA. The year-wise supply of mushrooms to USA from 1995–97 is given in Table 22.1.

Table 22.1: Mushroom quantity (M.T.) of Exports, by Leading Countries to USA

Countries	1995	1996	1997
China	34,198	31,840	33,557
Indonesia	16,826	14,210	16,595
Hong Kong	4,146	2,550	2,196
Chile	5.069	4.245	3,026
Taiwan	2,646	2,929	3,011
Mexico	1,754	2,597	2,666
Netherlands	2,464	702	600
India	2,749	2,015	4,566
Others	8,938	8,143	8,091
Total	78,790	69,229	74,310

Source: USDA april 1998 as presented in Mariental study, 1999.

Table 22.2: Estimates of Future Demand, Supply and Demand-Supply Gap for North America for the Years 1997–2006

Year	Demand (M.T.)	Suppy (M.T.)	Demand Supply Gap (M.T.)
1997	5,23,744	4.26,752	1,26,992
1998	5,43,123	4,32599	1,40,524
1999	5,63,218	438,526	1,54,693
2000	5,84,057	4,44,533	1,69,524
2001	6,05,667	4,50,625	1,85,044
2002	6,28,077	4,46,797	2,01,280
2003	6,51,316	4,63,055	2,18,261
2004	6,75,415	4,69,399	2,36,016
2005	7,00,405	4,75,830	2,54,575
2006	7,26,320	4,82,349	2,73,791

Source: Mariental study 1999.

The demand-supply gap for mushrooms in Northern America is likely to go up from 1,26,962 MT, 1997 to 2,73,971 MT by the year 2006. The unfortunate situation which can affect Indian supply to some extent is that the American consumers are switching-over from processed to fresh mushrooms, and it is difficult for India to export mushrooms in a fresh form because of the distance.

NATIONAL SCENARIO

Mushroom farming in India is hardly 3 decades old, and with an initial lag-phase it has started showing an upward trend. From a meagre 4000 ton production during 1985, India's total mushroom production reached 30,000 tones in 1995, which has further increased to 40,000 tons in 1996–97, and 52,620 tons in 2000–01. Similarly, its productivity has increased from 6kg/q compost in the year 1970 to 22kg/q compost during 1995. However, on some other fronts, like diversification of species, adoption of improved technology/strains and consumption of mushrooms by the masses, India has more or less remained static or has shown very sluggish growth. For example, even today, white button mushroom (*Agaricus bisporus*) comprises over 90% of the total mushroom production in our country. Although, artificial cultivation of paddy-straw mushrooms (*Volvariella* spp.) and oyster mushrooms (*Pleurotus* spp.) was standardised long ago, and a large part of the country having tropical/subtropical climate was more suitable for the cultivation of these two types, yet their adoption by the growers is yet to attain the desired levels. This is despite the fact that cultivation of these two types requires much less monetary inputs and their cultivation technology is much simpler than that of white button mushroom. In contrast, our neighbour China, managed to increases the production of oyster mushrooms by 400% during the late eighties, and even today China is the leading producer of that mushroom. India is yet to achieve such a landmark in case of any of these mushrooms, although it possesses all the necessary inputs for such a breakthrough, viz favourable climate, abundant agro-wastes, cheap man-power, a largely vegetarian population with rampant protein malnutrition, etc.

Export-oriented farms, improved/imported technology and strains with ample mechanisation are being adopted, while in seasonal farms, even the old and outdated technology is not being properly practiced. In between these two extremes, there are some small, climate controlled farms, which practice indigenous technology/strains with medium results. Obviously, there is considerable variation in the productivity, as well as quality of mushrooms being produced in these farms, which also reflects on the profitability of these farms. All these together have hampered the growth of the mushroom industry on a sound footing in our country, which in turn has arrested the increase in domestic consumption of this nutritious food thereby creating unnecessary marketing problems. Statewise estimates of mushroom production during 1996–97 and 2000–01 (Table 22.3) indicate that most of the states have ventured into this non-traditional farming activity, but no effort to develop the domestic marketing consumption of this highly perishable commodity has been made in any of these states, either by the private or public sector. Most of the big

Table 22.3: Statewise Production of Cultivated Mushrooms in India

Name of the State	Production (MT) 1996–97		Type of Mushroom 2000–2001
Jammu & Kashmir	400	400	*Agaricus*
Punjab	4000	13,000	*Agaricus*
Himachal	500	500	*Agaricus*
Haryna	4000	5000	*Agaricus*
Uttar Pradesh	4000	2000	*Agaricus/Pleurotus*
Uttaranchal	—	2640	*Agaricus*
Madhya Pradesh	2000	2500	*Agaricus/Pleurotus*
Rajasthan	50	80	*Pleurotus/Colocybe*
Maharashtra	7000	7000	*Agaricus/Pleurotus*
Goa	1000	1000	*Agaricus/Pleurotus*
Karnataka	600	700	*Agaricus/Pleurotus/ Volvariella*
Kerala	100	300	*Pleurotus/Volvarella*
Tamil Nadu	10,000	11,000	*Agaricus/Pleurotus/ Volvariella/Calocybe*
Andhra Pradesh	2500	3000	*Agaricus/Pleurotus*
Orrisa	700	700	*Pleurotus/Volvariella*
West Bengal	250	300	*Agaricus/Pleurotus*
Northeaster States (Including Sikkim)	1000	1200	*Agaricus/Pleurotus/ Volvariella/Lentinus*
Delhi/Pondichery/Chandigarh/Others	1000	1200	*Agaricus/Pleurotus*

units have therefore to depend entirely on the International market to sell their produce, at whatever price and terms they get from the foreign buyers.

MUSHROOM GROWING SYSTEMS

A. Marginal Scale production:
   - Crop Rooms (Huts): Made up of Sarkanda, Bamboo, Straw and grasses
   - Crop Room/Hut size: 40′ × 20′ × 12–14′ or 35′ × 18′ × 10–12′
   - Containers: Shelves or racks of Bamboo and Sarkanda
   - Composting: Long method
   - Yield: 10–12 kg/100kg compost in 8–10 weeks of harvesting

B. Small Scale production:
   - Crop Rooms: Conversion of old buildings into crop rooms or insulated crop rooms
   - Crop Room size: 40′ × 18′ × 12′or 50′ × 21′ × 12′
   - Containers: 3 to 5 tire bamboo shelves or metallic racks for 10 to 12 kg compost in bags
   - Composting: Long method/Short method
   - Yield: 10–15 kg/100kg compost in 8–10 weeks of harvesting

C. Industrial Scale production:
   - Crop Rooms: Insulated and controlled crop rooms
   - Crop Room: 48′–100′ × 18–27′ × 12–18′
   - Containers: Metallic racks for bags/ shelves
   - Composting: Short method
   - Yield: 18–22 kg/100kg compost in 5–6 weeks of harvesting

Export of mushrooms from India started looking up only in the nineties, due to the addition of preserved mushrooms in the export quota. Prior to that, mostly dried collections from forests comprising of morels/oyster were being exported from India. Table 22.4 shows that there was a tremendous increase in mushroom export from 1990–91 to 1997–98, and their value exceeded 50 crores of rupees. However, the major share (less than 80%) of the export has gone to USA and there is an urgent need to expand and diversify the foreign market for Indian mushrooms. In the international market, our main competitors have often been found to be lacking in quality, yet their cost of production being lower, their produce is proving more competitive in the world market. Indian mushroom exporters have therefore to face double challenge of maintaining good quality and keeping the cost of production competitive.

CHOOSING THE SPECIES OF MUSHROOMS FOR CULTIVATION

Several factors must be considered in choosing the types of mushroom to be cultivated. One important factor is the prevailing climatic conditions in the area where the mushroom farm is to be established. The different mushrooms that can now be cultivated artificially have different temperature requirements

Table 22.4: Mushroom Export from India –Status (1991–97)

	Indian Production –		Total 40,000 tons Button Mushroom Other Mushroom	(1996–97) 90% 10%
Years	Fresh/Chilled (MT)	Dried (MT)	Preserved (MT)	Total Value (Lakhs)
1991	—	—	128.00	821.65
1992	37.6	44.51	1174.76	1623.64
1993	110.53	49.17	3294.20	2065.90
1994	41.53	37.70	4811.48	2393.52
1995	1560.00	39.40	6883.00	5128.00
1996	2343.67	92.00	5119.88	4892.12
1997	5711.18	106.03	520.56	5697.57

To Countries: 80% to USA, Rest to Netherlands, Israel, Canada, Russia, South Africa, Italy, Kuwait, Philippines, Oman, France, UAE, Switzerland, Germany, Korea, Sweden, Australia, Brazil, Denmark and Others.
Source: Conference on Mushroom Industry in India, New Delhi, 1999.

Table 22.5: Temperature Requirement for Different Mushrooms

Common name	Scientific name	Temp.°C
Button	Agaricus bisporus	14–20
Dhingri	Pleurotus spp.	18–28
Shiitake	Lentinula edodes	14–22
Black ear	Auricularia spp.	18–34
Milky	Calocybe indica	25–35
Paddy straw	Volvariella volvacia	28–38

for mycelial growth and fruiting. Another factor to consider is the location of the mushroom farm, and related to this is the availability of the substrate for growing the chosen species. One should consider the distance between the farm and the place where the substrate will be obtained: the greater the distance, the higher will be the cost of transporting the substrate and the higher the cost of production. Other factors are the available technology and the acceptability of the specific mushroom to the local people or to the market for which it is intended.

The production technology for many species of mushrooms is undergoing dynamic changes. While the relatively rapid improvements in production technology may not be of much consequence to the small farmer, they are crucial to the larger commercial growers. Prior to choosing a particular species, it may be desirable to survey the available technology, to determine the most suitable type of production system. For example, when considering the cultivation of shiitake, (*Lentinula edodes*) consider whether the mushroom should be grown on supplemented sawdust or on natural logs. The state of development of the technology for each species should, therefore, be carefully considered before any decision is made.

SWOT ANALYSIS

Strengths, Weaknesses, Opportunities and Threats of Mushroom Production (SWOT)

A. Strengths:
- Abundant raw material availability.
- Enormous domestic market.
- Cheap manpower.
- Strong technical base.
- Strategic location.
- Climatic zones congenial for mushroom production.
- Govt. support.

B. Weakness:
- Much higher cost of bank borrowings.
- Costly packing material.
- Poor infrastructure ie inadequate research, training centres and spawn compost supply centres.
- Higher cost of energy.
- Inadequate procurement and processing units.
- High transportation costs.
- Improper marketing channels.

C. Opportunities:
- Increasing domestic demand.
- Technology absorption achieved.
- Production in other countries declining.
- International trade barriers breaking down.

D. Threats:
- China dumping mushrooms at cheap rates.
- Demand shifting towards fresh and speciality mushrooms.
- Crash of sea money market and imminent devaluation of Chinese currency.
- Increase in cost of production due to transportation of materials for installation of the production unit and raw materials for its operation.

(G) FUTURE STRATEGY

Key programmes and thrust areas that should be addressed:

A. Research

- Conservation and maintenance of genetic resources, germplasm characterisation and improvement.

- Improvement in technology suited to seasonal, as well as industrial production.
- Production technology of speciality mushrooms, including medicinal mushrooms.
- Newer substrates, packaging and production system for spawn.
- Integrated pest and disease management.
- Improvement in storage packaging and processing technologies, including diversification of value-added products.
- Basic studies on mushroom biology.

B. Development:

- Mushroom promotion campaign through transfer of technology by modern aids, including mass media.
- Mushroom training centres at district and state levels.
- Spawn and compost supply units, initially at districts and then at block levels.
- Procurement and processing units of cooperative societies, initially at districts and then at block levels.
- Interest rate on loan and electric rates, as in agricultural activities.
- Promotion of mushroom species as per available temperatures in different months.

C. Selection and development of mushroom villages for rural entrepreneurship:

- Marketing of fresh mushrooms through 'Milk Sangh' / 'Mandis'.
- Promotion of 100% E.O.U.
- Research/Scientific backup to the growers.

CONCLUSION

A number of marginal, small and industrial mushroom farm units have been established in the country. Marginal farmers are seasonal in nature and grow 1–3 crops in temporary huts or old buildings or in small constricted temporary crop rooms. Most of the industrial growers are growing mushrooms under controlled conditions round the year have 4–5 crops of *Agaricus bisporus*. As of today, *Agaricus bisporus* (button mushroom) is the only one grown on commercial lines in the country. In contrast, China, Japan and the European Union produce *Lentinula edodes, Pleurotus* spp., *Auricularia polytricha, Volvariella volvacia, Flamulina velutips, Trimella fuciformis, Pholiota nemiko.* Therefore, cultivation of other mushrooms, besides *Agaricus bisporus* needs to be undertaken in different parts of the country.

Our country offers varied climatic conditions, raw materials and manpower for the production of mushrooms. For development of mushroom cultivation in an industrialised manner, technical and financial backing is required. Good quality spawn and compost needs to be prepared and supplied to marginal/ small growers to enhance their yields and a proper marketing channel needs to be organised / opened for them. It is need of the hour to establish mushroom research, training and promotion centres, at State levels, and spawn production and composting units at district/block levels to provide the technical back-up, training, spawn and compost to the marginal and small mushroom growers.

Serious efforts should be made to collect, maintain, identify and domesticate the wild mushroom flora of different altitudes for inclusion of newer species of mushroom. There is a need to refine the available technology for seasonal as well as industrial scale growers of the country. Procurement and processing of mushrooms should be initiated through cooperative societies. To attract industrialists and small growers, interest rate on loans and electricity rates should be like in other agricultural activities.

REFERENCES

Apeda and MSI.1999. Conference on Mushroom Industry in India Page 1–39.

Cheng, S.C. 1979. Studies on the nutritive value of the straw mushroom, *Volvariella volvacea*. M. Phil. Thesis. The Chinese University of Hongkong.

Cochran, K.W. 1978. Medical effects. Pages 169–18. In: S.T. Chang and W.A. Hayes. The Biological and Cultivation of Edible Mushrooms. Acad. Press. New York. 819 p.

FAO. 1990. Technical guidelines for mushroom growing in the tropics. Organ. U.N. Rome.

Ikekawa, T., Uehara, N., Maeda, Y., Nakamishi, M. and Fukuoka, F. 1969. Anti-tumour activity of aqueous extracts of some edible mushrooms. *Cancer Research* 92, 734–735.

Mariental Study. 1999. Scope of Export of Mushrooms from India-potential markets. In Conference on Mushroom Industry in India Page 17–33.

Rai, R.D., Dhar, B.L. and Verma, R.N. 1997. Advances in Mushroom Biology and Production Page 1 –415.

Singh, R.P. and Chaubey, H.S. 1995. Mushroom production Technologies. Pb. Directorate of communication, Pantnagar pp. 104.

Singh, R.P. 1997. White Button Mushroom Cultivation and Centres of Production in Uttar Pradesh. Advances in Mushroom Biology and Production. Page 380–389.

23

Assessment of Fungal Diversity: A Molecular Approach

JYOTI SAXENA AND BHAKTI BAJPAI

Biosci. & Biotech Department., Banasthali Vidyapith, Banasthali–304 022, (Raj.), India

SUMMARY

DNA fingerprinting allows the rapid assessment of the genetic structure of complex communities in diverse environments and the extent of changes caused by environmental disturbances. Although community profiling techniques have only recently been applied to the studies of soil fungal ecology, their influence on the field has been considerable. Molecular methods promise to generate data that can fill the gaps observed in details concerning the roles and interactions of different fungal species and strains. Nevertheless, it must be emphasised that no single set of primers or community profiling techniques is optimal for the assessment of fungal diversity in all instances, rather they should be selected with the aim of the research, the target group of fungi and data analysis in mind.

Keywords: Fungal diversity, Molecular approach, PCR/ RAPD/ RFLP/ITS/ ARDRA/ DGGE/OFRG

There are about 2 million kinds of living organisms on the earth, to which fungi contribute approximately a hundred thousand species and many more still await discovery. Fungi are important components of the ecosystem and human civilization. No matter which aspect of fungi we look at, they are highly diverse and versatile organisms, adapted to all kinds of environments. They play roles in processes, such as soil formation, nutrient cycling, nutrient transportation to plant roots and the transformation of waste materials into useful commodities, such as compost. Fungi also represent a source of food, pharmaceuticals and biological control agents and many of them being important pathogens. The saprobic fungi represent the the largest proportion of fungal species in soil and they perform a crucial role in the decomposition of plant structural polymers, such as cellulose, hemicellulose and lignin, thus contributing to the maintenance of the global carbon cycle. In addition, these catabolic activities enable fungi to grow on inexpensive substrates. This property, coupled with their ability to produce commercially interesting organic molecules and enzymes, explains the significant interest in biotechnological utilisation of fungi.

Although considerable knowledge about fungi has been amassed, most of these organisms remain uncharacterised. Estimates suggest that there are 1.5 million species of fungi on earth; however, only approximately 70,000 species have been described, leaving 95% of the species undescribed (Hawksworth,1991; 2001; Hawksworth and Rossman, 1997). The reasons for this deficiency includes habitats that have not been well investigated, organisms that are difficult to culture axenically such as obligatory associated fungi and inaccurate identification of catalogued samples. As only a small fraction of fungi present in soils can be cultured, conventional microbiological techniques yield only limited information on the composition and dynamics of fungal communities in soil. On the contrary, DNA based methods do not depend on the culturability of the microorganisms and therefore, they offer an attractive alternative for the study of complex fungal community structures. Community finger printing techniques have commonly been applied to studies of bacterial ecology, and have significantly led to an understanding of the role of the diversity of bacteria in the soil (Ranjard *et al.*, 2000; Johnsen *et al.*, 2001; Kozdroj and van Elsas, 2001) In contrast to the large amount of data regarding bacterial communities, few reports are available on total fungal diversity in soils (Viaud *et al.*, 2000, Anderson and Cairney, 2004). The development and application of new methodologies to characterise, isolate and identify soil biota has indicated that we have only scratched the surface of soil biodiversity. The advent of major taxonomy projects, such as, the fungal specific endeavour known as Deep Hypha (http://ocid.NACSE.ORG/research/deephyphae/) and the Tree of Life (http://tolweb.org./tree/phylogeny.html) contribute to the need for an estimate of the total diversity of the fungal kingdom and the proportion of fungi that will be difficult to obtain in pure culture. These estimates are also important for assessing the potential of discovering novel fungi of value to biotechnology industry.

Some of the commonly used molecular methods to characterise fungal diversity are Random Amplified Polymorphic DNA (RAPD), Restriction

Fragment Length Polymorphism (RFLP), Terminal-Restriction Fragment Length Polymorphism (T-RFLP), 18S rDNA sequencing, Amplified Ribosomal DNA Restriction Analysis (ARDRA), Internal Transcribed Spacer (ITS) sequences, Intergenic Spacer sequences (IGS), Automated Ribosomal RNA Intergenic Sequence Analysis (ARISA), Denaturing Gradient Gel Electrophoresis (DGGE), and Oligonucleotide Fingerprinting of rRNA Genes (OFRG).

POLYMERASE CHAIN REACTION (PCR)

PCR is a technique for amplifying a specific region of DNA, defined by a set of two 'primers' at which DNA synthesis is initiated by a thermostable DNA polymerase. Usually, at least a million-fold increase of a specific section of a DNA molecule can be realised and the PCR product can be detected by gel electrophoresis. The regions amplified are usually between 150–3,000 base pairs in length.

RANDOM AMPLIFIED POLYMORPHIC DNA (RAPD)

RAPDs are DNA fragments amplified by the Polymerase Chain Reaction (PCR), using short (generally 10 bp) synthetic primers of random sequence. These oligonucleotides serve as both forward and reverse primers and are usually able to amplify fragments from 3–10 genomic sites, simultaneously. Amplified fragments (within the 0.5–5 kb range) are separated by gel-electrophoresis, and polymorphisms are detected as the presence or absence of a bands of particular size. These polymorphisms are primarily considered to be due to variations in the primer annealing sites. RAPDs have been used for many purposes, ranging from studies at the individual level (eg genetic identity), to studies involving closely related species. Due to their very high genomic abundance, RAPDs have also been applied in gene mapping studies (Williams *et al.*, 1990; Hadrys *et al.*, 1992).

RAPD analysis has been widely employed to study intraspecific genetic diversity. Groppe *et al.* (1995) used microsatellite containing loci as molecular markers for population studies of *Epichloe* species, whereas, intraspecific polymorphism within isolates of *Drechslera tritici-repentis* and *Biopolaris sorokiniana* was analysed by a few workers, using RAPD (Oliviera *et al.*, 2002; Santos *et al.*, 2002). Similarly, *Stemphylium solani* isolates were assessed for their genetic diversity on the basis of RAPD markers and ITS of rDNA. Isolates from potato and tomato were distinct from each other and fell into separate groups, which provides evidence that *S. solani* attacking cotton in Brazil belongs to a distinct genotype (Mehta, 2001).

Peltola *et al.* (2002) studied the toxigenic diversity of two different RAPD groups of *Stachybotrys chartarum* isolates for trichothecene production. These RAPD groups, obtained by unweighted pair group method of arithmetic averages (UPGMA) analysis, were found to be identical to the grouping obtained by Tri5 gene specific PCR.

Advantages

Very low quantity of template DNA is required (5–50 ng per reaction); no sequence data for primer construction is needed; the assay is easy and quick, amenable to automation and the cost involved is low. RAPD has very high genomic abundance, is randomly distributed throughout the genome and the technique generates multiple bands per reaction.

Disadvantages

Despite several advantages, this technique has many limitations such as it requires purified high molecular weight DNA; reproducibility of results is low, and great precautions are needed to avoid contamination of DNA as short random primers that can amplify DNA fragments in a variety of organisms are used. Other constraints are the need of highly standardised experimental procedures because of sensitivity to reaction conditions, and band profiles cannot be interpreted in terms of loci and alleles. Further, it can detect dominant alleles only, and similar sized fragments obtained may not be homologous.

Other related techniques of RAPD that are used to study microbial diversity are:

(i) AP-PCR (Arbitrarily Primed Polymerase Chain Reaction): uses longer arbitrary primers as compared to RAPDs.
(ii) DAF (DNA Amplification Fingerprinting): uses 5–8 bp primers to generate a larger number of fragments.
(iii) MAAP (Multiple Arbitrary Amplicon Profiling): collective term for techniques using single arbitrary primers.

RESTRICTION FRAGMENT LENGTH POLYMORPHISM (RFLP)

Restriction Fragment Length Polymorphism (RFLP) is a technique in which organisms may be differentiated by analysis of patterns derived from cleavage of their DNA. If two organisms differ in the distance between sites of cleavage of a particular restriction endonuclease, the length of the fragments produced will differ when the DNA is digested with a restriction enzyme. The similarity of the patterns generated can be used to differentiate species (and even strains) from one another.

Restriction Endonucleases

Restriction endonucleases are enzymes that cleave DNA molecules at specific nucleotide sequences, depending on the particular enzyme used. Enzyme recognition sites are usually 4 to 6 base pairs in length. Generally, the shorter the recognition sequence, the greater the number of fragments generated. If molecules differ in nucleotide sequence, fragments of different sizes may be generated. The fragments can be separated by gel electrophoresis.

PCR / RFLP

Isolation of sufficient DNA for RFLP analysis is time-consuming and labour intensive. However, PCR can be used to amplify very small amounts of DNA, usually in 2–3 hours, to the levels required for RFLP analysis. Therefore, more samples can be analysed in a shorter time.

Individual mycorrhizae, representing different morphotypes, were subjected to analysis of esterases (EST) isozyme profiling and RFLP of the ITS from fungal rRNA genes, following amplification using PCR. Ten morphophytes displayed different isozyme profiles (Timonen *et al.*, 1997). High genetic diversity in arbuscular mycorrhizal fungi has been studied by RFLP of a small subunit rDNA in the presence of different types of primers. Microsatellite primers were used by Vandenkoornhuyse *et al.* (2001) whereas, Wubet *et al.* (2003) used Glomeromycota specific primers for the same purpose.

Mochizuki *et al.* (2003) differentiated strains of *Trichophyton mentagrophytes* using RFLP-PCR amplified ITS regions of ribosomal DNA. RFLP analysis of ITS region has been suggested by several authors as a means of discriminating between ectomycorrhizal fungi at species level (Gardes *et al.*, 1991, Bruns *et al.*, 1991). At genus level, ITS -RFLP separated the 26 isolates into 8 genera with any of the enzymes used.

TERMINAL-RESTRICTION FRAGMENT LENGTH POLYMORPHISM (T-RFLP)

T-RFLP (Terminal-Restriction Fragment Length Polymorphism) involves the analysis of total soil DNA for the presence of rDNA genes from particular groups of microbes. This technique is useful since a vast majority of microbes are not culturable (Amann *et al.*, 1995) and culture enrichment often introduces bias or is not accurate. The TRF method employs PCR to amplify rDNA genes from soil DNA samples. The population of amplicons which are labelled with fluorescent tags on one primer are cut separately with several different restriction enzymes, and then separated using gel electrophoresis in polyacrylamide or capillary gel electrophoresis to resolve the different fragment lengths present. TRF profiles ie patterns from each enzyme used to generate a given sample, can then be used for comparative or identification purposes. T-RFLP is a very powerful technique, particularly for the identification of target organisms within a sample. However, the inability to generate data from T-RFLP peaks makes the identification of unknown OTU's difficult, and the applications to studies on total fungal diversity more problematic.

Cloned PCR amplicons that have been generated from environmental DNA have also been used for assessing soil fungal diversity (Chen and Caimey,

2002; Anderson *et al.*, 2003b; Jumponnen, 2003). Although clones can be screened using RFLP to group clones into OTUs before DNA sequencing, thus reducing the number of clones requiring sequencing, it is difficult to determine how many clones require analysis to have a fully sampled diversity contained within a single sample. A systematic evaluation of the value and the potential of T-RFLP analysis for the study of microbial community structure and dynamics had been undertaken by Osborne *et al.* (2000). They found the technique to be reproducible and robust for the assessment of environmental DNA samples isolated directly from polluted and pristine soil.

In a study for the identification and differentiation of dermatophyte fungi, Jackson *et al.* (1999) concluded that RFLP analysis of NTS (Non-Transcribed Spacer) and ITS regions of the rDNA repeat is a valuable technique, both for molecular strain differentiation of *Trichophyton rubrum* and for species' identification of common dermatophyte fungi. Both the 18S and 15S regions of the fungal rDNA gene complex have been evaluated for TRF efficacy, and ITS appear to provide an accurate picture of fungal species diversity (Lord *et al.*, 2002). T-RFLP analysis of the ITS region was also performed by Klammer *et al.* (2002) to determine the fungal community composition. They evaluated the specificities of different primer sets against plant and fungal species isolated from the coastal scrub oak forest soil. The species richness and community evenness of fresh water fungi were analysed, using both DGGE and T-RFLP; species richness based on DGGE was higher than those based on T-RFLP analysis (Nikolcheva *et al.*, 2003).

18S RRNA

Variations within a small subunit 18S rRNA gene sequence can be used to study the diversity and ecology of the soil fungal community. In contrast to bacteria, where 16S rRNA gene sequencing is successfully used for identification up to species level, taxonomic identification of fungi based on the sequence of 18S rRNA gene is more problematic with identification commonly limited to genus or family level. This is primarily due to the lack of variation within 18S rRNA gene, between closely related fungal species, as a result of the relatively short period of evolution of the kingdom fungi as compared with bacteria (Hegenholtz and Pace, 1996). Further, exhaustive database of fungal reference sequences is lacking.

One of the major limitations in investigating fungal diversity in soil is the availability of suitable PCR primers. White and colleagues (1990) designed the first PCR primer for amplification of fungal 18S rDNA and ITS regions from extracted DNA. Although these primers have proved to be very powerful tools and are still widely used today, the lack of specificity for fungal templates limits their usefulness in mixed DNA samples. To overcome the limitations of the original 18S rDNA primers (White *et al.*, 1990), several new primer pairs have been developed to amplify a broad taxonomic range of fungi (White *et al.*,

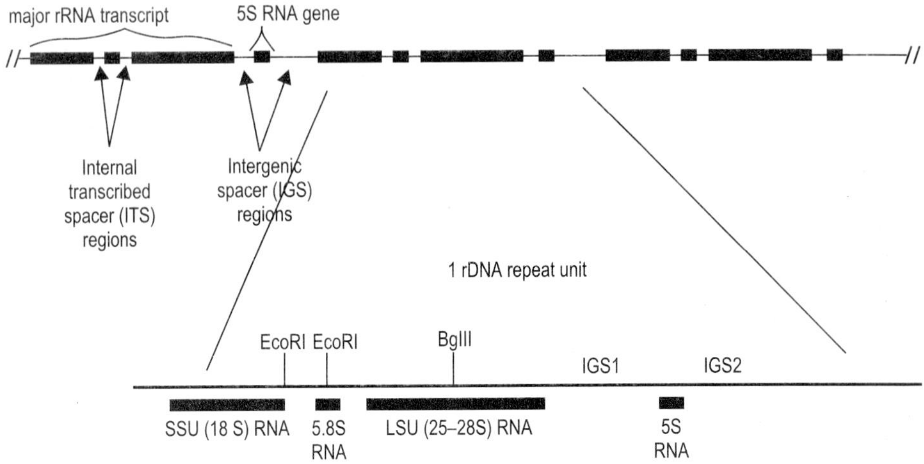

1990; Smit *et al.*, 1999; Borneman and Hartin, 2000; Vainio and Hantula, 2000), out of which some have been designed preferentially to amplify specific groups of fungi such as AM fungi (Helgason *et al.*, 1998). Based on 18S rRNA sequence variation, Wu *et al.* (2003) designed 108 oligonucleotide probes for 15 genera and 31 species. After homology searches and DNA hybridisation evaluation, 33 probes were verified as genus or species specific. These new probes can contribute to the molecular diagnostic research for environmental monitoring

Despite the limitations of 18S rRNA gene in fungi, the conserved and variable regions contained within the gene cluster have been used most widely for assessing fungal diversity (White *et al.*, 1990; Smit *et al.*, 1999; Borneman and Hartin, 2000; Hunt *et al.*, 2004). The early radiation of Glomeromycota within the kingdom has resulted in greater 18S rRNA variations belonging to this phylum, thus allowing discrimination between symbiotic AM fungal isolates to species, and sometimes below species level (Vandenkoornhuyse and leyval, 1998). Schabereiter–Gurtner *et.al.* (2001) analysed fungal communities from a historical church window glass by DGGE and phylogenetic 18S rDNA sequence analysis. Borneman and Hartin (2000) and Guo *et al.* (2001) detected endophytic fungi from frond tissue of *Livistona chinenesis* by amplification of 5.8S gene and flanking internal transcribed spacers, ITS1 and ITS2, using universal and fungal specific primers. Anderson *et al.* (2003) studied potential bias of fungal 18S rDNA and ITS-PCR primers for estimating fungal diversity in soil. Their work demonstrated the usefulness of 18S rDNA and ITS-PCR primers for assessing fungal diversity in environmental samples and also highlighted potential limitations to the approach with respect to PCR primer specificity and bias.

According to Millar *et al.* (2003), molecular identification should not be based solely on sequence analysis of the 18S rRNA gene, but should be confirmed by sequence analysis of an additional rRNA gene locus, such as the ITS region.

AMPLIFIED RIBOSOMAL DNA RESTRICTION ANALYSIS (ARDRA)

Another option for characterising a community of organisms is to use a technique called Amplified Ribosomal DNA Restriction Analysis (ARDRA). This method involves restriction enzyme digestion of amplified ribosomal DNA. The fragments are then visualised, using a polyacrylamide gel to produce a community specific pattern. While this process is more rapid than cloning, it is limited by several problems. This method is time consuming because it requires restriction enzyme digestion and gel electrophoresis of PCR-amplified products. Consequently, only a small number of samples can be assayed at a time. In addition, these as techniques are sensitive to contamination and are often inconclusive. Moreover, since ARDRA uses visual techniques, small fragments or small amounts of particular fragments may remain unresolved (SooKyung *et al.*, 2003).

ITS-ARDRA has been used by many workers to identify various wood decaying fungi (Annamalai *et al.* 1995; Schmidt and Moreth 1999; Jasalavich *et al.* 2000). The two ITS regions of rDNA can be PCR-amplified, using universal primers derived from the conserved regions (White *et al.* 1990), and subsequently analysed by restriction enzyme digestion (ARDRA-ITS) to generate fungal strain-specific restriction patterns that can be used as a marker for strain identification (Nho *et al.*, 1997). The fungal specific 18S and 28S primers FS38F (NS1; 5'-GTAGTCATATGCTTGTCTC-3') and FL2590R (ITS4; 5'-TCCTCCGCTTATTGATATGC-3') (*S. cerevisiae* numbering) were used to amplify most of the 18S rRNA genes and the entire ITS1/5.8S/ITS2 segment of the rRNA operon (Bruns *et al.*, 1991).

INTERNAL TRANSCRIBED SPACER SEQUENCE (ITS) / INTERGENIC SPACER SEQUENCE (IGS)

The ITS region is now perhaps the most widely sequenced DNA region in fungi. It has typically been most useful for molecular systematics at the species level, and even within species (eg, to identify geographic races). Because of its higher degree of variation than other genic regions of rDNA (SSU and LSU), variations among individual rDNA repeats can sometimes be observed within both the ITS and IGS regions. In addition to the standard ITS1+ITS4 primers used by most labs, several taxon-specific primers have been described that allow selective amplification of fungal sequences.

The nuclear-encoded ribosomal RNA genes (rDNA) of fungi exist as a multiple-copy gene family comprised of highly similar DNA sequences (typically from 8–12 kb each) arranged in a head-to-toe manner. Each repeat unit has coding regions for one major transcript (containing the primary rRNAs for a single ribosome), punctuated by one or more intergenic spacer (IGS) regions. In some groups (mostly basidiomycetes and some ascomycetous yeasts), each repeat also has a separately transcribed coding region for 5S RNA, whose position and direction of transcription may vary among groups. This technique is widely used for identification of fungi in clinical samples (Iwen *et al.*, 2002).

Molecular diversity of ericoid mycorrhizal fungi, using RFLP patterns from two ribosomal ITS and DNA sequences from ITS2, was characterised by Monreal *et al.* (1999). Thirty-four mycorrhizal isolates obtained from culture collection were divided into 16 groups, RFLP patterns were identical when fungal specific primers were used. The Phylogenetic and taxonomic heterogeneity of a rare opportunistic yeast pathogen, *Cryptococcus humicolus* was studied by Sujita *et al.* (2000) by sequence analysis of ITS region and 18S rDNA. *C. humicolus* strains were phylogenetically closely related to each other with the exception of one strain, and they clustered with *C. curvatus* and *Trichosporon* sp. with high boot strap values.

The greatest amount of sequence variation in rDNA exists within the IGS region (sometimes also known as the non-transcribed spacer or NTS region). The size of the IGS region may vary from 2 kb upwards. It is not unusual to find hypervariability for this region (necessitating cloning of individual repeat haplotypes).

Several patterns of organisation can be found in different groups of fungi:

1. Most filamentous ascomycetes have a single uninterrupted IGS region (between the end of the LSU and start of the next SSU sequence), which may vary in length from 2–5 kb or more. Amplification of the entire IGS region requires using primers anchored in the 3' end of the LSU gene (eg, LR12R) and 5' end of the SSU RNA gene (e.g., invSR1R).
2. In many ascomycetous yeasts and nearly all basidiomycetes, the IGS also contains a single coding region for the 5S RNA gene, which divides the IGS into two smaller regions that may be more easily amplified using primers. Depending on the orientation and position of the 5S RNA gene, the PCR may be used to sequentially amplify, either a portion of the intergenic spacer region (IGS) beyond the large subunit RNA coding region.

According to Fujita *et al.* (2001), multiplex PCR of ITS 1 and 2 regions, followed by electrophoresis, is a promising tool for the rapid identification of common and uncommon yeast strains. Earlier other workers (Gardes *et al.*, 1991; Henry *et al.*, 2000) have also used ITS 1 and 2 regions for identification of *Aspergillus* and ectomycorrhizal fungi. Specific primers targeting the conserved sequences of 5.8S and 28S ribosomal DNA, as well as those of 18S and 28S rDNAs, resulted in the respective amplification of the species specific ITS-2 region and of the ITS-1 and ITS–2 regions, which vary in amplicon length and sequence according to species.

AUTOMATED RIBOSOMAL RNA INTERGENIC SPACER ANALYSIS (ARISA)

rRNA intergenic spacer (RISA) is a simple and reliable method that exploits the variability in the length of the intergenic spacer (IGS) between the small (16S) and large (23S) subunit rRNA genes in the *rrn* operon. This approach has been used successfully to assess the structure of soil bacterial communities (Ranjard *et al.*, 2000).

Fisher and Triplett (1999) developed a method for ARISA involving a capillary electrophoresis system. This method provides a rapid and reproducible method for estimating the diversity and comparison of indigenous bacterial communities. Like terminal restriction fragment length polymorphism (T-RFLP) and length heterogeneity RFLP, ARISA involves the use of a fluorescence-tagged oligonucleotide primer for PCR amplification and subsequent electrophoresis in an automated system. Due to the high resolution of the gels and the high sensitivity of fluorescence detection, the number of peaks detected is much higher on ARISA profiles than on RISA profiles. Similarly, differences in the intensity of the bands can be estimated precisely, which allow a final comparison of the profile. However, this level of sensitivity might have some drawbacks because it may introduce a variability within profiles that have no biological origin. Another limitation of RISA is the availability of adequate spacer sequences in the databases.

Ranjard *et al.* (2001) standardised the conditions for the characterisation of bacterial and fungal communities by ARISA fingerprints. In fungal ARISA (F-ARISA) they exploited the length polymorphism of the nuclear ribosomal DNA (rDNA) region containing two internal transcribed spacers (ITS) and the 5.8S rRNA gene (ITS1-5.8S –ITS2).

Denaturing Gradient Gel Electrophoresis (DGGE)

Denaturing Gradient Gel Electrophoresis analysis (DGGE) was introduced by Muyzer *et al.* in 1993. The theory behind DGGE is very simple, the two strands of a DNA molecule separate, or melt, when heat or a chemical denaturant is applied. The temperature at which a DNA duplex melts is influenced by two factors:

(i) The hydrogen bonds formed between complimentary base pairs, GC rich regions melt at higher temperatures than regions that are AT rich.
(ii) The attraction between neighbouring bases of the same strand or 'stacking'.

Consequently, a DNA molecule may have several melting domains with characteristic melting temperatures (Tm), determined by the nucleotide sequence. DGGE exploits the fact that otherwise identical DNA molecules, which differ by only one nucleotide within a low melting domain, will have different melting temperatures. When separated by electrophoresis through a gradient of increasing chemical denaturant (usually formamide and urea), the mobility of the molecule is retarded at the concentration at which the DNA strands of low melt domain dissociate. The branched structure of the single stranded moiety of the molecule becomes entangled in the gel matrix and no further movement occurs. Complete strand separation is prevented by the presence of a high melting domain, which is usually artificially created at one end of the molecule by incorporation of a GC clamp. This is accomplished during PCR amplification, using a PCR primer with a 5' tail consisting of a sequence of 40 GC.

Advantages

The main advantage of DGGE is its high detection rate and sensitivity. The methodology is simple and a non-radioactive detection method is used. The PCR fragments may be isolated from the gel and used in sequencing reactions. This is an ideal method for the detection of mutations.

Disadvantages

The limitations of DGGE are that computer analysis and preliminary experiments are essential when setting up DGGE from scratch, so purchase of DGGE equipment may be required. Moreover, the primers are expensive because of the 40 bases of GC clamp. Additional primers may be required for sequencing. Another constraint is that an analysis of PCR fragments over 400bp is less successful and genes which are exceptionally GC rich are not easily analysed by DGGE. It also involves the use of formamide.

Variations of DGGE

Temperature Gradient Gel Electrophoresis (TGGE)

The chemical denaturant gradient is replaced by a gradient of increasing temperature down the gel.

Constant Denaturant Gel Electrophoresis (CDGE)

The chemical denaturant is at a constant concentration throughout the gel, equivalent to the melting temperature of the low melting domain. This approach requires different gel conditions for each PCR fragment to be analysed. The main application of CDGE is limited to the identification of known mutations.

Broad Range DGGE

One gel and a single set of conditions is used to screen all the exons of one gene. This approach is useful for identifying a mutation in a single patient and has been applied in the identification of mutations in the PAH gene, in patients with PKU.

DGGE / TGEE provides a rapid means of investigating soil fungal communities, particularly where the aim is to study shifts and changes in community composition (Smit *et al.*, 1999; van Elsas *et al.*, 2000; Anderson *et al.*, 2003b; Ascher *et al.*, 2004). Technique benefits from the ability to analyse and compare numerous samples on a single gel and allow rapid simultaneous comparison in between samples. Of late, fungal specific software package for statistical analysis of data has emerged, thereby, substantially improving our knowledge about the interpretation of ecological influence.

In general, shorter fragments of DNA result in better resolution between bands in a profile (<5000bp) thus, limiting the taxonomical information that can be obtained by the sequencing of excised bands, although some large products have also been used successfully in a few cases (Ranjard *et al.*, 2000, Landeweert *et al.*, 2004). Besides community profiling, DGGE has been used by some workers for the characterisation of fungi and species delineation. Kowalchuk *et al.* (1997) have used this method for the detection and characterisation of fungi infecting plant roots of *Ammophila arenaria*. Fagbola *et al.* (2004) distinguished *Colletotrichum circinans* from *C. coccodes* by DGGE analysis of 18s rDNA amplified fragments. The symbiotic AM and ectomycorrhizal fungal community composition from plant roots has been studied by amplification of a small subunit r RNA gene and DGGE (Opik *et al.*, 2003; Lendeweert *et al.*, 2003).

Nikolcheva *et a*l. (2003) compared the traditional and molecular approaches to study the diversity of aquatic Hyphomycete. They concluded that a combination of both the approaches are required to characterise the fungal community. According to them, both T-RFLP and DGGE are medium to high throughput techniques to analyse fungal community richness and evenness. Although both techniques require expensive equipment, many samples can be processed in a short time, and once the process is optimised, the processing costs per sample are low. One advantage of DGGE over T-RFLP in fungal community profiling is that it can also be used for the analysis of phylogenetic relationships among the species comprising the community. Moreover, bands excised from a denaturing gel can be used to design species—specific molecular probes to detect spatial patterns of fungal taxa on the plant substrate.

OLIGONUCLEOTIDE FINGERPRINTING OF RIBOSOMAL RNA GENES (OFRG)

The oligonucleotide fingerprinting of ribosomal RNA genes (OFRG) is a method that permits the identification of arrayed ribosomal RNA genes (rDNA) through a series of hybridisation experiments using small DNA probes. It provides a cost effective means to extensively analyse microbial communities and should have applications in medicine, biotechnology and ecosystem studies. Currently, most investigators analyse rDNA with methods such as denaturing gradient gel electrophoresis (DGGE) (Muyzer *et al.*, 1991), terminal random fragment length polymorphisms (T-RFLP) (liu *et al.*, 1997), and ribosomal intergenic spacer analysis (RISA) (Borneman and Triplett, 1997). Although these methods permit rapid analysis of numerous samples, they generate only superficial description of microbial community composition. Thorough depictions of community composition can be obtained by extensive sequence analysis of rDNA clone libraries. Yet, this approach is not commonly practiced because of the high costs associated with examining such diverse communities.

The recent development of array based methods, which permit thousands of hybridisation events to be examined in parallel, have brought great promise to the field of microbial ecology. In this approach, labelled rRNA or rDNA from environmental samples are analysed by their hybridisation to oligonucleotide probes, attached to a substrate. OFRG works by sorting arrayed rDNA clones into taxonomic clusters through a series of hybridisation experiments, using a single oligonucleotide probe. The OFRG process can be divided into six parts: (1) DNA extraction, (2) rDNA PCR, (3) Cloned library construction, (4) Array construction, (5) Array hybridisation, and (6) Data analysis. The arrays are printed by applying the PCR amplified rDNA fragments on nylon membranes, using a hand held 384-pin replicating tool. The most significant unresolved obstacle at this juncture appears to be the probe design. For this approach to work, each oligonucleotide probe must hybridise to a specific rDNA sequence or group of sequences. However, the development of such probes remains a significant challenge because of the highly conservative nature of rDNA and the extensive diversity of microbial life. An additional problem is the designing of probes for the multitude of microorganisms that have yet to be described. As with most PCR-based rDNA methods, OFRG does not provide quantitative population data. Thus, if firm conclusions concerning population trends are required, independent and quantitative analysis should be performed to verify the results obtained by OFRG.

Although the OFRG approach was originally developed to examine bacterial community composition, Borneman *et al.* (2001) used a simulated annealing algorithm to design probe sets that discriminated any clone type. Valinsky *et al.* (2002) developed an OFRG probe set for fungal rDNA and then demonstrated its utility by examining the fungal community composition of agriculture soil. The results indicated that the taxonomic identity generated by this new technique was consistent with those generated by a nucleotide sequence analysis.

SooKyung *et al.* (2003) identified wood decaying fungi by the hybridisation of immobilised sequence specific oligonucleotide probes with PCR amplified fungal rDNA-ITS spacers and found the method to be practical as well as robust.

REFERENCES

Anderson, I.C., Campbell, C.D. and Prosser, J.I. (2003a). Diveristy of fungi in organic soils under a Mooreland—Scots pine (Pinus sylvestris L.) gradient. *Emviron. Microbiol.* **5**:1121–1132.

Anderson, I.C., Campbell, C.D. and Prosser, J.I. (2003b). Potential bias of fungal 18S rDNA and internal transcribed spacer polymerase chain reaction primers for estimating fungal biodiversity in soil. *Environ Microbiol.*, **5**:36–47.

Anderson, I.C. and Caimey, J.W.G. (2004). Diversity and ecology of soil fungal communities: Increased understanding to the molecular techniques. *Environ. Microbiol.*, **6 (8)**: 769–779.

Annamalai, P.H., Ishii D.L. and Revathi R. (1995). Polymerase chain reaction and its applications in fungal disease diagnosis. *PflKrankh.*, **102**: 91 –104.

Ascher, J., Agnelli, A., Corti, G., Cecchernini, M. T., Namnipieri, P. and Pietramallara, G. (2004). Microbial biomass, soil respiration and DGGE –Analysis of total and extracellular DNA in a forest soil profile. Geophysical Research Abstarcts vol 6, 01846. sRef –ID 1607–7962/gra/EGU 04-A-01946.

Borneman, J. and Hartin, R.J.(2000). PCR primers that amplify fungal rRNA genes from environmental samples. *Appl. Environ. Microbiol.*, **66**: 4356–4360.

Borneman, J., and Triplett E. W. (1997). Molecular microbial diversity in soils from Eastern Amazonia: Evidence for unusual microorganisms and microbial population shifts associated with deforestation. *Appl. Environ. Microbiol.*, **63**: 2647–2653.

Borneman, J., Chrobak, M., Vedova, G.D., Figueroa, A. and Jiang, T. (2001). Probe selection algorithm with applications in analysis of microbial communities. *Bioinform.*, **17(1)**: S39–S48.

Bruns, T.D., White, T.J., Taylor, J.W. (1991). Fungal molecular systematics. *Ann. Rev. Ecol. Systematics*, **22**: 525–564.

Chen, D.M. and Caimey, J.W.G. (2002). Investigation of the influence of prescribed burning on ITS profiles of ectomycorrhizal and other soil fungi at three Australian sclerophyll forest sites. *Mycol. Res.*, **106**: 532–540.

van Elsas, J.D., Duarte, G.F., Keijzer-Wolters, A. and Smit, E. (2000). Analysis of the dynamics of fungal communities in soil via fungal specific PCR of sol DNA followed by denaturing gradient gel electrphoresis. *J. Microbiol. Methods*, **43**: 133–151.

Fagbola, O. and Abang, M.M. (2004). *Colletotirchum ciricinans* and *C, coccodes* can be distinguished by DGGE analysis of PCR-amplification of 18S rDNA fragments. *African J. Biotechnol.*, **3(3)**: 195–198.

Fisher, M.M. and Triplett, E.W. (1999). Automated approach for ribosomal intergenic spacer analysis of microbial diversity and its applications to fresh water bacterial communities. *Appl. Environ Microbiol.*, **65**: 4630–4636.

Fujita. S., Senda, Y., Nakaguchi, S. and Hashimoto, T.(2001). Multiplex PCR using internal transcribed spacer 1 and 2 regions for rapid detection and identification of yeast strains. *J. Clin. Micorbiol.*, **39(10)**: 3617–3622.

Gardes, M.., White, T.J., Fortin, J.A., Bruns, T.D. and Taylor, J.W. (1991). Identification of indigenous and introduced symbiotic fungi in ectomycorrhizae by amplification of nuclear and mitochondrial ribosomal DNA. *Can. J. Bot.*, **69**: 180–190.

Gomes, E.A., Kasuya, M.C.M., de Barros, E.G., Borges, A.C. and Areujo, E.F. (2002). Polymorphism in the internal transcribed spacer (ITS) of the ribosomal DNA of 26 isolates of ectomycorrhizal fungi *Genet. Mol. Biol.*, **25(4)**: 477–483.

Groppe K., I. Sanders, Wiemken, A. and Boller, T. (1995). A microsatellite marker for studying the ecology and diversity of fungal endophytes (*Epichloe* spp.) in grasses. *Appl. Envion. Microbiol.*, **61**: 3443–3949.

Guo, L.D., Hyde, K.D. and Liew, E.C. (2001). Detection and taxonomic placement of endophytic fungi within frond tissues of Livistona chinenesis based on rDNA sequences. *Mol. Phylogenet. Evol.*, **20(1)**: 1–13.

Hadrys, H., Ballick M., and Schierwater, B. (1992). Applications of random amplified polymorphic DNA (RAPD) in molecular ecology. *Mol. Ecol.*, **1**: 55–63;

Hawksworth, D.L. (1991). The fungal dimension of biodiversity: Magnitude, significance and conservation. *Mycol. Res.*, **95**: 641–655.

Hawksworth, D. L. (2001). The magnitude of fungal diversity: The 1.5 million species estimate revisited. *Mycol. Res.*, **105**: 1422–1432.

Hawksworth, D. L and Rossman, A.Y. (1997). Where are all the undescribed fungi? *Phytopathol.*, **87**: 888–891.

Helgason, T., Daniell, T.J., Husband, R., Fitler, A.H. and Young, J.P.W. (1998). Ploughing up the wood wide web? *Nature*, **394**: 431.

Henry, T., Iwen, P. C. and Hinrichs, S. H. (2000). Identification of *Aspergillus* species using internal transcribed spacer regions 1 and 2. *J. Clin. Microbiol.*, **38**:1510–1515.

Hugenholtz, P. and Pace, N. R. (1996). Identifying microbial diversity in the natural environment: A molecular phylogenetic approach. *Trends Biotechnol.*, **14**: 190–197.

Hunt, J., Boddy, L., Randerson, P. F. and Rogers, H. J. (2004). An evaluation of 18S rDNA approaches for the study of fungal diversity in grassland soils. *Microb. Ecol.* (In press).

Iwen, P. C., Hinrichs S. H. and Rupp, M. E. (2002). Utilisation of the internal transcribed spacer regions as molecular targets to detect and identify human fungal pathogens. *Med. Mycol.*, **40 (1)**: 87–109.

Jackson, C. J., Barton, R. C. and Evans, E.G.V. (1999). Species identification and strain differentiation of dermatophyte fungi by analysis of ribosomal-DNA intergenic spacer region. *J. Clin. Microbiol.*, **37 (4)**: 931–936.

Jasalavich, C.A., Ostrofsky A. and Jellison J. (2000). Detection and identification of decay fungi in spruce wood by restriction fragment length polymorphism analysis of amplified genes encoding rDNA. *Appl. Environ. Microbiol.*, **66**: 4725–4734

Johnsen, K., Jacobsen, C.S., Torsvik, V. and Sorensen, J. (2001). Pesticide effect on bacterial diversity in agricultural soils—A review. *Biol. Fertil. Soils.*, **33**: 443–453.

Jumpponen, A. (2003). Soil fungal community assembly in a primary successional glacier forefront ecosystem as inferred from rDNA sequence analysis. *New Phytol.*, **158**: 569–578.

Klamer, M., Roberts, M.S., Levine, L.H., Drake, B.G. and Garland J.L. (2002). Influence of elevated CO_2 on the fungal community in a coastal scrub oak forest soil investigated with terminal restriction fragment length polymorphism analysis. *Appl. Environ. Microbiol.*, **68(9)**: 4370–4376.

Kowalchuk, G. A., Gerands, S. and Woldendorp, J. W. (1997). Detection and characterisation of fungal infections of *Ammophila arenaria* (marram grass) roots by denaturing gradient gel electrophoresis of specifically amplified 18s rDNA. *Appl. Environ. Microbiol.*, **63(10)**: 3858–3865.

Kozdroj, J. and van Elsas, J. D. (2001). Structural diversity of microorganisms in chemically perturbed soil assessed by molecular and cytochemical approaches. *J. Micorbiol. Methods*, **43**: 197–212

Landeweert, R., Leeflang, P. Smit, E. and Kuyper, T. W. (2003). Molecular identification of ectomycorrhizal mycelium in soil horizons. *Appl. Environ. Microbiol.*, **69**: 327–333.

Landeweert, R., Leeflang, P., Smit, E. and Kuyper, T. W. (2004). Diversity of an ectomycorrhizal fungal community studied by root tip and total soil DNA approach.. *Mycorrhiza* (in press).

Liu, W.T., Marsh, T.L., Cheng, H., and Forney, L.J. (1997). Characterisation of microbial diversity by determining terminal restriction fragment length polymorphisms of genes encoding 16S rRNA. *Appl. Environ. Microbiol.*, **63**: 4516–4522.

Ludwig, R.I. and Schleifer, K.H. (1995). Phylogenetic identification and in situ detection of individual microbial cell without cultivation. *Micorbial Rev.*, **59**: 143–169.

Mehta, V.R. (2001). Genetic diversity among isolates of *Stemphylium solani* from cotton. *Fitopathologia Brasileira*, **26**: 703–709.

Michaelides K, Schwaab R, lalloz M.R.A., Schmidt W., Tuddenham E.G.D.(1995). Mutational analysis: New mutations. In: McPherson MJ, Hames BD, Taylor, GR, editors. PCR 2 A practical approach. Oxford: IRL Press at Oxford University Press; pp.255–287

Millar, B.C., Jiru, X., Walker, N. J., Evans, J. P. and Moore, J. E. (2003). False identification of *Coccidioides immitis*: Do molecular methods always get it right? *J. Clin. Microbiol.* **41 (12)**: 5778–5780.

Mochizuki, T., Ishizaki, H., Barton, R.C., Moore, M.K., Jackson, C.J., Kelly, S.L. and Evans, E.G. (2003). Restriction fragment length polymorphism analysis of ribosomal DNA intergenic regions is useful for differentiating strains of *Trichophyton mentagrophytes J. Clin. Microbiol.*, **41(10)**: 4583–4588.

Monreal, M., Berch, S.M. and Berbee, M. (1999). Molecular diversity of ericoid mycorrhizal fungi. *Can. J.* Bot., **77:** 1580–1594.

Muyzer, G., De Waal, E.C. and Uitterlinden, A.G. (1993). Profiling of complex microbial populations by denaturing gradient gel electrophoresis analysis of polymerase chain reaction –amplified genes coding for 16s rRNA. *Appl. Environ. Microbiol.*, **59**: 695–700.

Nho, S., Anderson M., Moore C. and Denning D. (1997). Species differentiation by internally transcribed spacer PCR and Hha I digestion of Fluconazole-resistant *Candida krusei, Candida inconspicua*, and *Candida norvegensis* strains. *J. Clin. Microbiol.*, **35**, 1036 –1039

Nikolcheva, C.G., Cockshutt, A.M. and Barlocher, F. (2003). Determining diversity of freshwater fungi on decaying leaves: Comparison of traditional and molecular approaches. *Appl. Environ. Microibol.*, **69(5)**: 2548–2554.

Oliviera de A.M.R., Matsumara, A.T.S., Prestes, A.M. and vander Sand, S.T. (2002). Intraspecific variability of *Bipolaris sorokiniana* isolates determined by random –amplified polymorphic DNA (RAPD). *Genet. Mol. Res.*, **1(4)**: 350–358.

Opik, M., Moora, M., Liira, J., Kdjalg, U., Zobel, M. and Sen, R. (2003). Divergent arbuscular mycorrhizal fungal communities colonised roots of *Pulsatilla* spp. in boreal Scots pine forest and grassland soils. *New Phytol.*, **160(3)**: 581–593.

Osborn, A. M., Moore, E. R. and Timmis, K. N. (2000). An evaluation of terminal restriction fragment length polymorphism (T-RFLP) analysis for the study of microbial community structure and dynamics. *Environ. Microbiol.*, **2(1)**: 39–50.

Peltola J., Niessen L., Nielsen K.F., Jarvis B.B., Anderson B. , Salkinoja-Salonen M. and Moller E. M (2002). Toxigenic diversity of two different RAPD groups of *Stachybotrys chartarum* isolates analysed by potential for trichothecene production and for boar sperm cell motility inhibition. *Can. J. Microbiol.*, **48(11)**: 1017–1029.

Ranjard, L., Poly, F. and Nazaret, S. (2000). Monitoring complex bacterial community using culture-independent molecular techniques: application to soil environment. *Res. Microbiol.*, **151**: 167–177.

Ranjard, L., Poly F., Combrisson J., Gourbiere F., Richaume A., Thiolouse J. and Nazaret S. (2000). Heterogeneous cell density and genetic structure of bacterial pools associated with various soil microenvironments as determined by enumeration and DNA fingerprinting approach (RISA). *Microb. Ecol.*, **39**: 263–272.

Ranjard, L., Poly, F., Lata, J.C., Mougel, C., Thioulouse, J. and Nazaret, S. (2001). Characterisation of bacterial and fungal soil communities by automated ribosomal intergenic spacer analysis fingerprints: biological and methodological variability. *Appl. Environ. Microbiol.*, **67(10)**: 4479–4487.

Santos, A.M.P.V., Matsumura, A.T.S. and Sand, S.T.D. (2002). Intraspecific genetic diversity of *Drechslera tritici-repentis* as detected by random amplified polymorphic DNA analysis. *Genet. Mol. Bot.*, **25(2)**: 243–250.

Schabereiter-Gurtner C., Pinar, G., Lubitz, W. and Rolleke, S. (2001). Analysis of fungal communities on historical church window glass by denaturing gradient gel electrophoresis and phylogenetic 18S rDNA sequence ananlysis. *J. Microbiol Methods,* **47(3)**: 345–354.

Schmidt, O. and Moreth, U. (1999). Identification of the dry rot fungus, *Serpula lacrymans* and the wild *Merulius, S. himantioides,* by amplified ribosomal DNA restriction analysis (ARDRA). *Holzforschung,* **53**: 123 –128.

Smit, E., Leeflang, P., Glandorf, B., van Elsas, J. D. and Wernars, K. (1999). Analysis of fungal diversity in wheat rhizosphere by sequencing of cloned PCR amplified genes encoding 18s rRNA and temperature gradient gel electrophoresis. *Appl. Environ Microbiol.* **65**: 2614–2621.

SooKyung, Oh. D., Kamden, P., Keathley, D. E. and Kyung-Hwan Han (2003). Detection and species identification of wood decaying fungi by hybridisation of immobilised sequence specific oligonucleotide probes with PCR amplified fungal ribosomal DNA internal transcribed spacers. *Holzforschung,* **57**: 346–352.

Sujita, T., Takashima, M., Ikeda, R., Nakase, T. and Shinoda, T. (2000). Phylogenetic and taxonomic heterogeneity of *Cryptococcus humicolus* by ananlysis of the sequences of the internal transcribed spacer regions and 18S rDNA, and the phylogenetic relationships of *C. humicolous, C. curvatus* and the genus. *Trichosporon. Microbiol Immunol.*, **44(6)**: 455–461.

Timonen, S., Tammi, H. and Sen, R. (1997). Characterisation of the host genotype and fungal diversity in Scots pine ectomycorrhiza from natural humus microsome using isozyme and PCR-RFLP analysis. *New Phytol.*, **135(2)**: 313.

Valinsky, L., Vedova, G. D., Jiang, T. and Borneman, J. (2002). Oligonucleotide fingerprinting of rRNA genes for analysis of fungal community composition. *Appl. Environ. Micorbiol.*, **68(12)**: 5999–6004.

Vainio, E.J. and Hantula, J. (2000). Direct analysis of wood inhabiting fungi using denaturing gradient gel electrophoresis of amplified ribosomal DNA. *Mycol. Res.*, **104**: 927–936.

Vandenkoornhuyse, P., Leyval, C. and Boninin, I. (2001). High genetic diversity in arbuscular mycorrhizal fungi: evidence for recombination events. *Heredity* **87**, 243–253.

Vandenkoornhuyse, P. and Leyval, C. (1998). SSU rDNA sequencing and PCR fingerprinting reveal genetic variation within *Glomus mosseae*. *Mycologia* **90**: 791–797.

Viaud, M., Pasquier, A. and Brygoo, Y. (2000). Diveristy of soil fungi studied by PCR-RFLP of ITS. *Mycol. Res.*, **104**: 1027–1032.

White, T.J., Bruns, T.D., Lee, S. and Taylor, J. (1990). Analysis of phylogenetic relationships by amplification and direct sequencing of ribosomal RNA genes, In PCR Protocols: A Guide to Methods and Applications. Inns. M.A. Gefland, D.H. Sninsky, J.J. and White, T.J. (eds) New York, Academic Press, pp. 315–322.

White, T.J., Bruns T., Lee S. and Taylor J. (1990). Amplification and direct sequencing of fungal ribosomal RNA genes for phylogenetics. *In*: PCR Protocol: A Guide to Methods and Applications. Eds. M.A. Innis, D.H. Gellfand, J.J. Sninisky, T.J. White. Academic Press, New York., pp. 315 –322.

Williams, J.G.K., Kubelik A.R., Livak K.J., Rafalski J.A., and Tingey S.V.. (1990). DNA polymorphisms amplified by arbitrary primers are useful as genetic markers. *Nucleic Acids Res.*, **18**: 6531–6535.

Wu, Z., Tsumura, Y., Blomquist, G and Wang, X. R. (2003). 18S rRNA gene variation among common airborne fungi, and development of specific oligonucleotide probes for the detection of fungal isolates. *Appl. Environ. Microbiol.*, **69(9)**: 5389–5397.

Wubet, T., Weib, M., Kottke, I. and Oberwinkler, F. (2003). Morphology and molecular diversity of arbuscular mycorrhizal fungi in wild and cultivated *Taxus baccata Can. J. Bot.*, **81(3)**: 255–266.

24

Mushrooms of Medicinal Importance and Their Components

PRATIBHA BHATT, S.C. SATI* AND R.P. SINGH

Mushroom Research and Training Centre, G.B. Pant Univ. of Agriculture and Technology, Pantnagar-263145 (Uttaranchal), India
Department of Botany, Kumaun University, Nainital-263002, (Uttaranchal), India

SUMMARY

Mushrooms have a long history of use in traditional Chinese and Japanese medicines. Recent studies suggest that mushrooms are Probiotic. They help our body strengthen itself and fight against illness. The present paper deals with some mushrooms such as *Ganoderma lucidum, Coriolus versicolour, Lentinula edodes, Grifola frondosa* and species of *Crodyceps, Agaricus, Pleurotus, Auricularia, Volvoriella, Armillaria, Tremella* which have been found useful. Beside having antibacterial, antitumour anticancer and antiinformatory properties, they help in reducing HIV-T deeper cells in AIDS. They also lower blood poressure and cholestrol in the body.

Keywords: Mushrooms, Medicinal use, Human health

The discovery of the so called 'wonder drug' penicillin was a landmark in the field of medicinal use of fungi. Since then several fungi have been well recognised for their antifungal, antibacterial, antiviral, antitumour and many other such properties of pharmacological value. In the recent past, a variety of medicinal preparations have been produced and marketed in the form of tablets and capsules from the mushrooms extracts. Mushrooms are perhaps the only fungi deliberately and knowingly consumed by human beings and they complement and supplement the human diet with various ingredients not encountered in or deficient in food items of plant and animal origin. Moreover, their unique chemical composition makes them suitable for specific groups suffering from certain physiological disorders or ailments. Mushrooms are regarded as an ultimate health food (King, 1993), low in calories due to the presence of good amounts of quality proteins, iron, zinc, vitamins, minerals and dietary fibres, which protect from digestive ailments and help in strengthening the human immune system.

Recent investigations have proved the empirical observations of the oriental herbalists that certain mushrooms possess very useful medicinal attributes. In 1991, the value of world medicinal crops was estimated at 8.5 billion dollars and in the same year, 1.2 billion dollars was estimated to have been generated from medicinal products from mushrooms. This was based on the sale value of products from *Coriolus, Ganoderma, Lentinula, Schizophyllum* and five other mushrooms (Chang, 1993).

Although mushrooms have traditionally been used for their gastronomic and nutritional appeal, there has always been great interest in certain mushrooms for their medicinal attributes. Production of medicinal mushrooms is increasing world wide, nowadays. In the present era, a variety of proprietary products based on mushroom nutriceuticals and pharmaceuticals have already been produced and marketed. Various mushrooms and their metabolic extracts have been reported to protect against cancer, tumour and pathogenic microorganisms. It is suggested that regular consumption of different mushroom varieties not only protects humans from heart trouble, but it also has medicinal potential for curing certain ailments.

Mushroom nutriceutical, as defined by Chang and Buswell (1996), is a new class of compounds that have been extracted from either mushrooms or the vegetative mycelium of mushroom species. It has potential therapeutic application, increased resistance to diseases and causes regression in diseases.

MEDICINALLY IMPORTANT MUSHROOMS

Mushrooms have a long history of use in traditional Chinese medicine. In fact, it is estimated that in China more than 270 species of mushrooms are believed to have medicinal properties, with 25% of them having antitumour properties. Few of the edible mushrooms have also gained importance in modern medicine for their pharmacological value. The details of medicinally important mushrooms are summarised hereunder.

Ganoderma lucidum (Reishi mushroom)

Ganoderma lucidum (Ling Zhi or Reishi in Chinese, Saruno, Koshikake or Mannendake in Japanese), belonging to higher Basidiomycetes has been used extensively as 'The mushroom of immortality' in China and other Asian countries for 2000 years.

It is full of medicinal compounds and is said to promote a healthy heart and liver. Reishi is also believed to be a powerful anticancer agent. It is used in Chinese medicine for its immunogenic, antibacterial and anti-inflammatory properties. It demonstrates its antitumour and immunomodulatory activities by increasing bodily resistance against the growth of tumours. It can also act as a metabolic regulator.

Two main classes ie Triterpenes and Polysaccharides, belonging to compounds present in *Ganoderma*, have been found to have pharmacological properties.

The antitumour effects of *G. lucidum* are due to polysaccharide (referred to as GP, Ganoderma polysaccharide). These polysaccharides, especially the B glucans fractions, have been found to have broad stimulatory effects on white blood cells. These effects may ultimately lead to the release of cytokines and lymphokines which exhibit antitumour properties (Chang, 1994). It has been shown that high molecular weight ß1-3, ß1-6 D glucans exhibit antitumour activity. Studies by Wang *et al.*, (1994) have shown that *Ganoderma* polysaccharide can potentiate the immune function by stimulating the production and release of lymphokines and monokines from T-cells and macrophages. Elevated levels of cytokines are capable of inhibiting tumour growth. It is also believed that Reishi can induce the production of chemotherapy agents such as interferon , a protein that is produced inside cells to fight viral infection and interleukin-1 (Jia *et al.*, 1993) and Interleukin-2 (Zhang *et al.*,1993 and Shiuh-Sheng-Lee *et al.*, 2003). The polysaccharides from *G. lucidum* are potent hepatoprotective agents b-D glucans appear to be one of the major contributors to the multiple biological activity of *G. lucidum* (Yihuai *et al.*, 2003).

The triterpines have been tested and found to lower blood pressure, and are anti-inflammatory and antiviral. Nature health practitioners in the West are beginning to use Reishi for its sleep promoting effects . Compounds called triterpenes are thought to be responsible for producing a calming effect on the nervous system. Triterpenes also show a variety of responses like platelet aggregation (Wang *et al.*, 1991). Ling Zhi-8 (LZ-8), a potent biological response modifier is a small protein (110 amino acids). The gene for coding LZ-8 has been cloned (Murasugi *et al.*, 1991).

Ganoderma research is now beginning to examine the specific mode of action of various compounds isolated from fruit bodies as well as mycelium. In Chinese history, it has been depicted as a symbol of longevity in artistic works. In recent years, this mushroom has achieved immense fame due to its wide medicinal usage. It is cultivated in China and Japan and is also found throughout India as a stem parasite of several trees.

Coriolus versicolour (Turkey tail)

This fungus is strongly called 'turkey tail' in English. Probably, because it has multi-coloured fruiting bodies which grow in overlapping clusters on rotten wood of broad leaf trees and occasionally on the trunks of pine. It is known in Chinese as Yun-zhi. Traditionally, it was used as a tea or a dried powder to treat infection or inflammation of the respiratory, urinary and digestive tracts, liver diseases, general weakness and tumours. *Coriolus versicolour*, also known as *Trametes versicolour*, is a common non-edible polypore which has been found to produce polysaccharides. Commercially known as Krestin or PSK an anti-tumour agent. It is a protein-bound polysaccharide that is obtained as hot water extract from the mycelium of *Coriolus versicolour* (Sakagami and Takeda, 1993).

This protein-bound polysaccharide (PSK or PSP) is a Biological Response Modifier (BRM), which is capable of modifying the host's biological response by stimulating the immune system and thus, exhibiting diverse therapeutic effects, particularly antitumor activity. The main component of the polysaccharide moiety of Krestin is b-glucanes. The main chain consists of a b 1–4 glucose polymer with branches at b1–3 end and b1–6 positions. It has minor components of galactose, mannose, xylose and fucose. Another biological response modifier substance derived from the mycelium of *Coriolus versicolour* is PSP (Polysaccharide peptide). In many respects, the properties of PSP are similar to PSK, but PSP contains small amounts of galactose, mannose, xylose, arabinose and rhamnose. PSP does not contain fucose which is present in PSK, and PSK lacks arabinose and rhamnose which are present in PSP (Yang *et al*, 1993)

Krestin or PSK, extracted from mycelium of *C. versicolar*, is prescribed for cancer of the digestive system, breast cancer and pulmonary cancer in Japan (Fukushima, 1989). PSK has been shown to activate functional maturation of macrophages and scavenging of oxygen radical.

Lentinula edodes (Shiitake)

L. edodes can be considered as a leader of mushrooms as is used, both for edible and medicinal purposes. It is known as Xiang gu in Chinese, which means fragrant mushroom, probably due to its characteristic flavour, and shiitake in Japan, which derives its association with the shiia tree. The shiitake (*Lentinula*) is the most popular edible mushroom in Japan. The Asians have used it for thousands of years as a tonic and stimulant to increase vitality, prevent cerebral haemorrhage strokes as well as to improved circulation.

L. edodes is source of the well known antitumour polysaccharide lentinan from the fruiting body or mycelia (Chihara, 1993). Lentinan is a highly purified, high molecular weight polysaccharide which does not attack cancer cells directly but produces its antitumour effect by activating different immune responses in the host.

In *lentinula edodes*, an active hypolipidemic substance, identified as Eritadenine, is found to exhibit a general response, affecting cholesterol, triglyceride and phospholipid levels.

Therefore, Lentinan is being studied not only for its antitumour properties, but also for its ability to lower cholesterol and blood pressure. This lowering of the serum cholesterol level, hypocholesteromia, has been attributed to Tokuda and Kaneda (1976).

Grifola frondosa (Maitake)

This species is found in eastern US, Europe and Asia. In Japan, it is known as Maitake, which means 'dancing mushroom'. *G. frondosa* is another mushroom which is highly effective in controlling blood pressure, diabetes and constipation. Preliminary research in United States at the National Cancer Institute indicates that Maitake prevents the destruction of HIV-T-Helper cells. A patient's T-helper cells count is measured to trace the progression of HIV to AIDS. Initial animal studies show that this mushroom may also be able to inhibit the escalation of blood glucose level, thus indicating anti-diabetic possibilities (Nanba 1993, King 1993).

Cordyceps species (Worm grass)

Cordyceps, an extremely rare mushroom, is a native of the high Himalayan mountains. It can be found for a short time, each summer, growing on its natural host—a caterpillar. Recent evidence has shown that the extract of *C. sinesis* has immunoregulator properties. One of the most valued traditional Chinese medicines is used commonly for the replenishment of the body health. It consists of the dried fungus *C. sinesis* growing on Lepidopteran (caterpillar) larvae. For medication, the fruiting body (fungus) and the worm (caterpiller) are used together.

The fruiting body and the caterpillar host of *C. sinensis* show close resemblance in their main constituents and antioxidation activity, and the result suggests that the function of the worm in *Cordyceps* is to provide a growth medium for the fruiting body (Li-sp *et al.*, 2002). The methanol extract of the fruit body of *C. sinensis* shows immunomodulatory response (Kuo-yuhchi *etal*, 1996). Its polysaccharide also shows hypoglycemic activity. *C. cicadae* is involved in the regulation of interleukin-2 (IL-2) and interferon-gamma (IFN-g) in human mononuclear cells (Weng-shu-cheng *et al.*, 2002).

Pleurotus species (Oyster or Dhingri)

Oyster mushroom is the third most popularly grown mushroom in the world, ranking second in India. In India, *Pleurotus sajor-caju* is one of the commonly growing species. *Pleurotus sajor-caju* has been shown to exhibit hypotensive properties and reduce the rate of nephron deterioration, which may be useful for chronic renal failure patients. (Tam *et al.*,1986). The oyster mushroom (*Pleurotus ostreatus*) has also been found to have hypocholesterolemic effects on rats (Bobek *et al.*, 1990). *Pleurotus tuberregium* also shows some antimicrobial activity. PSPC (Polysccharide Protein Complex) a polysccharide protein has been isolated and characterised from *Tricholoma giganteum* as it shows antitumour activity (Liu *et al.*, 1993, Ooi-vec, 2001). It was obtained from culture filtrate of the mycelium of *T. giganteum.*

SOME OTHER COMMON MUSHROOMS

A few other edible mushrooms have also gained importance in modern medicine for their various pharmacological values. An antitumour substance, namely 'Calvacin' has been obtained from the giant puffball *Calvatia gigantea* (Beneke E.A., 1963). Benedict and Brady (1972) emphasised that polyacetylenes are common in a number of genera of higher Basidiomycetes, including edible genera like *Pleurotus* and *Tricholoma.*

Clavatis gigantula, Auricularia polytricha, Flammulina velutipes and *Agaricus bisporus* are reported to be effective in lowering blood cholesterol (Kaneda, 1964). *A. bisporus* is the most commonly grown mushroom in the world as well as in India. Regular consumption of this mushroom makes the digestive system strong and induces the immune system against diseases. 'Retene', an antitumour substance, has also been reported to be present in this mushroom (Vogel *et al.*, 1974). Antitumour activity of *Flammulina* is due to polysaccharides and a low molecular weight protein, 'Proflamin' from *F. velutipes.* (Ikekawa, 2001).

Auricularia, the first mushroom cultivated by man has for long been a favourite in Chinese cuisine and is also used for its medicinal properties. It is used for the treatment of piles, stomach ailments, sore throat and tumours. Markhija and Bailley (1981) recorded antiplatelet aggregation activity in *Auricularia polytricha* due to adenosine.

Volvariella volvacea is an important mushroom of the tropics. The cardiotoxic protein (Volvatoxin-A) protects the body from cancer by preventing the respiration of tumour cells.

Armillaria mellea (Honey mushroom) is reported to be an excellent purgative medicine for stimulating the secretion of bile, and for alleviating dizziness and headache. It is also used for improving vision and counteracting opthalmia and night blindness.

Tremella aurantia (Golden ear mushroom) and *Tremella fuciformis* (Silver ear mushroom) are extensively used in Japan for preparing tonic, for strengthening health and resisting diseases in Japan beverages.

The prominent medicinal ingredients of mushroom are mostly polysaccharides which are different in their composition, mode of action, degree of ramification and polymerisation etc.

CONCLUSION

Medicinal fungi have become increasingly popular and are widely used as a traditional medicinal material in the Far East. For example, the premier

Fig. Plate 24.1. A. *Ganoderma* sp.1 (Reishi mushroom); B. *Agaricus bisporus* (Button mushroom) C. *Auricularia* (Ear mushroom); D. *Pleurotus sajor caju* (Oyster/ Dhingri); E. *Lantinula edodes* (Shitake); F. *Ganoderma* sp. 2; G. *Cordyceps sinensis* (Worm grass); and H. *Calocybe indica*

Ganoderma lucidum (ling-zhi) has been used for a broad spectrum of health benefits, ranging from preventive measures and maintenance of health to the regulation or treatment of chronic, as well as acute life-threatening illness. In recent years extensive research on mushroom polysaccharides has contributed to a wider application of these active bio-polymers in medicine

Medicinal mushrooms are being used in ayurvedic medicines since ancient times. This means that thoughtless research has been done in this field; recent research has proved beyond doubt that some mushroom possess pharmacological properties. The pharmacological uses of mushrooms demand further intensive research to screen other mushrooms, wild as well as cultivated, for their medicinal attributes. Mushrooms also have a long history of use in traditional Chinese medicine. Their legendary effects on promoting good health and vitality and increasing the body's adaptive abilities have been supported by recent studies. These studies suggest that mushrooms are 'Probiotic'. They help our body strengthen itself and fight illness by maintaining physiological homeostasis and by restoring the body's balance and natural resistance to diseases. The compounds they contain have been classified as Host Defense Potentiators (HDP), which can have immune system enhancement properties (boost immune system). That is one of the reasons why they are currently used as adjuncts to cancer treatment in Japan and China. It is of paramount importance that more research efforts is invested to characterise and standardise the bioactive ingredients of mushroom products for earning enduring public credibility and securing an expensive world market in the future.

REFERENCES

Beneke, E.A.1963 . Clavatia, Clavacin and cancet. *Mycologia* **55**: 227–270.

Benedict, R.G. and Brady, L.R. 1972. Antimicrobial activity of mushroom metabolites ,*Journal of Pharmaceutical Sciences*, **61**:1820–1822.

Bobek,P.Ginter.L,Ozdin and Cerven, J. 1990. Hypocholesterolemic effect of oyster mushroom (*Pleurotus ostreatus*) in rat with hereditary increased sensitivity to dietary cholesterol. *Biologia (Bratislava)*, **45**: 961–966.

Chang, S.T .1993. Mushroom Biology: The impact on mushroom production and mushroom products, in *Mushroom biology and Mushroom Products* ,Ed. by S.T.Chang, J.A.Buswell and Chiu, S.W. The Chinese University press, Hong Kong,3–20

Chang, S.T and Buswell, J.A.1996. Mushroom nutriceutical. *World Journal of Microbiology Biotechnolo.*, **12**:473–476.

Chang, R.,1994. Effective dose of *Ganoderma* in human Proc. contributred symposia 59 a, b. 5th International Myco. Cong. (Ed.by Buchanan, P.K., Hseu R.S. and Moncalvo, J.m.), 117–121

Chihara, G .1990. Lentinan and its related polysaccharides as host defence potentitor: their application to infectious disease and cancer. *'Immunotherapeuctic Prospects of Infectious Diseases'* (Ed. by K. Noel Masihi and W. Lange, Springer-Verlag, Heidelberg 1990). (Cited from Chihara, G. 1993)

Chihara, G. 1993. Medical aspects of lentinan isolated from *lentinus edodes* (Berk) Sing in *Mushroom Biology and Mushroom Products*. Ed. by Chang, S.T., Buswell, J.A. and Chiu, S.W. The Chinese University Press. Hong-Kong. 261–266.

Fukushima, M. 1989.The overdose of drugs in Japan. *Nature* **342**: 850–851.

Ikekawa,T. 2001.Beneficial effects of edible and medicinal mushrooms on health care; International conference " Perspectives of medicinal mushrooms in health care and nutrition in the 21st century" ,Kiew ,Ukraine, *International Journal of Medicinal Mushrooms,* **3** (4),291–298.

Jia, Y.P., XU, W.M., Ren, J., yin, X. and Zhang, L. 1993. Effect of Ling Zhi on production of Interleukin 1 (IL–1) immunopharmacological study. In the Research on *Ganoderma* (Part-1): (Ed.by. Zhu, S and Mori, M.) Shanghai Medical University Press, 254–258.

Kaneda, T. Arai, K. and Tokuda.1964 .The Effect of dried mushroom, *Cortinellus shiitake* on cholesterol metabolism in rats.*J. Jpn. Soc. Food Nutri.*, **16**:106–108.

King, T.A. 1993.Mushrooms, the ultimate health food, but little research in US to prove it. *Mushroom News*, **41**(2): 46–29.

Kuo-Yuhchi, Tsai-Weijern, Shiao-Mingshi, Chen-Chieh Fu, Lin-ChingYuang, KuoYC, Tsai-WJ, Shiao-MS, Chen-CF. and Lin-CY. 1996. *Cordyceps sinensis* as an immunomodulatory agent. *American Journal of Chinese Medicine,* **24** (2): 111-125.

Li-SP. Su-ZR. Dong-TTX. and Tsim-kwk.2002.The fruiting body and its caterpillar host of *Cordyceps sinensis* show close resemblance in main constituents and anti-oxidation activity. *Phytomedicine,* **9** (4):319–324.

Liu, W.K., Ooi, V.E.C., Ng, T.B. and Chang, S.T. 1993. Immunomodulatory activities of mushroom mycelial extracts, in *Mushroom Biology and Mushroom Products*,Ed. By S.T. Chang, J.A. Buswell, and S.W.Chiu, The Chinese University press. Hong Kong, 285–291.

Markhija, A.N. and Bailley, J.M.1981.Identification of the antiplatelet substance in Chinese Black Tree fungus. *New England Journal of Medicine,* **304**:175.

Murasugi, A., Tanak, S., Komiyama, N., Iwata, N., Kino, K., Tsunoo, H. and Sakuma, S. 1991. Molecular cloning of a cDNA and a gene encoding an immunomodulatory protein Ling Zhi -8, from a fungus *Ganoderma lucidum. J. boil. Chem.*, **266**:2486–2493.

Nanba, H.1993. Maitake mushroom—the king mushroom. *Mushroom News,* **41**(2):22–25.

Ooi,V.E.C. 2001. Pharmacological studies on certain mushrooms from China. *International Journal of Medicinal Mushrooms.,* **3**(4) :341–354.

Shiuh-Sheng lee, Po-Lei Lee, Chieh –Fu Chen, Sheng-Yuan Wang and Kuang-Yao Chen. 2003. Antitumor Effects of Polysaccharides of *Ganoderma lucidum*.(Curt.:Fr.) P.Karst. (Ling Zhi,Reishi Mushroom) (Aphyllophoromycetideae). *International Journal of Medicinal Mushrooms,* **5**: 1–16.

Tam, SC., Yip, KP., Fund, KP. and Chang, ST. 1986. Hypotensive and renal effect of an extract of the edible mushroom *Pleurotus sajor caju. Life Sciences,* **38**:1155–1161.

Tokuda, S and Kaneda, T. 1976. Reducing mechanism of plasma cholesterol by shiitake. *Mushroom Sci.,* **9**:445–462.

Sakagami, H. and Takeda, M. 1993. Diverse biological activicity of PSK (Krestin), a protein–bound polysaccharide from *Coriolus versicolor* (Fr.) Quel. In *Mushroom Biology and Mushroom Products* (Ed.by S.T. Chang ,J.A .Buswell and S. W. Chiu) . The Chinese University Press, Hong Kong, 237–245.

Vogel, F.S., McGarry, S.J. Kamper, L.A.K. and Graham, D.G. 1974. Bacteriocidal properties of a class of quinoid compounds related to the sporulation of mushroom, *Agaricus bisporus. Am. J. Pathol* **76**:165–174.

Weng-Shu-Cheng, Chou-Cheng-Jen, Lin-Lie-Chwen, Tsai-Wei-Jenn, and Kuo-Yuh-Chi. 2002. Immunomodulatory functions of extracts from the chinese medicinal fungus *Cordyceps cicadae..J.of Ethnopharmacology,* **83**:79–85.

Wang, C.N., Chen, J.C., Shiao, M.S. and Wang, C.T. 1991.The inhibition of human platelet functions by *Ganodermic* acid. *S. Biochem.* J., **277**:189–197.

Wang, S.Y., Hsu, M.L., Lee, S.S., Hong, C.Y.B, and Shiao, M.S. 1994. The role of *Ganoderma lucidum* in immunopotentiation: effect on cytokine release from human macrophages and T-lymphocytes, in Abstract of 94 symposium on *Ganodrma research,* Beijing Medical University. 56–57.

Yang, Q.I, Hu, Y.J., Li, X.L., Yang, S.X., Liu, J.X., Liu, T.F., Xu, G.M., and Liao, M.L. 1993. A new biological response modifier substance PSP. In *Mushroom Biology and Mushroom Products,*(Eds. By Chang, S.T., Buswell, J.A. and Chiu, S.W.). The Chinese University Press. Hong Kong, 247–259.

Yihuai Gao, Zhi–Bin Lin, Min Huang and Shufeng Zhou. Hepatoprotective activity and the mechanisms of action of *Ganoderma lucidum.* (Curt.:Fr.) Karst, P. (Ling Zhi,Reishi Mushroom) (Aphyllophoromycetideae) *International Journal of Medicinal Mushrooms,* **5**: 111–131

Zang, LX., Meng, H. and Zhon, ZB. 1993a. Effect of Ling Zhi on the production of interleukin-2 (IL-2) immunopharmacological study (7): In the *Research on Ganoderma* part 1 (Ed. by Zhu, S and Mori, M) Shanghai Medical University press, Shanghai, 259–265.

25

Ecological Study of Epiphytic Lichens on *Cupressus torulosa* in Nainital

NEERJA PANDE, S.C. SATI AND PRACHI JOSHI

Department of Botany, Kumaun University Nainital-263002, India

SUMMARY

Four lichen species were found on *Cupressus torulosa* stems of three girth classes (30 –70 cm; B = 70 –100 cm and C = more than 100 cm) at CBH in two sites viz, Hanumangarh and Ramsay Hospital. *Parmelia cristifera* was the dominant lichen species on the basis of frequency, density, biomass and Importance Value Index (IVI) at both the sites. *Ramalina fraxinia* was the co-dominant species in all the girth Classes, excepting girth class (ie, more than 100 cm) at site I, where it dominated. The growth patterns of *Parmelia* and *Ramalina* were strikingly remarkable, though both the species shared the same habitat. Decline in biomass of one species was accompanied by an increase in dry matter of the other species which may be due to the competition between the species for nutrients, water and space on the same substratum.

Keywords: Lichen biomass, *Cupressus torulosa*, Importance Value Index (IVI)

The survival of epiphytic vegetation (lichens and bryophytes) is mainly dependent on the atmospheric sources of water and nutrients. Owing to this, the study of lichen communities may be used to determine the changes in atmospheric conditions (Barkman, 1969). So far, several workers have studied the distribution of epiphytic lichens, in relation to substrate specificity (Hale, 1955, Culberson, 1955b), and lichen communities of various epiphytes like

Pseudotsuga menzeisii, Abies lasiocarpa (Gough, 1975); *Pinus ponderosa* (Eversman, 1982), *Populus termuloides* (Jonescu, 1970; Case, 1977), *Larix lyallii* and *Pinus albicaulis* (Kalguatkar and Bird, 1969). Compared to such diverse studies, the lichen flora of this area is relatively less known. D.D. Awasthi and his group at Lucknow did extensive taxonomical and distributional studies of lichens from different habitats, including lichen as epiphytes, but the ecological study was totally neglected. Therefore, the objective of this study was to analyse the lichen communities present on *Cupressus torulosa* trees in Nainital, as a base line information, and to compare the *C. torulosa* epiphytes with those of other dominant trees of this area.

MATERIALS AND METHODS

The study sites ie, *Cupressus torulosa* stands were located at an altitude of 1938 m asl (29° 22′ –24′ N latitude to 79° 28′ –29′ E longitude) in Nainital hills. The mean maximum temperature of the area ranged from 23.7°C (June) to 12.0°C (December) and the mean minimum temperature from 18.1°C (June) to 4.0°C (December). The mean annual rainfall was 191.3 mm (data recorded from State Observatory, Nainital).

Two pure stands of *C. torulosa* were selected at Hanumangarh (Site I, young stand) and Ramsay Hospital (Site II, mature stand) for the phytosociological study of epiphytic lichens. At each site, trees having good cover of lichens were selected and divided into three girth classes at CBH viz, girth Class A (30 –70 cm), girth Class B (70 –100 cm) and girth Class C (more than 100 cm). Lichen vegetation on the selected trees (3 in number for each girth class) was quantitatively estimated during the study period (1986 –1987), following the standard methods of Misra (1968), by placing 3 quadrats of 10 × 10 cm size on the stems of each tree of *C. torulosa*. The quadrat data thus collected were analysed for species composition, frequency, density, Importance Value Index (IVI) and biomass (Mishra, 1968). The data were also computed for Similarity Index (SI): by using the following formula:

SI = 2c / a+b; where c= number of species common to both sites, a= number of species at site 1, and b= number of species at site II.

Lichens were identified in the laboratory by following pertinent monographs and flora. Various characteristics of bark samples like texture, colour, pH, moisture content and moisture holding capacity (Brodo, 1974) were also analysed, following Mishra (1968).

RESULTS

Texture, colour, pH and moisture holding capacity of the bark samples are given in Table 25.1. Four species of lichens, namely *Parmelia cristifera, Ramalina fraxinia, Phaeophyscia hispidula* and *Graphina acharii* were encountered as epiphytes on the *Cupressus torulosa* stems. At site II ie, Ramsay Hospital area, only two species *Parmelia cristifera* and *Phaeophyscia hispidula* formed the community (Table 25.2). Based on frequency, density and IVI, *Parmelia cristifera* was found dominant in girth class A and B at site I and site II, *Ramalina fraxinia* at girth class C at site I and *Phaeophyscia hispidula* at girth class C at site II (Table 25.2) .

Total community biomass ranged from 42.6 gm^{-2} (girth class C) to 36.5 gm^{-2} (girth class A) at site I and from 5.5 gm^{-2} (girth class C) to 1.5 gm^{-2} (girth class A) at Site II. At site I, *Parmelia cristifera* contributed maximal biomass of 26.7gm^{-2} at girth class A, while *Ramalina fraxinia* added 20.8gm^{-2} at girth class C. Across the season at site I, maximum per cent contribution to biomass was 52 by *Parmelia cristifera* in summer, while the co-dominant species *Ramalina fraxinia* contributed only 41% to the community. Compared to this, *Phaeophyscia hispidula* added 69 per cent to the community biomass during the rainy season at site II. Across the girth class, at site I, *Parmelia cristifera* contributed maximum 73% (girth class A) and 77% biomass (girth class B) at site II (Table 25.4).

Table 25.1: Certain Features of the Bark of *Cupressus torulosa* Bark at Two Sites in Nainital

Parameters	Site I	Site II
Texture	Smooth, soft and shredding type	Relatively smooth, soft shredding type, flakes smaller in size
Colour	greyish brown to dark--brown	greyish brown
PH		
(rainy)	6.7	6.6
(winter)	6.2	6.5
(summer)	6.4	–
Water holding capacity		
percent vol. (cc)	75.9	75.1
per unit dry (g)	177.1	176.1

DISCUSSION

Morphologically, all the three forms viz, crustose, foliose and fruticose lichens formed the community at site I whereas only foliose form constituted the community at site II. At both the sites, there was not much difference in species composition of lichens with increasing girth class. This is in accordance with the opinion of Hawksworth and Hill (1984) that a specific substratum tend to bear the uniform and remarkable assemblage of lichen species under the influence of similar environmental conditions. Kershaw (1964) opined that the zonation of lichens of young trees was similar to that of mature trees, but more compressed. In our study, at site I, the number of species was similar at all

Table 25.2: Across the girth class, Vegetational Analysis of Lichen Communities Growing on *Cupressus torulosa* Trees at Two Sites in Nainital (F = Frequency(%); D = Density (individuals m^{-2}); B = Biomass (g m^{-2}); IVI = Importance Value Index)

Species	Girth Class A				Girth Class B				Girth Class C			
	F	D	B	IVI	F	D	B	IVI	F	D	B	IVI
Site I												
Parmelia cristifera	209	209	26.7	172.3	143	143	18.9	110.9	110	110	11.9	82.0
Ramalina fraxinia	132	138	6.2	81.0	121	121	9.8	79.1	198	198	20.8	145.9
Graphina acharii	33	33	1.9	20.9	88	99	10.6	45.8	44	44	4.3	31.8
Phaeophyscia hispidula	44	44	1.7	25.5	77	77	2.6	41.6	55	55	5.6	40.1
SITE II												
Parmelia cristifera	121	154	2.9	180.5	77	77	2.7	154.5	110	55	0.7	119.6
Phaeophyscia hispidula	77	77	2.6	119.3	121	121	0.8	145.3	55	99	0.8	125.2

Table 25.3: Across the Season, Per cent Biomass of Lichen Vegetation Growing on *Cupressus torulosa* bark at two sites in Nainital

Species	Rainy	Winter	Summer
Site I			
Parmelia cristifera	42.7	50.5	52.0
Ramalina fraxinia	32.3	32.6	41.1
Graphina acharii	10.4	24.3	2.2
Phaeophyscia hispidula	14.5	4.9	4.5
SITE II			
Parmelia cristifera	31.1	58.6	–
Phaeophyscia hispidula	68.9	41.3	–

Table 25.4: Across the Girth Class Per cent Biomass of Lichen Vegetation Growing on *Cupressus torulosa* Bark at Two Sites in Nainital.

Species	Girth class A	Girth class B	Girth class C
Site I			
Parmelia cristifera	73.1	45.1	27.98
Ramalina fraxinia	16.98	23.4	48.7
Graphina acharii	5.35	25.3	10.2
Phaeophyscia hispidula	4.6	6.2	13.1
Site II			
Parmelia cristifera	52.8	76.9	47.0
Phaeophyscia hispidula	47.2	23.1	52.9

girth classes, but the thalii were more compressed at girth class C (eg, the dominant species, *Parmelia cristifera* at site I had maximum density at girth class A and B during rainy season (Table 25.2). On the basis of IVI, in the present study, at site I two communities (eg *Parmelia–Ramalina* at girth class A and B and *Ramalina –Parmelia* at girth class C) of lichens were observed.

Density data indicated that with the increase in diameter of the tree, the number of thalii decreased eg, *Parmelia*, the dominant species, had 209 thalii at girth class A, followed by 143 and 110 thalii at girth class B and C, respectively. This may be due to the fact that at lower girth classes, the size of the thalii were smaller than those present at higher girth class, therefore more thalli occupied the similar area. However, in other species, no regular pattern was observed. This is in agreement with the study of Eversman (1982) on the lichen vegetation of *Pinus ponderesa*. Possibly, at higher girth class, the lichen vegetation was lost due to the flaking off of the bark, and the gradual replacement of lichens by the slow and steady colonisation of bryophytes, because of the congenial habitat conditions created by the lichens for the successful establishment of bryophyte propagules (Barkman, 1958).

In terms of biomass, the communities growing on the trees at site I had greater accumulation of dry matter as compared to site II (site I = 102.7 gm^{-2};

Table 25.5: Lichen Vegetation Growing on Different Habitats of the Study Area

Lichen Species	Ct	Cd	Pr	Qf	Ql	Psp.	Ba	Pc	Eh
Anaptychia hypoluca	–	–	–	–	–	–	–	–	+
A. firmula	–	–	–	–	–	–	–	–	+
A. speciosa	–	–	–	–	–	–	–	–	+
Anthroracuthecium variolosum	–	–	–	–	+	–	–	–	–
Cladonia pityrea	–	–	–	–	–	–	–	–	+
Collema flaccidum	+	–	–	+	+	–	–	–	+
Dermatocarpon vellereum	–	–	–	–	–	–	–	–	+
Graphina acharii	+	–	–	–	–	–	–	–	–
Heterodermia himalayensis	–	–	–	–	–	–	+	+	–
Lecania caesia	–	–	–	–	–	–	–	–	+
Lecidia subkochiana	–	–	–	–	–	–	–	–	+
Leptogium pedicilatum	+	–	–	+	+	–	–	–	+
Lepraria sp.	–	–	–	–	–	–	–	–	+
Lobaria pulmonaria	–	+	–	–	–	–	–	–	–
Parmelia apicola	–	+	+	–	–	–	+	–	–
P. cristifera	+	+	+	+	+	+	+	+	–
Parmelia latissima	–	–	–	–	–	–	–	–	+
Phaeophysica hispidula	+	–	–	+	+	+	–	–	–
Physcia dilatata	–	–	–	–	–	–	–	–	+
Physcia stellaris	–	–	–	–	–	–	–	–	+
Placodium tetrastichium	–	–	–	–	–	–	–	–	+
Placodium vitellinulum	–	–	–	–	–	–	–	–	–
Ramalina fraxinia	+	–	–	+	+	–	–	–	–
Ramalina farinaceae	–	–	–	+	+	–	–	–	–
Usnea orientalis	–	–	–	+	+	–	+	+	–

Ct = *Cupressus torulosa;* **Cd**=*Cedrus deodara;* **Pr**=*Pinus roxburghii;* **Qt**=*Quercus floribunda;* **Qf**=*Quercus leucotricophora;* **P sp** =*Prunus sp.;*
Ba = *Berberis asiatica;* **Pc** = Pyrenthus crenulatus and **Eh**=Epilithic habitats

site II = 11.6 gm^{-2}). This may be perhaps due to the presence of more species, larger size of thallus and less competition with bryophytes at site I, compared to site II where bryophyte predominated. Across the girth class, mean annual biomass was maximum for *Ramalina –Parmelia* community (girth class C = 42.6 gm^{-2}) at site I, and minimum for *Parmelia* community at site II (girth class B = 3.51 gm^{-2}).

The growth behaviour of two species ie *Parmelia* and *Ramalina* at Site I showed a characteristic pattern. Although the two species shared the same habitat, yet the biomass was quite different, increase in biomass of one species was accompanied by a decline in the dry matter of the other species (Table 25.2). This may be due to competition between the two species for nutrients, water and space in the same substratum.

Table 25.6: Similarity (%) of Lichen Vegetation Growing on *Cupressus torulosa* with Other Habitats (epiphytic and epilithic)

Habitats	*Cupressus torulosa*
Cupressus torulosa	100
Cedrus deodara	22.2
Pinus roxburghii	44.4
Quercus floribunda	77.0
Quercus leucotrichophora	71.0
Prunus sp.	12.5
Berberis asiatica	20.0
Pyrenthus crenulatus	11.0
Rocks	10.7

The variations in the species number and biomass, at two sites on the same phorophyte, was largely due to the differences in the nature of the stand (young and mature), degree of mesicness and texture of the underlying habitat, mist, humidity and temperature. Further, perhaps the propagules have not chanced to land on favourable places on the tree surfaces at the right time for their successful establishment (Phillips, 1951).

COMPARISON WITH OTHER HABITATS

Three lichen species conspicuously present on *Cupressus torulosa* trees were also consistent components of *Quercus floribunda, Q. leucotrichophora,* rocks and stone walls of this locality (Table 25.5): *Collema flaccidum, Leptogium pedicilatum* and *Phaeophyscia hispidula.*

Parmelia cristifera, together with *Ramalina fraxinia,* exhibited the widest environmental tolerance in this area. This species grew in different vegetation types of this region, eg, on the barks of *Cupressus torulosa, Cedrus deodara, Quercus leucotrichophora, Q. floribunda, Pinus roxburghii,* shrubs like *Pyrenthus crenulatus, Prunus* sp. and *Berberis asiatica,* but was totally absent on epilithic habitats. However, *Parmelia latssima,* another species of *Parmelia* was collected from the rocks of this locality.

Maximum 77 per cent similarity of the lichen vegetation growing on *Cupressus torulosa* was observed with *Quercus floribunda,* followed by *Quercus leucotrichiphora* (Table 25.6). Low similarity values were found with the lichens growing on shrubs (*B. asiatica* SI=20%) and on rocks (SI = 10.7%).

REFERENCES

Barkman, J.J. (1958). Phytosociology and Ecology of Cryptogamic Epiphytes. Van Gorcum. Assen.

Barkman, J.J. (1969). The influence of air pollution on bryophytes and lichens. Air Pollution Proceedings, European Congress on Influence of Air Pollution on Plants and Animals, Ist. 1968 pp. 197–209.

Brodo, I.M. (1974). Substrate Ecology. In Ahmadjian, V. and Hale, M.E. eds. The Lichens. Academic Press, New York and London. 401–439

Case, J.W. (1977). Lichens on *Populus tremuloides* in Western Central Alberta, Canada. The Bryologist., 80: 48–70.

Culberson (1955b). The corticolous communities of lichens and bryophytes in the upland forests of Northern Wisconsin. Ecol. Mongr., 25: 215–23.

Eversman, S. (1982). Epiphytic lichens of a Ponderosa Pine forest in South Eastern Montana. The Bryologist, 83 : 204–213.

Gough, L.P. (1975). Cryptogam distributions on *Pseudotsuga menzeisii* and *Abeis lasiocarpa* in the Front Range, Boulder County, Colorado. The Bryologist, 78: 124–145.

Hale, M.E. (1955) Phytosociology of corticolous cryptogams in the upland forests of southern Wisconsin. Ecology, 36: 45–63.

Hawksworth, D.L. and Hill, D.J. (1984). The Lichen –Forming Fungi, Blackie, Glasgow and London, 158 pp.

Jonescu, M.E. (1970). Lichens on *Populus tremuloides* in west-central Canada. The Bryologist 78: 557–578.

Kalgutkar, R.M. and Bird, C.D. (1969). Lichens found on *Larix lyallii* and *Pinus albicaulis* in south western Alberta, Canada. Canadian Journal of Botany, 47: 627–648.

Kershaw, K.A. (1964). Preliminary observations on the distribution and ecology of epiphytic lichens in Wales. Lichenologist, 2 : 263–276.

Mishra, R. (1968). Ecology Workbook. Oxford and IBH Publishing House, New Delhi, 244 p.

Phillips, E. A. (1951). The associations of bark inhabiting bryophytes in Michigan. Ecol. Mongr., 21: 301–316.

26

Perception of Elicitors of Arbuscular Mycorrhiza and Nitrogen-fixing Bacteria: Signal Transduction

MOHD. W. ANSARI, R. SRIVASTAVA,
V. TEWARI AND A.K. SHARMA

Department of Biological Sciences, C.B.S.H., G.B. Pant University of Agric. and Tech., Pantnagar-263 145, Uttaranchal, India

SUMMARY

Roots of plant form intricate intracellular symbiotic association with fungi and bacteria, resulting in arbuscular mycorrhiza and nitrogen-fixing root nodules. Because of the nutritional and protective benefits that AMF confer to their hosts, there is considerable interest in understanding the ecological, physiological, biochemical and molecular interactions involved in arbuscular mycorrhiza with land plants, and in root nodule symbiosis with nitrogen-fixing bacteria. It is widely believed that in both the cases the interaction is mediated by signal molecules (elicitors) produced by the arbuscular mycorrhiza and rhizobacteria and the most complementary receptor molecules. Signals and receptors are thought to interact at the contact surface between the two organisms to start a signal transduction pathway leading to the activation of various responses. The present review is an attempt to synthesise information on the interactions of AM-fungi and rhizobacteria with early land plants, suggested to be mediated through signal transduction that might be helpful in exploring the current knowledge.

Keywords: Arbuscular mycorrhiza, N_2-fixing bacteria, Signal transduction

The Arbuscular Mycorrhizal (AM) symbiosis is the association between fungi of order Glomales (Zygomycetes) and the roots of terrestrial plants (Harley and Smith, 1983). This type of symbiosis dates back to the Devonian era and it affects approximately 90% of plant taxa (Remy *et al.*, 1994). In this type of association, the plant supplies the fungus with carbon (from its fixed photosynthates), while the fungus assists the plant in its uptake of phosphate and other mineral nutrients from the soil (Smith and Read, 1997). Because of this ability, they have proved to be potentially useful, as biofertilisers, in increasing the yield of crops and tree species. This bidirectional association takes place through the formation of extensively branched haustoria, arbuscules. This AM symbiosis also helps the plants to increase resistance to root pathogens and tolerate drought stress (Hwang *et al.*, 1992). A major challenge for the mycorrhizologists is to understand the AM fungus-host signalling mechanisms and the colonisation processes. This symbiotic relationship is obligately biotropic in nature (Williams, 1992). In all of the current methods of cultivating AM fungi, the presence of the host plants is indispensable. Recently, there was considerable interest in understanding the ecology, physiology and molecular interactions involved in this symbiosis which AMF confer on their host. Microorganisms inhabit the rhizosphere, the 'hot spot', and the beneficial rhizobacteria may also enhance plant growth and nutrient uptake. In contrast to AM formation, only a few plant species have the ability to interact symbiotically with the bacteria of the genera *Azorhizobium, Bradyrhizobium, Rhizobium* and *Sinorhizobium* (Symbiotic nitrogen-fixers). This interaction is almost completely restricted to leguminous plants and results in the formation of a completely new organ, the root nodule (Mylona *et al.*, 1995). In addition, this association also enhances the infection of mycorrhiza with the plant roots by producing volatile chemicals that trigger spore germination (Azcon, 1987; Mugnier and Mosse, 1987) or stimulate root colonisation by increasing root cell permeability (Bowen, 1980). Free-living and symbiotic nitrogen-fixers also contribute the most important nutritional input, nitrogen and thus, they play an important part in the biogeochemical cycle by fixing the atmospheric nitrogen. Their interaction with AM fungi which mobilises several nutrients, particularly P, shows nutritional as well as non-nutritional, host mediated effects.

Plant cell system is capable maintaining the compatibility itself by means of the perception of mycorrhizal elicitors (recognition). After the perception of signal molecules (elicitors), the plant cell, at the site of recognition, mediates the signals through complementary receptor molecules and transmits the information inside the cell, across the plasma membrane as well as neighbouring cells. Before the trigger of signal transduction pathway, organisms first maintain the basic compatibility, either by biochemical means (Bais *et al.*, 2004) or by susceptible genes. In the light of these facts, this review is an attempt to explore the current information on the interaction of AM fungi and nitrogen-fixing bacteria with early land plants, thought to be mediated through signal transduction.

EVALUATION OF SIGNAL TRANSDUCTION IN INTRACELLULAR SYMBIOSIS

Rhizospheric interactions involve complex exchanges between plant roots and inhabiting microbes. These interactions in the rhizosphere were highly influenced by edaphic factors; however, the below ground diversity depends upon the exudation. Plant roots form intracellular symbiotic association with fungi and bacteria, resulting in arbuscular mycorrhiza and nitrogen-fixing root nodules, respectively. This association includes the trafficking of signals in between the neighbouring plant roots and the roots and soil microbes. The signalling event between host root and AM fungi, before and after colonisation, is not yet fully understood. But the association can be classified as 'host-dependent' and 'host-independent'. Experimental evidence suggests that the quality and source of the exudates plays an important role in triggering germination. For example, the exudates from the non-host plants, such as *Brassica* spp. and *Lupinus* spp., do not stimulate germination (Giovannetti *et al.*, 1993). A novel receptor-like kinase was reported to be responsible for the transduction of both bacterial and fungal symbiotic signals. This kinase defines an ancient signalling pathway that probably evolved in the context of arbuscular mycorrhiza and has been recruited subsequently for endosymbiosis with bacteria (Kistner and Parniske, 2002). AM and root nodule symbioses are endosymbioses which enter the host intracellularly. Rhizobia are recognized by hosts via Nod factors, extracellular polysaccharides, lipopolysaccharides and secreted proteins, while equivalent signal molecules have not yet been described for AM symbioses. Investigations of both AM and RNS (Root Nodule Symbioses) in legume species have revealed a genetic overlap between the two types of symbiosis, supporting the idea that during the evolution of RNS, parts of older evolutionary programmes were recruited from AM (Harrison, 1999; Duc *et al.*, 1989).

INDUCTION OF A SIGNALLING PATHWAY IN RESPONSE TO SYMBIOTIC MICROORGANISMS

The intracellular infection process in the root endosymbiosis is regulated by the plant. In infection thread symbiosis, the microbe gains an entry through the plant cell wall and lives in the infected cell. Then they move on for infecting the adjoining cells. The induction of the accommodation of the intracellular residing microorganisms is mediated by several signalling pathways which act as keys required for opening a lock. To understand the signalling pathways of AM fungal symbiosis, the plant mutants serve as the key tool for the genetic dissection of AM development. By far, the largest number of AM-mutants, however, have been isolated from legumes, by examining the ability of the mutants, that have defects in the nitrogen-fixing root nodule symbiosis (RNS), to develop AM (Table 26.1).

The mutants isolated this way are defective in both bacterial and fungal symbiosis (Marsh and Schultze, 2001; Stougaard, 2001). These 'impaired' genes have been collectively referred, hereafter, as SYM genes. The evolutionary implication is that the younger RNS has recruited functions from the ancient AM symbiosis (LaRue and Weeden, 1994; Oldroyd and Downie, 2004). The common *SYM* genes are therefore at the centre of interest from both the mechanistic and evolutionary view point. The mutants isolated thereafter, were defective in both bacterial and fungal symbiosis. The list of these genes also

Table 26.1: Orthologous Relationship Between Some Legume Symbiosis Genes

Lotus japonicus		*Pisum sativum*	*Medicago truncatula*	Predicted Function	Reference(s)
(a) Cloned Nodulation Genes that are Required for Early Stages of RNS					
Cloned gene	Initial designation				
NFR1	*LjSYM1*		*LYK3*	Nod-factor receptor kinase	Madsen *et al*, 2003 Radutoiu *et al*, 2003
NFR5	*LjSYM5*	*PsSYM10*	*NFP*		
NIN		*PsSYM35*		Membrane protein with DNA binding domain	Schauser *et al*, 1999 Borisov *et al*, 2003
(b) Common SYM Genes Required for AM and RNS					
		PsSYM8	*DMI1*	Ion Channel	Ane *et al*., 2004
SYMRK	*LjSYM2*	*PsSYM19*	*DMI2* *(MsNORK)*	Symbiosis receptor kinase	Stracke *et al*, 2002 Endre *et al*, 2002
		PsSYM9/30	*DMI3*	Calcium-and calmodulin-dependent kinase	Levy *et al*, 2004 Mitra *et al*, 2004
	LjSYM3				Schauser *et al*, 1998
	LjSYM4				Schauser *et al*, 1998
	LjSYM15				Schauser *et al*, 1998
	LjSYM23				Szczyglowski *et al*, 1998
	LjSYM30				Szczyglowski *et al*, 1998

comprises three genes from *Melilotus alba* (Ma) *MaSYM1*, *MaSYM3* and *MaSYM5* and one gene each from alfalfa (*Medicago sativa*), bean (*Phaseolus vulgaris*) and faba bean (*Vicia faba*) (Hirsch *et al.*, 2001; Marsh and Schultze, 2001). Gene expression that the affected studies and physiological investigations (Walker *et al.*, 2000) indicate that genes are involved in perception and/or transduction of microbial signals. Whether the mycorrhizal-defective mutants are isolated from defects in orthologous common *SYM* genes or in mycorrhiza-specific genes need to be investigated further.

PERCEPTION OF FUNGAL AND BACTERIAL SIGNALS, A SPECIFICITY PROBLEM

A plant receptor-like kinase is required for symbiosis to occur between plants and microorganisms. This was achieved by the map-based cloning of one of the common *SYM* genes, *NORK* (Nodulation Receptor Kinase)/ *SYMRK* (Symbiosis Receptor-Like Kinase). The discovery that one of the common *SYM* genes does encode a signaling component was an important confirmation for the prediction based on phenotypic observations (Stracke *et al.*, 2002). *SYMRK/NORK* carries the hallmarks of the Receptor-Like Kinase (RLK); a predicted signal peptide, three extracellular leucine-rich-repeats (LRRs), a transmembrane domain and an intracellular protein kinase domain. The overall structure suggests the role of Leucine-Rich Receptor-Like Kinase (*LRR-RLK*) and its involvement in the perception of microbial signal molecules, either as a receptor or as a part of the receptor complex. The signal transduction pathway of the highly promiscuous AM fungi was found to be mediated by chitin, an oligomer, of the fungal cell wall (Kistner and Parinske, 2002). A high concentration of chitin oligomers induced calcium spiking in peas, although it was quantitatively different from that of the NF-induced one. Calcium spiking in *PsSym10* mutants, which form mycorrhizal association but do not support bacterial infection, was not induced by NF or chitin oligomers. This can be inferred as some other Fungal Factor (FF), instead of NF or chitin oligmers which are required to establish a symbiotic interaction with AM-fungi, which is quite different from that. Thus, a fungal factor (FF) feeding in common SYM pathway is likely to be different from both NF and chitin oligomers (Fig. 26.1). The question arises as to how *SYMRK/NORK* can integrate presumably different fungal and bacterial signals. Although the possibility exists, that its extracellular domain itself binds a variety of different molecules, however, genetic evidence suggests the involvement of different additional protein molecules that mediate the specificity of the recognition of NF. Genetic pathways are required for symbiosis related to gene expression (Albrecht *et al.*, 1998; Stracke *et al.*, 2002) and symbiotic infection (Strouggard, 2001; Hirsch *et al.*, 2001; Catoria *et al.*, 2000; Geurts and Bisseling, 2002). Genes required for root nodules and arbuscular mycorrhizal symbiosis define the common *SYM* pathway (Fig. 26.1). A key symptom of early symbiotic signalling in root hairs is calcium spiking, an induction of repetitive calcium oscillations. Calcium spiking is the only phenotypic marker that discriminates between common *SYM* genes (Table 26.1), and has been used to establish a genetic hierarchy. Like *NORK/MtDM12* from (*Medicago tranculata*)/ *PsSYM19* (from *Pisum sativum*) and, by analogy, *SYMRK*, the genes *PsSYM8*, *LjSYM4* (from *Lotus japonicus*) and *MtDM1* are required for calcium spiking, whereas *PsSYM9/PsSYM30* and *MtDM13* are not (Walker *et al.*, 2000; Wais *et al.*, 2000; Ehrhardt *et al.*, 1996). The non-nodulating *Ljsym1*, *Ljsym5*, *Pssym10* and *Mtnfp* mutants do form mycorrhizal association (Stouggard, 2001), so these genes do not affect signaling mediated by, as yet, hypothetical fungal signalling factor(s). The lack of any morphological root hair responses to the Nod-Factor (NF) in *Ljsym1*, *Ljsym5* and *Mtnfp* mutants and the absence of calcium-spiking responses in *Pssym10* and *Mtnfp* mutants led to the placement of the affected genes upstream of the common pathway. Whether the genes responsible for such specificity in AM exist or not, is not known. It is likely that the common *SYM* pathway is as old as AM and might be present in all AM forming plants, whereas the genes specific for NF recognition are possible key inventions for the process of bacterial infection in the Fabales, Fagales, Cucurbitales and Rosales. Other genes in the common pathway have been identified in *Melilotus alba* and *Lotus japonicus* and not been placed relative to calcium-spiking (Strouggard, 2001; Walker *et al.*, 2000). We expect the majority of plant genes involved in the perception and transduction of bacterial and fungal signals to be cloned over the next few years.

PERCEPTION OF BIOCHEMICAL SIGNALS

The rhizosphere is the densely populated area in which the roots must compete with the invading root systems of neighbouring plant species for space, water, and mineral nutrients, and with soil-borne microorganisms, including bacteria and fungi feeding on the abundant source of organic material (Ryan and Delhaize, 2001). In addition to the classical role of providing mechanical support and allowing water/nutrient uptake, roots also perform certain specialised roles, including the ability to synthesise, accumulate, and secrete a diverse array of compounds eg flavonoids, phytoalexins, antimicrobial compounds (Rosmarinic Acid and Naphthoquinones) (Flores *et al.*, 1999). The complex array of the inhabitants of the underground world and their diversity suggests that the roots also serve as the targets for soil organisms. In fact, the compounds secreted by plant roots serve an active and relatively well-documented role in the regulation of symbiotic and protective interaction with microbes. The unexplored chemodiversty of root exudates is an obvious place to search for biologically novel active compounds, including antimicrobials. Rosamarinic Acid (RA) secreted from basil (*Ocimum basilicum*) and Naphthoquinones from *Lithospermum erythrorhizon* roots, were demonstrated to show potent

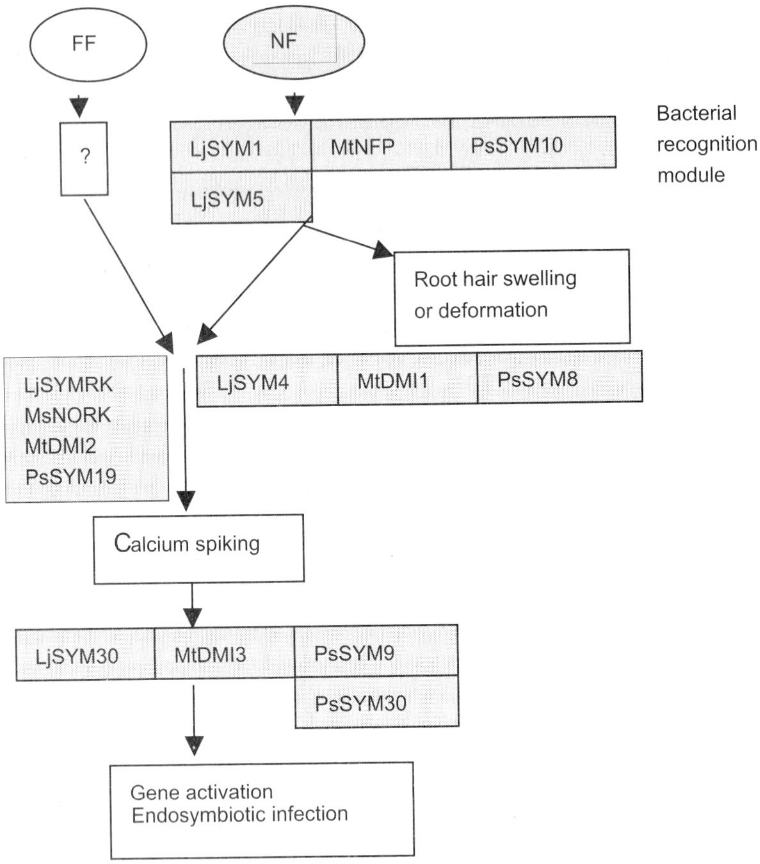

Fig. 26.1. Pathway Defined by Plant Genes Required for Both Fungal and Bacterial Symbiosis.

antimicrobial activity against an array of soil-borne microorganisms including *Pseudomonas aeruginosa* (Bais *et al.*, 2002). Through exudation of wide varieties of compounds, roots may regulate the soil microbial community in their immediate vicinity and encourage beneficial symbioses (Hirsch *et al.*, 2003; Bais *et al.*, 2001; Buee *et al.*, 2000; Jones *et al.*, 2003), the converse is also true; many bacteria and fungi are dependent on their association with plants, that are often regulated by root exudates (eg, flavonoids) of legumes that activate *Rhizobium meliloti* genes, responsible for the nodulation process (Peters *et al.*, 1986). A chemotactic response towards root-secreted organic and amino acids is the first step in root colonisation (Zheng and Sinclair, 1996). Isoflavonoids and flavonoids present in the root exudates of a variety of leguminous plants activate the *Rhizobium* genes responsible for the nodulation process and might be responsible for arbuscular mycorrhizal colonisation (Trieu *et al.*, 1997; Becard *et al.*, 1995). Border cells exude a diverse array of biochemical substances that influence the behaviour of fungi and bacteria (Gunawardena and Hawes, 2002).

Border cells released in advance of the vulnerable root tip might protect plant health by inhibiting tip infection, by pathogens, or by stimulating the development of beneficial associations.

Under controlled conditions, border cells and their associated products can contribute upto 98% of the carbon rich material that is released by plants as root exudates, so the potential impact on root-microbe interaction is large (Griffin *et al.*, 1976). A report on isoflavonoid (genistein) exudation from the roots of plants indicates that a specific elicitation regime is crucial for the enhanced exudation of genistein (Kneer *et al.*, 1999). Research has shown that chemical or biological elicitors, acting as natural stresses, stimulate roots to exude an assay of compounds (Walker *et al.*, 2003; Bais *et al.*, 2002). An elicitor triggers the production of different compounds in different plant species eg, Acylated Homo-Ser Lactones (AHLs) for Gram-negative bacteria and peptide-signalling molecules for Gram-positive bacteria (Walker *et al.*, 2003). Elicitation can drastically increase the quantities of certain continuously expressed compounds in the exudates (Walker *et al.*, 2003; Bais *et al.*, 2002). Because many of the compounds and a battery of proteins found in root exudates display both antifungal and antibacterial activity, it is likely that several secreted compounds act synergistically, accounting for the overall antimicrobial activity of root exudates. The presence of antimicrobial activity in the compounds, secreted by the inhabiting microflora in the rhizospheric zone enables them protect themselves from the other non-symbionts. Roots can selectively sense harmful pathogens, other than symbionts to ward off the pathogens in the vicinity (Bais *et al.*, 2002; Walker *et al.*, 2003; Wu *et al.*, 2000 & 2001; Park *et al.*, 2002; Flores *et al.*, 1999).

CONCLUSION

Where Do We Move From Here Now?

The most exciting prospects in the field of symbiotic signalling are the imminent results from plant genetics. We expect the majority of plant genes involoved in the perception and transduction of bacterial and fungal signals to be cloned over the next few years. With a more complete understanding of early signalling pathways, it should become apparent which symbiosis genes are missing from crop plants, and do not form symbiosis with mycorrhiza and nitrogen-fixing bacteria. Our rapidly expanding understanding of the biology of root exudation should contribute to new strategies for improving the fitness of plants used by humans.

REFERENCES

Albrecht, C., Geurts, R., Lapeyrie, F. and Bisseling, T. 1998. Endo-mycorrhizae and rhizobial Nod factors both require SYM8 to induce the expression of the early nodulin genes *PsENDO5* and *PsENDO12A*. *Plant J.*, 15, 605–614.

Ane, J.M., Kiss, G.B., Riely, B.K., Penmesta, R.V., Oldroyd, G.E., Ayax, C., Levy, J., Debelle, F., Baek, J.M., Kalo, P., Rosenberg, C., Roe, B.A., Long, S.R., Denarie, J. and Cook, D.R. 2004. *Medicago trancatula DMI1* required for bacterial and fungal symbiosis in legumes. *Science*, 303, 1364–1367.

Azcon, R. 1987. Germination and hyphal growth of *G. mosseae*: *In vitro* effects of rhizosphere bacteria and cell free culture media. *Soil Biol. Biochem.*, 8, 135–138.

Bais H.P., Walker T.S., Stermitz F.R., Hufbauer R.S., Vivanco J.M. 2002. Enantiomeric dependent phytotoxic and antimicrobial activity of (±)–catechin; a rhizosecreted racemic mixture from *Centaurea maculosa* (spotted knapweed). *Plant Physiol.*, 128, 1173-1179.

Bais, H.P., Loyola-Vargas, V.M., Flores, H.E. and Vivanco, J.M. 2001. Root-specific metabolism: the biology and biochemistry of underground organs. *In Vitro Cell. Dev.-PLANT*, 37, 730–741.

Bais, H.P., Park, S.W., Weir, T.L., Callaway, R.M. and Vivanco, J.M.. 2004. How plants communicate using the underground information superhighway? *Tr. Plant Sci.*, 9, 26–32.

Becard, G., Taylor, L.P., Douds, D.D., Pfeffer, P.E., Doner, L.W. 1995. Flavonoids are not necessary plant signal compounds in arbuscular mycorrhizal symbiosis. *Mol. Plant-Microbe Interact.*, 8, 252–258.

Borisov, A.Y., Madsen, L.H., Tsyganov, V.E., Umehara, Y., Voroshilova, V.A., Batagov, A.O., Sandal, N., Mortensen, A., Sshauser, L., Ellis, N., Tikhonovich, I.A. and Stougaard, J. 2003. The *sym35* gene required for root nodule development in pea is an ortholog of *nin* from *Lotus japonicus*. *Plant Physiol.*, 131, 1009–1017.

Bowen, G.D. 1980. Misconceptions, concepts and approaches in rhizosphere biology. *In*: *Contemporary Microbial Ecology*. Ellwood, D.C., Latham, M.J., Iledger, J.N., Lynch, J.M. and Slater, J.M. (eds.). Academic Press, London. Pp. 283–304.

Buee, M., Rossignol, M., Jauneau, A., Ranjeva, R., Becard, G. 2000. The pre-symbiotic growth of arbuscular mycorrhizal fungi is induced by a branching factor partially purified from plant root exudates. *Mol. Plant-Microbe Interact.*, 13, 693–698.

Catoria, R., Galera, C., De Billy, F., Penmesta, R.V., Journet, E.P.,Maillet, F., Rosenberg, C, Cook, D., Gough, C., J. 2000. Four genes of *Medicago truncatula* controlling components of a Nod factor transduction pathway. *Plant Cell*, 12, 1647–1666.

Duc, G, Trouvelot, A., Gianinazzi-Pearson, V. and Gianinazzi, S. 1989. First report of non-mycorrhizal plant mutants (myc$^-$) obtained in pea (*Pisum sativum* L.) and faba bean (*Vicia faba* L.). *Plant Sci.*, 60, 215–222.

Ehrhardt, D.W., Wais, R. and Long, S.R. 1996. Calcium spiking in plant root hairs responding to Rhizobium nodulation signals. *Cell*, 85, 673–681.

Endre, G., Kereszt, A., Kevei, A., Mihaceae, S., Kalo, P. and Kiss, G.B. 2002. A receptor kinase gene regulating symbiotic nodule development. *Nature*, 417, 962–966.

Flores, H.E., Vivanco, J.M. and Loyola–Vargas, V.M. 1999. "Radicle" biochemistry: The biology of root-specific metabolism. *Tr. Plant Sci.*, 4, 220–226.

Geurts, R. and Bisseling, T. 2002. Rhizobium Nod factor perception and signalling. *Plant Cell*, 14, S239–S249.

Giovannetti , M., Sbrana, C., Avio, L., Citernesi, A.S. and Logi, C. 1993. Differential hyphal morphogenesis in arbuscular mycorrhizal fungi during pre-infection stage. *New Phytol.*, 125, 587–593.

Griffin, G.J., Hale, M.G. and Shay, F.J. 1976. Nature and quantity of sloughed organic matter produced by roots of axenic peanut plants. *Soil Biol. Biochem.*, 8, 29–32.

Gunawardena, U. and Hawes, M.C. 2002. Tissue specific localisation of root infection by fungal pathogens: Role of root border cells. *Mol. Plant-Microbe Interact.*, 15, 1128–1136.

Harley, J.L. and Smith, S.E. 1983. Mycorrhizal Symbiosis. Academic Press, London.

Harrison, M.J. 1999. Molecular and cellular aspects of the arbuscular mycorrhizal symbiosis. *Annu. Rev. Plant Physiol. Plant Mol. Biol.*, 50, 361–389.

Hirsch, A.M., Bauer, W.D., Bird, D.M., Cullimore, J., Tyler, B., Yoder, J.I. 2003. Molecular signals and receptors: controlling rhizosphere interactions between plants and other organisms. *Ecology*, 84, 858–868.

Hirsch, A.M., Lum, M.R. and Downie, J.A. 2001. What makes the rhizobia-legume symbiosis so special? *Plant Physiol.*, 127, 1484–1492.

Hwang, S.F., Chang, K.F. and Chakaravaty, P. 1992. Effects of vesicular-arbuscular mycorrhizal fungi on the development of *Verticillium* and *Fusarium* wilts of alfalfa. *Plant Dis.*, 76, 239–243.

Jones, D.J., Farrar, J. and Giller K. E. 2003. Associative nitrogen fixation and root exudation—what is theoretically possible in the rhizosphere? *Symbiosis* 35, 19–38.

Kistner, C. and Parniske, M. 2002. Evolution of signal transduction in intracellular symbiosis. *Tr. Plant Sci.*, 7, 511–518.

Kneer, R., Poulev, A., Olesinski, A. and Raskin, I. 1999. Characterisation of the elicitor-induced biosynthesis and secretion of genistein from roots of *Lupinus luteus* L. *J. Exp. Bot.*, 50, 1553–1559.

LaRue, T.A. and Weeden, N.F. 1994. The symbiosis genes of the host. *In*: Proceedings of the 1st European Nitrogen-Fixation Conference. Kiss, G.B.; Endre, G. and Officina (Eds.), 147–151.

Levy, J., Bres, C., Geurts, R., Chalhoub, B., Kulikova, O., Duc, G., Journet, E.P., Ane, J.M., Lauber, E., Bisseling, T. and Denarie, C. 2004. A putative Ca^{2+}/Calmodulin-dependent protein kinase required for bacterial and fungal symbiosis. *Science*, 303, 1361–1364.

Madsen, E.B., Madsen, L.H., Radutoiu, S., Olbryt, M., Rakwalska, M., Szczyglowski, K., Sato, S., Kaneko, T., Tabata, S., Sandal, N. and Stougaard, J. 2003. A receptor kinase gene of the LysM type is involved in legume perception of rhizobial signals. *Nature*, 425, 637–640.

Marsh, J.F. and Schultze, M. 2001. Analysis of arbuscular mycorrhizas using symbiosis-defective plant mutants. *New Phytol.*, 150, 525–532.

Mitra, R.M., Gleason, C.A., Edwards, A., Hadfield, J., Downie, J.A., Oldroyd, G.E. and Long, S.R. 2004. A Ca^{2+}/Calmodulin-dependent protein kinase required for symbiotic nodule development gene identification by transcript-based cloning. *PNAS,USA*, 101, 4701–4750.

Mugnier, J. and Mosse, B. 1987. Spore germination and viability of a vesicular-arbuscular mycorrhizal fungus, *Glomus mosseae*. *Trans. Br. Mycol. Soc.*, 88, 411–413.

Mylona, P., Pawlowski, K. and Bisseling, T. 1995. Symbiosis nitrogen-fixation. *Plant Cell*, 7, 869–885.

Oldroyd, G.E.D. and Downie, J.A. 2004. Calcium, kinases and nodulation signalling in legumes. *Nat. Rev. Mol. Cell Biol.*, 5, 566–576.

Park, S.W., Lawrence, C.B., Linden, J.C., and Vivanco, J.M. 2002. Isolation and Characterisation of a Novel Ribosome-Inactivating Protein from Root Cultures of Pokeweed and Its Mechanism of Secretion from Roots *Plant Physiol*, 130, 164–178.

Parniske, M. 2004. Molecular genetics of the arbuscular mycorrhizal symbiosis. *Curr. Opin. Plant Biol.*, 7, 414–421

Peters, N.K., Frost, J.W. and Long, S.R. 1986. A plant flavone, luteolin, induces expression of Rhizobium meliloti nodulation genes. *Science*, 233, 977–980.

Radutoiu, S., Madsen, L.H., Madsen, E.B., Felle, H.H., Umehara, Y., Gronlund, M., Sato, S., Nakamura, Y., Tabata, S., Sandal, N. and Stougaard, J. 2003. Plant recognition of symbiotic bacteria requires two LysM receptor-like kinases. *Nature*, 425, 585–592.

Remy, W., Taylor, T.N., Hass, H. and Kerp, H. 1994. Four hundred million year old vesicular arbuscular mycorrhizae. *PNAS, USA*, 91, 11841–11843.

Ryan, P.R. and Delhaize, E. 2001.Function and mechanism of organic anion exudation from plant roots. *Annu. Rev. Plant Physiol. Mol. Biol.*, 52, 527–560.

Schauser, L., Handberg, K., Sandal, N., Stiller, J., Thykjaer, T., Pajuelo, E., Nielsen, A. and Stougaard, J. 1998. Symbiotic mutants deficient in nodule establishment identified after T-DNA transformation of *Lotus japonicus*. *Mol. Gen. Genet.*, 259, 414–423.

Smith, S.E. and Read, D.J. 1997. Mycorrhizal Symbiosis, Ed. 2 Academic Press, San Diego.

Stougaard, J. 2001. Genetics and genomics of root symbiosis. *Curr. Opin. Plant Biol.*, 4, 328–335.

Stracke, S., Kistner, C., Yoshida, S., Mulder, L., Sato, S., Kaneko, T., Tabata, S., Sandal, N., Stougaard, J., Szczyglowski, K. and Parniske, M. 2002. A plant receptor-like kinase required for both fungal and bacterial symbiosis. *Nature*, 417, 959–962.

Szczyglowski, K., Shaw, R.S., Wopereis, J., Copeland, S., Hamburger, D., Kasiborski, B., Dazzo, F.B. and de Bruijn, F.J. 1998. Nodule organogenesis and symbiotic mutants of the model legume *Lotus japonicus*. *Mol. Plant Microbe Interact.*, 11, 684–697.

Trieu, A.T., Van Buuren, M. L. and Harrison, M. J. 1997. Gene expression in mycorrhizal roots of *Medicago truncatula*. *In*: Radical Biology: Advances and Perspectives on the Function of Plant Roots. H.E. Flores, J.P. Lynch, D. Eissentat, (eds.), American Society of Plant Physiologists, Rockville, MD, pp 498–500.

Wais, R.J., Galera, C., Oldroyd, G., Catoria, R., Penmetsa, R.V., Cook, D., Gough, C., De'narie, J. and Long, S.R. 2000. Genetic analysis of calcium spiking responses in nodulation mutants of Medicago truncatula. *PNAS, USA*, 97, 13407–13412.

Walker, S.A., Viprey, V. and Downie, J.A. 2000. Dissection of nodulation signalling using pea mutants defective for calcium-spiking, induced by Nod-factors and chitin oligomers. *PNAS, USA*, 97, 13413–13418.

Walker, T.S., Bais, H.P., Halligan, K.M., Stermitz, F.R. and Vivanco, J.M. 2003. Metabolic Profiling of Root Exudates of *Arabidopsis thaliana*. *J. Agric. Food Chem.*, 51, 2548–2554.

Williams, P.G. 1992. Axenic culture of arbuscular mycorrhizal fungi. *Methods Microbiol.*, 24, 203–220.

Wu, H., Haig, T., Pratley, J., Lemerle, D. and An, M. 2001. Allelochemicals in wheat (Triticum aestivum L.): Cultivar difference in the exudation of phenolic acids. *J. Agric. Food Chem.*, 49, 3742–3745.

Wu, H., Pratley, J., Haig, T., Lemerle, D. and An, M. 2000. Allelochemicals in wheat (Triticum aestivum L.): Variation of phenolic acids in root tissues. *J. Agric. Food Chem.*, 48, 5321–5325.

Zheng, X.Y. and Sinclair, J.B. 1996. Chemotactic response of *Bacillus megaterium* strain B153-2-2 to soybean root and seed exudates. *Physiol. Mol. Plant Pathol.*, 48, 21–35.

Index